MATLAB for Brain and Cognitive Scientists

MATLAB for Brain and Cognitive Scientists

Mike X Cohen

The MIT Press
Cambridge, Massachusetts
London, England

This book was set in Stone Sans and Stone Serif by Toppan Best-set Premedia Limited. Printed and bound in the United States of America.

Library of Congress Cataloging-in-Publication Data

Names: Cohen, Mike X., 1979- author.
Title: MATLAB for brain and cognitive scientists / Mike X. Cohen.
Description: Cambridge, MA : MIT Press, [2017] | Includes bibliographical
  references and index.
Identifiers: LCCN 2016033649 | ISBN 9780262035828 (h : alk. paper)
Subjects: LCSH: MATLAB. | Neurosciences--Data processing. | Cognitive
  science--Data processing.
Classification: LCC QP357.5 .C56 2017 | DDC 612.8--dc23 LC record available at
  https://lccn.loc.gov/2016033649

10 9 8 7 6 5

Contents

List of Interviews

Preface

MATLAB changed my life. It wasn't the first programming language I learned (that was Basic) nor was it the only one (C++, HTML/CSS, and a few others not worth mentioning). But somehow MATLAB has the right balance of usability, visualization, and widespread use to make it one of the most powerful tools in a scientist's toolbox. I started learning MATLAB in the context of computational modeling, and the MATLAB environment allowed me to understand and to visualize models and to explore parameter spaces when staring at equations gave me nothing but a vague and superficial feeling (coupled with a stronger feeling of anxiety and fear that I was in way over my head).

When I started using MATLAB for data analysis, the ease of inspecting, implementing, and modifying code and the ease of plotting data at each step of the analysis gave me the deep and satisfying feeling of comprehension that reading online tutorials and methods papers could not provide. In fairness, MATLAB is not the only programming language or environment that can be used to understand and to implement data analyses, and I have too little expertise in other languages to claim unequivocally that MATLAB is The Best. It became my go-to tool because so many neuroscience toolboxes and analysis scripts are already written in MATLAB, and because almost everyone around me was using MATLAB. But I have yet to encounter a data analysis or visualization problem I could not solve in MATLAB, so my motivation to gain expertise in other languages is fairly low.

I wrote this book because I want you to be able to use MATLAB the way I use MATLAB—as a means to two ends: (1) to understand data analyses and (2) to analyze data. The Internet is ripe with MATLAB introductions and tutorials, many of which are excellent, and I encourage you to find and go through them. But most of them will guide you gently and slowly from total novice to dipping your toes into the water, and then promptly drop you in the middle of the ocean. What we as a psychology/neuroscience community are lacking is a corpus of resources that will help you gain real expertise in MATLAB with a focus on the applications that you will actually use in your scientific career. I hope this book is a useful contribution to building such an educational corpus. Above all else, I hope that this book helps you to develop and sharpen your skills as a scientist. Good luck and have fun!

I   Introductions

1 What Is MATLAB and Why Use It?

Welcome to your new life as a MATLAB programmer and data analyzer.

If you ask your colleagues what MATLAB is, they might give you some of the following answers:

- It's a platform for data analysis.
- It's a high-level programming language.
- It's a fancy calculator.
- It's that thing you open to use analysis toolboxes.
- It's software that lets you write and customize data analysis tools.
- It's a program to make nicer plots than Excel or SPSS.
- It's an unceasing source of suffering, frustration, and red font.

These points are all true (except the last; the frustration does eventually diminish), but they are not the whole truth. MATLAB is a culture, a way of thinking, a language that cuts across countries, and a way of sharing and collaborating with other scientists regardless of whether they are in your research lab or even scientific field. The purpose of this book is to teach you how to program in MATLAB, with a focus on applications most commonly used in neuroscience and psychology.

1.1 "I Want to Be a Scientist; Do I Also Need to Be a Good Programmer?"

There are at least three reasons why being a good programmer is increasingly becoming part of being a good scientist. First, the brain is really complex, and technology for measuring it is getting really sophisticated. For better or worse, simple univariate tests (e.g., the one-sample t-test) are becoming less and less insightful for studying the brain. A lot of the simple questions in neuroscience are answered ("Does activity in the visual cortex increase when the light turns on?"), and thus more nuanced and specific questions are becoming important ("Do neurons fire in a specific temporal sequence and does this depend on stimulus features?"). Neuroscience data analyses are moving away from simple one-size-fits-all analyses toward custom-tailored analyses. And custom-tailored analyses means custom-tailored code.

Second, and related to the previous point, even if you want to be a scientist and not a programmer, your programming skills are a bottleneck for your research. As experiments and analyses get more complicated, point-and-click software tools will impose stronger limits on the experiments you can do and the analyses you can perform. This is not good. The bottleneck in science should be our neuroscientific theories and our understanding of biological processes, not your MATLAB skills.

Third, learning to program is learning a skill. Programming is problem solving. To program, you must think about the big-picture problem, figure out how to break down that problem into manageable chunks, and then figure out how to translate those chunks into discrete lines of code. This same skill is also important for science: You start with the big-picture question ("How does the brain work?"), break that down into something more manageable ("How do we remember to buy toothpaste on the way home from work?"), and break that down into a set of hypotheses, experiments, and targeted data analyses. Thus, there are overlapping skills between learning to program and learning to do science.

1.2 Octave

Octave is a free software that emulates many of MATLAB's capabilities. There are continual developments that make Octave an attractive alternative to MATLAB, particularly with the release of version 4, which includes a graphical interface similar to that of MATLAB. Octave can interpret nearly all MATLAB functions and works in all major operating systems (Windows, Linux, and even Mac).

The main advantage of Octave is the price (zero units of any currency). The main disadvantage of Octave is speed—tests generally show that Octave is several times slower than MATLAB, and MATLAB already is not known for its computation speed. If MATLAB is available, use it. If costs or licensing are limiting factors, use Octave.

Sometimes systems administrators, particularly when residing over large compute clusters, prefer Octave over MATLAB to reduce the number of MATLAB licenses that are required. If MATLAB licenses are limited, and if some computational time can be sacrificed, then it's a good idea to check that your code runs in Octave. If you do the kinds of off-line data analyses where you can start the script at night and look at the results the next morning or the next week, then the slight decrease in computation time is not prohibitive, and using Octave will free up MATLAB licenses.

All of the code in this book was tested in Octave. Occasionally, the Octave code looks slightly different than the MATLAB code or the MATLAB code does not work in Octave; this is noted in the code. Because nearly everything in this book can be implemented in Octave with little or no modification, the term "MATLAB" for the rest of this book really means "MATLAB and probably also Octave."

1.3 Python, Julia, C, R, SPSS, HTML, and So Forth

In terms of neuroscience data analysis, there really is no viable alternative to MATLAB. True, many or perhaps all of what can be accomplished in MATLAB can be accomplished in Python or C++ or Julia or several other programming languages, but this does not make them viable alternatives. For one thing, despite some hype about other languages, MATLAB remains a dominant language for data analysis in psychology, neuroscience, cognitive neuroscience, and many other branches of science. In part this is because MATLAB is specifically designed to work with multidimensional matrices, and neuroscience data nearly always reside in matrix form.

To be clear, there is nothing wrong with these other programming languages. There will be situations where other languages are superior to MATLAB and should be used. R and SPSS, for example, are well suited for analyses of multifactor parametric statistical models such as mixed-effects linear modeling. Python is arguably better than MATLAB at searching through online text databases to locate specific entries and associate them with information from other databases. C is generally much faster than MATLAB. But within the context of neuroscience data analysis, these are the exceptions. By far, the majority of neuroscience data analyses are done in MATLAB.

In practice, you will use several programming languages in your scientific career. It would behoove you to gain some familiarity with other languages, but if the topics listed in the table of contents of this book fit the description of the kinds of analyses you will be doing, MATLAB is the best option. For example, in addition to MATLAB, I occasionally use R, SPSS, Python, Presentation, and hoc, but I use these relatively infrequently and am no expert in these other languages.

Non-MATLAB languages are also not viable alternatives simply because so few people use them. Many research labs, researchers, and neuroscience data analysis toolboxes use MATLAB. That is why MATLAB is also a culture and a means of collaborating and sharing data and data analysis tools.

1.4   How Long Does It Take to Become a Good Programmer?

That depends on how "good" a programmer you want to be. It can take only a few weeks to gain a basic working knowledge of MATLAB. Obviously, the more time you spend working in MATLAB, the better programmer you will become.

Psychology research shows that what is important in skill acquisition is not just the amount of time spent on the activity but the amount of *focused, attentive time* dedicated to acquiring the new skill (Ericsson, Krampe, and Tesch-Römer 1993). If you try to learn how to program while watching television and eating, your progress will be really slow. If you dedicate 45 minutes a day to learning to program while doing nothing else (not even Facebook!), you will learn much faster. Programming is no easy business, but every hour you spend learning how to program is an hour invested in your future.

1.5   How to Learn How to Program

Programming languages are languages. They have vocabulary and syntax, they have sentences (lines of code), paragraphs (sections of code), and discourse, and they have styles and ways of thinking.

The only way to learn to speak a human language is by speaking. And the only way to learn how to program is by programming. Looking at someone else's code will help you learn how to program in the same way that looking at a Russian newspaper will help you learn to speak Russian. The chapters in this book close with exercises; take them seriously. It's often possible to cheat and to find the solution to the exercises somewhere else in the book code. Some people might argue that this isn't "cheating" because part of learning to program is learning to spot the right code and paste it in the right place. I don't entirely disagree, and knowing which code to copy and how to modify it is a useful skill. The question is whether you want to learn a little bit and be an okay programmer or whether you want to learn a lot and be a good programmer.

In the words of Cohen (2014): "No one is born a programmer. The difference between a good programmer and a bad programmer is that a good programmer spends years learning from his or her mistakes, and a bad programmer thinks that good programmers never make mistakes."

1.6 The Three Steps of Programming

Step 1: *Think*. Writing code starts in your head with the big-picture idea. What exactly do you want the code to do? What are the steps to accomplish this goal? I find that thinking about pictures helps—what will the data look like in the beginning, middle, and end of the script? Even if the script does not do any plotting, it's still useful to think about what plots would look like.

Now turn to MATLAB and open a new blank script. Don't write any actual code yet. Instead, make an outline using comments (comments, as will be explained later, are not interpreted by MATLAB and are indicated with %) of what the code should do. These comments should be descriptions of what needs to be done in what order. For example, at the end of the first step of programming, your script might look like this:

```
%% define parameters
% here I need to define frequencies of interest
% and the time windows to export to a .mat file
% also initialize the output matrices
%% load and process data
% find the right file, maybe with a simple GUI?
% load the file, checking whether it contains
% the variable "raw_data" with 3 dimensions
%% extract frequencies
% this part needs to loop over channels
% run fft on the data
% there are many trials, maybe one fft?
% extract just the frequencies of interest
%% save data to output file
% create a filename
% check to see whether this file already exists
% save the data to the .mat file
```

This is the hardest but most fun part of programming. And it feels like real science while you're doing it.

Step 2: *Write the code*. This involves translating what you specified in English in the comments into the language of MATLAB. The more detailed your comments, the easier this step is.

Is it Anglo-centric to suggest that comments be written in English? Yes, but in the same way that it is Anglo-centric to write this book in

English. One of the advantages of MATLAB is that code can be shared across countries. English is, at present and for the foreseeable future, the lingua franca of science. International scientific journals, conferences, websites, and e-mail lists are nearly always in English. If you want other people to be able to use your code, it's useful to write the comments in a language that they are likely to understand. If English is not your native language, then don't worry; grammar and spelling are less important here than in formal scientific communications. Just make sure the comments are comprehensible.

This step of programming is either fun or painful, depending on your level of programming skills and on how much you enjoy programming. This step is a real part of science because you are forced to convert ideas and plans that might initially be vague and underspecified into concrete and specific steps.

Step 3: *Debug*. Yes, MATLAB will print a lot of red font indicating errors and warnings, so prepare yourself. Some errors are easy to find and fix, like simple typos. Other errors are easy to find but harder to fix.

The worst and most dangerous errors happen when MATLAB does not issue a warning or error message because you technically did nothing illegal, but the result is wrong. There is really only one way to find these errors: Plot as much data as you can, in several different ways, and think carefully about what the data *should* look like, and whether the data *actually* look like what you expect. If you expect to see local field potential (LFP) traces but see a line of all zeros, something went wrong. Go back to previous lines, and keep plotting the data until it looks like something you'd expect.

The process of thinking about what a result should be and then checking or plotting to confirm is called "sanity checking." The importance of sanity checks cannot be understated. This book will offer many suggestions for sanity checks, and many exercises contain code with errors that you need to find and fix.

This step of programming is not much fun for anyone. You probably won't really feel like a scientist while doing this. But it is necessary and it must be done. The good news is that you will experience a deep sense of satisfaction after finally fixing all errors and resolving all warnings.

1.7 How Best to Learn from This Book

1. First the obvious points: Read the book, look at the MATLAB code, and run the code on your computer.

2. Slightly less obvious: Don't just run the entire script and then look at the plots. Run the code line by line. If there are multiple functions or variables on a single line, run each one in turn. Try to predict what each line does before running it. For example, if you see the following line of code,

```
plot(max(data,[],2), 'o')
```

don't run the line and passively wait for something to happen. Instead, first say to yourself: "This line will take the two-dimensional matrix called data, return the maximum value from the second dimension of this matrix, and plot it as a series of circles. I expect there will be 40 points because the size of the variable data is 40×100. This matrix contains non-normalized spike counts, which can only be positive, so if I see any negative values in the plot, something must be wrong. And if I see any zeros, then it could be okay but I'll be suspicious and investigate." And *then* run the line. If you get it right, congratulations. If the result was different from what you expected, figure out where you got it wrong and learn from it.

3. The code in this book is written to be a tool to help you learn, not as a set of "black-box" scripts with which to analyze your data. After running the code and figuring out how it works, make more plots and different kinds of plots. Change parameter values to see what the effects are. Start with code that works, change something in the code to make MATLAB produce an error, and then figure out why it gave an error. See if you can write new code such that the input data are the same, the resulting outputs are the same, but your code looks and works slightly differently from the book code. Study the code for a bit, then open a blank script and see how much of the original code you can rewrite from scratch.

4. Integrate across chapters. For example, in chapter 11 you will learn about the Fourier transform, and in chapter 31 you will learn about classification. Try performing the Fourier transform on some data and then classifying the power values from the Fourier coefficients.

5. In the medical world they have a saying: "See one, do one, teach one." In pedagogics they have a similar saying: "To teach is to learn." Work with other people while going through the book. Not too many other people—if there are more people than lines of code, that's a party, not a MATLAB learning session. Work on code together with one or two other people, and meet in groups to discuss programming or mathematical or conceptual problems. It's likely that someone

else can explain something you struggle with, and that you can explain something that someone else struggles with. If you are the better programmer in the group, let the less experienced person do the typing while you act as "backseat driver" (remember to be nice and patient).

6. Use the index. There are a lot of tips and tricks in this book, but the book is organized according to analysis topic, not according to MATLAB function. That was intentional, to make the content more approachable to a reader interested in neuroscience or psychology. But it means that some MATLAB-specific functions may be difficult to find from the table of contents. I tried to make the index as detailed and helpful as possible.

1.8 Exercises and Their Solutions

Chapters end with exercises that test and develop the skills taught in each chapter. You should do the exercises. Really. You will develop your MATLAB programming skills much more from the exercises than from reading the text and looking at my code. To encourage you to do the exercises, I occasionally put material in the exercises that is not presented in the chapter.

You should try to start and complete the exercises from scratch. If you need a hint, download the exercise starter-kit from my website (www.mikexcohen.com). This code is not a complete set of solutions, but it provides some skeleton scripts and hints to help you along. Also available online are screenshots of plots that correct solutions might produce. If your code produces a plot that looks like the ones online (random numbers notwithstanding), it's likely you got the correct solution (there are usually several correct ways to solve a MATLAB or data analysis problem).

If you really want to maximize your knowledge, complete the assignments multiple times. I'm not kidding. Come back to the same problems after a few days or weeks, and solve them again. Each time you re-solve the same problem, you'll find different or more efficient solutions. And you will start being able to recognize the types of problems that come up over and over again.

1.9 Written Interviews

I thought it would be insightful and encouraging for you to read interviews with some of the top MATLAB programmers in the cognitive and neuroscience fields. The interviews are placed throughout the book; all interviewees were asked the same questions, and their answers reflect their own opinions. Interviewees were selected because they have made contributions to the scientific field in part through programming in MATLAB. Of course, many people—from students to postdocs to professors to full-time programmers—contribute to their scientific field in part through MATLAB programming; please know that your efforts are highly appreciated even if you were not interviewed. The interviews are thought provoking, and I hope you enjoy them.

1.10 Where Is All the Code?

There are more than 10,000 lines of code that accompany this book (I know that sounds really impressive, but there are a few redundant lines). Printing every line of code in the book would greatly increase the page count and would require a lot of manual typing on your part. Therefore, the most important pieces of code are printed in the book, and all of the code can be downloaded from www.mikexcohen.com.

The code printed in the book does not always exactly match the online code. In some cases, variable names are shortened for formatting purposes; in some cases, the online code provides additional options or examples; and in some cases, the book shows additional code to facilitate comprehension. Although this requires you to spend more time going back and forth between the book and the online code, the extra effort forces you to appreciate that there are multiple ways to accomplish the same goal when programming. I want you to become a good and flexible programmer, not to memorize a few elementary programming sentences.

1.11 Can I Use the Code in This Book for Real Data Analyses?

Yes you can. The code here is valid for application to real data analysis. That said, the primary goal here is to *learn* how to use MATLAB—the focus is more on the learning aspect than on the applications aspect. The code is not a cookbook for data analysis, and you definitely should not simply copy and paste the code into your analysis scripts without understanding what the code does (in part because the code occasionally contains intentional errors that you need to find!). Yet the main purpose of using MATLAB in neuroscience is to analyze neuroscience data, and it doesn't make sense to learn programming without learning some analyses. Thus each chapter deals with one specific aspect of data analysis, but the focus is more on the programming implementations, tips and tricks, and potential mistakes to avoid, rather than on mathematical derivations and proofs.

Because of this, the analysis methods presented in this book are not necessarily the most cutting-edge methods that reflect the state of the art in neuroscience. Instead, they are the analyses that are commonly used in the literature and that provide a good framework for learning how to program in MATLAB. This separates *MATLAB for Brain and Cognitive Scientists* from the book *Analyzing Neural Time Series Data*, which goes into great detail about how to analyze electrophysiologic data while providing some instruction about MATLAB programming. *MATLAB for Brain and Cognitive Scientists* goes into great detail about MATLAB programming while providing only the necessary background details about the intricacies of data analysis, statistics, and interpretation.

1.12   Is This Book Right for You?

This book is written specifically for those studying or considering the fields of neuroscience, psychology, and cognitive neuroscience. The level is intended for advanced undergraduates up to professors, but probably master's students, PhD students, and postdocs will benefit the most. The book starts with the most elementary introduction to MATLAB but then quickly progresses to more medium-level and advanced material. This is intentional—there are many excellent resources for beginner-level introductions to MATLAB, but there are very few structured resources that can guide you from beginner to moderate level. If your goal is to spend a year slowly working your way up to writing a for-loop, then this book is probably not for you. If you have limited time, a positive attitude, and want to learn a lot of MATLAB quickly so you can start your data analyses before your hair turns gray, I hope you find this book to be the right resource.

1.13   Are You Excited?

If you are patient and motivated, learning how to program in MATLAB will change your life. It will open new possibilities for scientific discovery, it will make you more independent, and it will make you more competitive for your next job as a student, postdoc, or professor. If you are exploring a career outside science, programming skills will likely be important for any job in our increasingly digitized and programmed world. I don't know if it's good or bad that programmers are taking over human civilization, but hey, if you can't beat 'em, join 'em.

I tried to write this book to be approachable and encouraging, with a few subtle jokes here and there to keep you engaged. While writing, I tried to imagine that I'm sitting next to you, talking to you, and helping you each step along the way (don't worry, in my imagination I brushed my teeth and hadn't eaten canned tuna fish in several days). When I write "you" I am speaking directly to you, the reader. When I write "we," it's not the Royal We; I am imagining that you and I are sitting together working through MATLAB, and I'm trying to make you feel better by giving the impression that I'm in the process of figuring this stuff out along with you.

So, turn the page and let's begin!

2 The Philosophy of Data Analysis

Do no harm.
—Often misattributed to the Hippocratic oath, according to Wikipedia

Just because you can, doesn't mean you should.
—Common saying

When it comes to analyzing your data, MATLAB is simultaneously an amazing resource that gives you the freedom you need as a scientist and the biggest danger to doing good science.

"Philosophy of data analysis" is not about analyses per se; it is about how to think about data analysis, a set of guiding principles that you should keep in mind when analyzing your data.

2.1 Keep It Simple

Data analyses should be as simple as possible, and you should use more complicated analyses only when necessary. Don't run an independent components analysis (ICA) when an average will suffice. Don't run a mixed-effects hierarchical linear model with 15 covariates and all interaction terms using Monte Carlo simulations to estimate *a posteriori* distributions of the width of possible parameter estimates when a one-sample *t*-test will sufficiently provide evidence for your hypothesis.

This advice is difficult to give and difficult to take. Analyses are fun, and the brain is complicated. Many neuroscientists believe (unfortunately, despite its probable truth) that more fancy analyses will make a manuscript more likely to be accepted, particularly at high-impact-factor journals.

Complicated data require complicated analyses, and the brain is certainly complicated. The argument here is not that you should avoid complicated analyses. Rather, always start with simple analyses and move to more complicated analyses only when the simple analyses are insufficient to provide evidence for or against the hypotheses.

2.2   Stay Close to the Data

Imagine this situation: You start with raw data. Then you filter the data. Then you apply a regression model in which each time point of the data is predicted by two independent variables and their interaction. Next, you filter the regression weights for the interaction term. Then you convolve those filtered regression weights with a kernel defined by the output of a computational model. Then you compute the Fourier transform of the convolved signal, take a weighted-by-distance average of several frequency components, and perform a k-means clustering to separate the averaged Fourier coefficients into three groups. Then you compute the Euclidean distance from each point to the center of mass of each cluster. Finally, you run a one-factor analysis of variance (ANOVA) to determine whether the point-to-center distances within those three clusters are different between the Alzheimer-model mice and the control mice.

How can we relate the final ANOVA results back to the raw data (i.e., the activity of the brain that you measured)? Probably your answer is, "I have no idea." This is not a situation you want to be in with your scientific research.

Each of the steps described above gets further away from the data. Steps away from the data are nearly always necessary, partly because most data contain noise, and partly because multiple signals can be embedded in the data simultaneously. The purpose of data analysis is to recover what cannot be obviously visually observed in the data, and this necessarily involves moving away from the raw data. Multivariate or multidimensional data entail even bigger difficulties, because the raw data may be too complex to use or even visualize directly.

There are two primary dangers of letting your analyses get too far away from the raw data. First, errors, suboptimal parameters, or poor fits of models to data carry forward to all future steps. These issues will compound as you get further away from the data, meaning that a small error early on can produce large errors later on. Second, the more steps between the data and the final results, the more difficult it becomes to interpret those results or to link the results to other findings, theories, and so forth.

This is not to say that you should avoid any analyses or interpretations of parameters and meta-parameters. Rather, each step away from the raw data should be made with increasing caution and should be done only when it can be justified.

2.3   Understand Your Analyses

"Understand" can be interpreted on many levels. It is ideal to have a deep mathematical understanding of everything that happens to your data. One might interpret this to mean that you should be able to program any analysis yourself from scratch. This level of understanding is a noble goal that you should work toward, at least for the analyses you use most often.

But limiting yourself to analyses that you fully mathematically understand involves excessive constraints, particularly for those without a strong formal mathematical background or those relatively new to programming.

A more realistic interpretation of "understand your analyses" is that for any analysis you want to apply to your data, you should be able to explain

1. generally how the analysis works;
2. how to interpret the results;
3. what the effects of different parameter settings are; and
4. how to determine if something went wrong.

If you want to use a data analysis method, try to be comfortable with the four points listed above before publishing or presenting the results in a formal scientific setting. Being a good programmer helps enormously for understanding analyses, because you will be able to write code to test and inspect results after testing many different analysis parameters, input data, and so on.

2.4   Use Simulations, but Trust Real Data

The main advantage of simulated data is that you get to define the ground truth, which means you know what the outcome of the analysis should be. If you are trying to understand what a data analysis method does or are developing a new or modified analysis method, simulated data is a great place to start. In this book, you will learn several techniques for simulating data to evaluate data analysis methods.

The main disadvantage of simulated data is that simulated data often lack characteristics of real data. Methods might work well in simulated data and then fail to produce sensible results in real data. Or you might fine-tune an analysis method to capture a simulated dynamic that does not exist in real data.

When testing new methods, the best kind of empirical data to use are data where there is a simple and large effect that you can anticipate *a priori*. For example, if you are testing a novel data analysis method for fMRI data, evaluate the method on a task in which visual stimuli are presented to the left or right of fixation, and the human research participants make responses with their left or right hand. Experiments like this may not be very interesting from a neurocognitive perspective, but they are a good model for analysis methods because you know what the results should be.

2.5 Beware the Paralysis of Analysis

Because analyzing data is fun, because the parameter space for complex analyses tends to be large, and because most neuroscience data sets are multidimensional, it is easy to get caught in a loop where you keep reanalyzing the same data using different methods or different parameters. To make matters worse, there are many ways to analyze data, and there are new publications each year describing even more ways to analyze data or to improve existing analysis methods.

It is tempting to think that there is a really amazing and novel finding in your data, and you just need to keep searching for the right analysis method to get that result to come out. When you are stuck reanalyzing the same data without moving forward, this is called the paralysis of analysis. There are two major problems with the paralysis of analysis.

1. You increase the risk of overfitting and explaining noise rather than or in addition to signal. This point is discussed in more detail in the next section.
2. No data set is perfect, and repeatedly reanalyzing the same data increases the risk of nonreplicable results (this point is discussed in more detail in the next chapter). Progress in science is more likely to come from independent replications than from obsessive reanalyses of the same data set, particularly if that data set contains a limited amount of data.

At some point you hit the data analysis version of the law of diminishing returns: More time and energy goes into the analyses, but no new results come out. This is when you need to stop analyzing and move on. The difficult part is knowing when to stop. Ideally, you will decide the end point of the analyses before you even start. In practice, I suspect this rarely happens.

Here are my criteria for stopping data analysis: I stop analyzing data when different parameters or different analysis methods produce (a) the same pattern of results (including null results) or (b) completely different results. If different analyses produce the same results, I trust that there is a real finding and that it is robust. If different analyses produce qualitatively different results, I become suspicious that the effect is either very weak or is an artifact that some analyses are sensitive to. If the effect is theoretically relevant, I would report it but interpret it cautiously.

2.6 · Be Careful of Overfitting

Probably you've seen a graph like figure 2.1 before. If you haven't, the idea is fairly straightforward: Data contain signal and noise, and you want to fit only the signal and not the noise.

There are both qualitative and quantitative methods to avoid overfitting data. Qualitatively, you should begin with models that have few parameters, and then increase the number of parameters only when necessary. Visual inspection of data and model fits is sometimes sufficient to determine whether the model is more complex than necessary (discussed in chapters 28 and 29). Quantitative methods include formal model

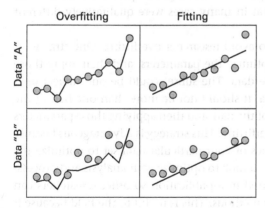

Figure 2.1
An illustration of overfitting. Two data sets were generated using the same linear trend plus independent random noise. The top left panel illustrates overfitting by using a 10-parameter polynomial model, and the top right panel illustrates properly fitting using a two-parameter linear (intercept and slope) model (you will learn more about these kinds of models in chapter 28). One of the problems of overfitting is that while the model fits data A, it cannot generalize to new data B.

comparisons, in which the fit of various models to the data (as measured, e.g., through log likelihood) are compared; the number of model parameters is used to penalize more complicated models. Akaike and Bayes information criteria are perhaps the most commonly used model comparison methods.

This is the easy-to-understand, textbook example of overfitting data. There is another, and more insidious, form of overfitting, which is more problematic and probably more widespread.

This can be called "researcher overfitting" and results from the large number of choices that the data analyzer has during all the steps of processing and analyzing the data. Each time you re-run the analysis using a different time window, different frequency band, different filter cutoff, different smoothing kernel, different data rejection criteria, and so forth, you are refitting the data with a slightly different "model."

If you keep changing these parameters and reanalyzing the data, there is a danger that you will be optimizing the data processing pipeline in part to capture noise or at least an idiosyncratic effect that might not replicate in an independent sample. A striking illustration of this possibility is a study showing that different combinations of processing steps in an fMRI study (processing steps included temporal filtering, slice-time correction, spatial normalization, and spatial smoothing) can produce a wide range of supra-threshold results that in many cases were qualitatively different (Carp 2012).

There are two strategies to avoid researcher overfitting. One strategy is to use a subset of data to optimize the parameters, and then apply those parameters to the rest of the data. The subset could be one subject or it could be one half of the data. It should not be more than one half of the data, because fitting to 99% of the data and then applying those parameters to 1% still runs the risk of overfitting. This strategy is advantageous because you can use the characteristics of your particular data set to optimize the analysis pipeline. The steps you took to optimize your analysis and processing pipeline should be reported in a publication so other researchers can evaluate and reproduce your methods. This is useful to the field because it will minimize the time other researchers will need to spend on processing steps unlikely to be successful.

A second strategy is to find other published studies that use similar analysis methods and use their processing and analysis protocols. This is a completely unbiased approach and runs zero risk of overfitting your data. But it comes with the risk that the processing pipeline might be suboptimal for your particular experiment. For example, imagine that a published study on the same topic and using a similar experimental design used a time window of 400–700 milliseconds for averaging action potentials. Inspecting your data, however, reveals that the spiking activity peaks earlier, say from 300 to 600 milliseconds. A window of 400–700 milliseconds might yield weak or nonsignificant results. This difference in activation timing might be meaningful (e.g., if there are differences in motivation or age) or it might simply reflect natural sampling variability.

In practice, a balance between these two strategies is probably the best way to go. Many research labs have standard in-house processing and analysis protocols that generally work well and that they use for most of their research. The important aspects are to detail all of your processing steps so they can be evaluated and replicated, and to apply the same processing steps to all subjects, data sets, and conditions. To the extent that there are biases, those biases should at least be consistently applied to all of the data.

2.7   Noise in Neuroscience Data

It is important to realize and to accept that data contain noise. Thinking that fancy filtering techniques can completely denoise data is dangerous: Signal and noise are often coupled, and filtering out too much noise typically entails filtering out some signal. Relatedly, some noise is easy to remove, while other noise is difficult or impossible to remove. There are no cookbook procedures that blindly and successfully denoise any data set.

The good news is that a lot of noise can be successfully removed from data, and the more you know about your noise and how to isolate it in time, frequency, or space, the cleaner your data can become. Still, try to avoid excessively pre-processing your data. With each filter or processing step, the risk of losing signal increases. In other words, use a chisel, not a sledgehammer.

Of course, the easiest and best way to remove noise is to avoid it during recording. If you see excessive noise in the data, try as best you can to find the part of your experimental setup that produces the noise. Noisy data cause major headaches during analyses, and a few extra days/weeks/months hunting down and eradicating experimental noise sources will improve the rest of your research.

Finally, before trying to denoise your data, think about whether it's even necessary. If some vibrating equipment causes an artifact at 748 Hz but your analyses will focus on activity below 40 Hz, it might be unnecessary to worry about removing this high-frequency artifact.

2.8   Avoid Circular Inference

Imagine generating 100 random numbers from a normal distribution (mean of zero and variance of one). Now pick the 20 largest numbers and use a *t*-test to determine whether the collection of those numbers is statistically significantly greater than zero. Would you be surprised if the *p*-value associated with that *t*-test is less than 0.05? Of course not—you *selected* numbers *because* they were larger than zero.

This is an example of circular inference (also sometimes called "double-dipping"). Circular inference means that a subset of data is selected in a way that is biased toward obtaining a specific result, even if there is no true result. This issue arises often in neuroscience, because data sets tend to be large and thus data selection is often necessary. To avoid circular inference, the method of selecting data must be independent of the pattern of results that the statistical analyses are designed to test.

In many cases, circular inference is easy to detect by critical thinking. The key here is to repeat the thought experiment described above: What would happen if you generated random numbers and applied the same data selection procedure and statistical analysis? Would you expect statistically significant results? If you select neurons that show increased firing in condition A compared to condition B, testing whether the firing rate differs between conditions A and B is a biased test. However, selecting neurons based on A versus B, but then testing the firing rate differences between conditions C and D, is not circular inference (assuming C and D are independent of A and B).

In some cases, circular inference is more difficult to detect. When in doubt, ask a colleague to evaluate your data selection and statistical procedures. Better to be a little confused in front of a colleague than to be embarrassed about a published result.

Circular inference is not illegal per se. There are situations where biased statistics can be informative when interpreted correctly and in the context of other analyses. In these situations, it is important to clarify explicitly which results are based on a biased selection and which are not. For example, if you select neurons based on a difference between conditions A and B, and then enter the data from those neurons into an ANOVA with factors *condition* (A vs. B) and *state* (anesthetized vs. awake), the statistically significant main effect of *condition* is biased and must be reported as being biased by the selection procedure, but the main effect of *state* and the interaction between the two factors are not biased and can be safely interpreted. To read more about circular inference and how to avoid it, see Kriegeskorte et al. (2009).

2.9   Get Free Data

If you want to analyze neuroscience data but don't have your own data, there are several online repositories where you can download published data sets. Occasionally, data sets are incomplete, undocumented, or otherwise unusable, but most publicly available data sets have passed through some quality-control check. There was a special issue in the journal *Neuro-Image* in 2015 on data repositories (http://www.sciencedirect.com/science/journal/10538119/124/supp/PB). This is not a complete list; two other repositories, for example, are CRCNS (https://crcns.org) and modelDB (http://senselab.med.yale.edu/ModelDB/). Open-access data is becoming increasingly popular in neuroscience, and new repositories are continually being developed. Printing an exhaustive list makes little sense because the list would be outdated by the time you read this. But searching the Internet will reveal extant repositories.

3 Do Replicable Research

Many fields in biology, including psychology and neuroscience, have been experiencing a sometimes-ignored but unavoidable crisis of nonreproducible findings. One alarming study estimated that somewhere between 30% and perhaps up to 90% of findings in scientific papers are not reproducible (see, e.g., the recent special issue in *Nature* on this topic: http://www.nature.com/news/reproducibility-1.17552).

Who is "to blame" for this? We can blame the researchers for rushing through experiments to publish more papers faster. We can blame the university departments that evaluate researchers on the basis of the number of their publications and the impact factors of the journals in which those publications appear. We can blame funding agencies for preferring to fund novel, high-risk projects instead of more trustworthy, incremental, replication-based science. We can blame the editors of high-impact-factor (sometimes called "luxury") journals for promoting novel and surprising findings (and therefore necessarily more likely to be statistical flukes) over more methodologically sound research that is more likely to reflect the true state of the world.

But I prefer a more positive outlook. Yes, we all could and should be trying to do better. But above all else, I blame Mother Nature for being fickle and creating a world full of incredible diversity and complexity that permeates every aspect of biological systems on our lovely green planet Earth. H. G. Wells wrote more mildly about human attempts to understand and control our environment in *The Time Machine* (1895): "nature is shy and slow in our clumsy hands." This incredible diversity produces biological systems that are complex, that change over time, and that can be highly sensitive to even minor fluctuations in the environment. This makes measuring and studying those systems difficult and dirty, even when the measurement devices (and the scientists using them) are nearly perfect. In other words, no matter how hard we might try to control everything and produce replicable research, nature will find a way to rebel.

We should all be concerned about findings that do not replicate. But we should not be so quick to assume that nonreplicable findings are the product of rushed, lazy, or unethical behavior. Certainly there are cases of outright fraud and scientific misconduct (see, e.g., http://retractionwatch. com), but I believe that most researchers try to do the best they can, given the constraints of limited time and budget resources and the pressure to publish in order to survive the competitive job market of scientific research.

There are many reasons why findings may fail to replicate. There could be statistical flukes (type I errors), small effect sizes, effects that are highly sensitive to minor experimental manipulations, seasonal or time-of-day effects, cultural or linguistic differences, developmental differences, and so on. There are also honest mistakes in experimental design or data analyses. Some of these factors are beyond our control. But there are strategies to improve research and produce findings that are more likely to replicate. The rest of this chapter provides a nonexhaustive list of tips that should help you do replicable research.

This discussion may seem out of context in a book on MATLAB programming, but I believe that striving for solid, replicable research is an important topic that should permeate our conversations, our research, and our education. It should not be relegated to occasional special issues of scientific journals and vague complaints in online forums.

3.1 Avoid Mistakes in Data Analysis

This is easier said than done, of course. And some errors are more likely to be detected than other errors. In particular, errors that produce strange or null effects are more likely to be found than errors that produce positive or plausible effects.

There are three ways to help prevent—or find and fix—mistakes in data analyses. The first is to use simulated data to confirm that the analyses can reveal the true result. The second is to perform many sanity checks on the MATLAB code by examining and plotting the data at each step of the analysis. Throughout this book, you will have many opportunities to learn how to sanity-check code and results.

The third way to prevent mistakes in data analysis is to keep detailed records of what analysis steps were applied, in what order, and using what parameters. This is not only for the sake of other people to follow your analysis; you'd be surprised how quickly analysis details are forgotten, so detailed notes allow you to reproduce your own analyses after months or years. One advantage of programming your analyses is that the code, along with comments, provides an unambiguous list of what happened in what order.

3.2 Have a "Large Enough" N

Sample size is important to make sure you have sufficient statistical power, that the findings are generalizable to the population from which they are drawn, and that your analyses are not overly sensitive to outliers or extreme data values.

There is no magic number that makes a sample size "large enough." It depends on the method, the experimental paradigm, the effect sizes, the quality of the data, and so on. For human scalp EEG, somewhere around $N = 20$ subjects is often a sufficient number, but this is just a rule of thumb. Some of the factors that influence whether a sample size is large enough cannot be precisely determined in advance, such as the effect size and the quality of the data.

To estimate the sample size that is likely to be sufficient, you can use statistical power calculators (e.g., Faul et al. 2007). You can also report effect sizes and *post hoc* power analyses in publications. For most kinds of neuroscience research, it is wise to collect at least 10% more data than you think you need, because data are often discarded due to artifacts, technical problems, attrition (meaning the subject does not complete the experiment or dies before the data are collected), or other factors.

3.3 Maximize Level 1 Data Count

There is a distinction between level 1 data and level 2 data. Level 1 data are the lowest level of observations. For example, in a cognitive task in which the subject repeats many trials of some condition, level 1 would correspond to trials in each condition. For neurophysiology experiments, this might correspond to neurons within an animal or within an *in vitro* slice preparation. For individual differences research, level 1 might correspond to the entire animal or research participant. Level 2 data are averages of level 1 data. And level 3 data would be one step higher than that.

Here is a brief example of the different levels. Imagine a study comparing teaching styles in different classroom settings. Level 1 might be the test performance of each student. Level 2 might be the entire classroom comprising a dozen students, where each classroom uses one of several teaching strategies. Level 3 might be classrooms in different neighborhoods to compare the effects of socioeconomic status. Level 4 could be different countries to compare cultural effects.

In psychology and in neuroscience, most experiments comprise level 1 and level 2. Generally, the idea of these two levels is that level 1 averaging and statistical procedures are designed to estimate the direction of effects within each individual, while level 2 averaging and statistical procedures are designed to determine whether the effects are likely to be observed in many individuals across the population from which the data were sampled.

The amount of data acquired at these two levels differs markedly. In neuroscience research, it is typical to acquire hundreds or thousands of trials within each individual, while only 10–20 individuals are tested, and sometimes fewer. Having more within-subject trials (or neurons, or whatever is the level 1 variable) is important for ensuring stability of the results during level 2 (group) analyses. Therefore, you should try to maximize the amount of level 1 data you can acquire.

3.4 Try Different Analysis Parameters, and Trust Analytic Convergence

A real finding should be robust to a reasonable range of analysis parameters. If you get a significant effect when using a filter that has a 6-dB roll-off but not when using a filter that has an 8-dB roll-off, this is not an effect that should inspire much confidence. However, even real effects with large effect sizes can be obliterated by extreme or inappropriate analysis parameters, so you need to know what constitutes a "reasonable range" of parameters for each type of analysis (see chapter 2.3).

Similarly, a real finding should be observed when using different analysis approaches. For example, if you observe prestimulus alpha power when using wavelet convolution, you should observe the same pattern of findings when using the short-time Fourier transform.

Any time you change an analysis or an analysis parameter, the results will necessarily change at least a bit. The important question is whether the results remain stable enough to lead you to the same conclusion about the findings. If your interpretation of the results changes with each minor modification to the analysis procedure, then you should be suspicious of those results.

3.5 Don't Be Afraid to Report Small or Null Effects, but Be Honest About Them

If you have an effect that borders on statistical significance, for example depending on the filter characteristics, particularly if it is a theoretically relevant finding, it is okay to report the finding. But you should be explicit and honest about the effect size. You could write something like "this finding was relatively small and dependent on analysis parameters. It is thus an interesting possibility, but must be confirmed in future experiments." Sometimes, effects are small because the study was designed to maximize other effects. Reporting small effects also facilitates future meta-analyses.

3.6 Do Split-Half Replication

Split-half replication is a good way to demonstrate that your findings are robust and a good way to avoid overfitting. The idea is very simple: Perform all of the analyses on one half of your data, then repeat the analyses using the same parameters and procedures on the other half of the data.

The data could be split according to subjects, or according to trials (e.g., even trials vs. odd trials), or according to blocks of trials. You should make sure that the way you split the data preserves the global characteristics across the two subgroups. For example, you should not split the data by gender, and you should not separate the first half versus the second half of trials if the outcome measure is likely to differ early versus late in the experiment.

The ability to perform split-half replication requires having a sufficient amount of data to perform statistics on each half of the sample. This is another good motivation for collecting a sufficient amount of data.

3.7 Independent Replications

Of course, the best way to determine whether a finding is replicable is to try to replicate it. The best kind of replications are from independent research groups using different equipment. A real finding shouldn't depend on whether the data were collected with Brand X or Brand Y equipment, and many fundamental brain processes shouldn't depend on whether the data were collected by Scientist M in the Netherlands or by Scientist R in Lithuania (unless, of course, the research involves human cultural or linguistic processes, in which case one would predict that the results differ meaningfully across countries).

But independent replications don't always happen, in part because different research groups do different research. Thus, you should try to replicate your own findings in subsequent experiments. This does not need to slow down your progress or scientific output, but you should try when possible to incorporate replications of previous findings into new experiments. This will happen naturally if your research follows a programmatic line of investigation.

3.8 Write a Clear Methods Section

The purpose of a methods section in a scientific publication is to provide sufficient detail for other researchers to replicate your scientific methods. Unfortunately, methods sections are too often written in a vague manner with few details, and interested researchers would be unable to replicate the exact procedures. Occasionally, the methods section is so poorly written that it's not even clear what was done in that experiment. Please do the world a favor and make sure your methods section is complete and comprehensible. Most journals allow supplemental online sections, so there is rarely an excuse of insufficient space. It's better to err on the side of providing too much rather than too little detail.

If you use toolboxes or software packages, include the names of the functions and procedures that you used, and specify any nondefault parameters.

3.9 Make Your Analysis Code or Data Available

Making your analysis code freely available has at least three benefits. First, other people will be able to learn from and use your code, which benefits the scientific community. Second, and relatedly, if someone uses your code in their publication, they will cite your paper, which increases the visibility of your research. Third, even if no one ever uses or even looks at your code, just knowing that your code will be available for the world to inspect will help motivate you to write error-free and well-commented code, which can have only positive effects on the quality of your research.

There are many ways to make your code available. You can put it on your own website or your lab website, upload it to https://github.com or http://code.google.com (or any other related website), or you can simply note in the publication that code is available upon request (this is the least preferred option).

4 The MATLAB Program

4.1 The MATLAB Program Graphical User Interface

When you start the MATLAB program, one or several windows will open. It might look something like figure 4.1. The MATLAB program has around six subwindows that show relevant information that you can optionally show or hide. The purpose of this chapter is to introduce you to the MATLAB environment. If you are completely new to MATLAB, this chapter is a prerequisite for every other chapter in this book. If you already have some experience with MATLAB, it would behoove you to skim through this chapter just in case there are a few things you don't already know.

All of the programming and executing of MATLAB scripts happens in the MATLAB graphical user interface (GUI). It is possible to batch MATLAB jobs, meaning you tell MATLAB which script to run, and that script runs on a server with no interface. But most of the time, you will run MATLAB in the so-called interactive mode.

It is possible to have multiple instances of MATLAB running on the same computer. That is, you can open MATLAB four times and run four completely different analysis jobs. If you have a multicore computer, it is good to have maximum $N - 1$ MATLAB instances, so that one core can be kept free for other computing needs. Multiple MATLAB instances is an easy way to parallelize your work: You can analyze half of the data sets in one MATLAB instance and half of the data sets in another MATLAB instance. If you are sharing computing resources with other people, don't use all the cores for your MATLAB instances (more on this in the MATLAB etiquette section at the end of this chapter).

The MATLAB GUI changes somewhat with each version. This chapter is based on the R2015a version. Other versions may have slightly different layouts, but the differences will be minor. If you have a version of MATLAB

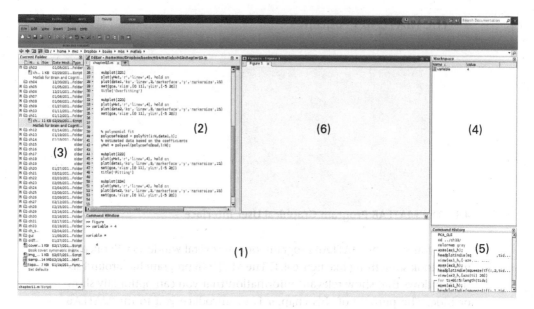

Figure 4.1
When you open MATLAB for the first time, it might look something like this.

older than R2007, you should consider upgrading, because several important MATLAB functions have changed since then.

4.2 Layouts and Visual Preferences

The MATLAB window is highly customizable. Each MATLAB user (i.e., you) has personal preferences for layout and color schemes, so you should feel free to modify the layout to be most comfortable for you. Windows can be made visible or hidden by selecting *Home, Layout* from the main *Menu* bar. The important windows include the following (these numbers correspond to numbers in figure 4.1).

1. *Command window.* This is where you interact with MATLAB. All code is evaluated here, and all of the variables and stored data are accessible and viewable here. Your MATLAB experience will be extremely limited without the Command window being visible.

2. *Editor.* This is where you view and edit MATLAB scripts and functions. "Scripts" and "functions" are text files that contain code and comments. It is likely that you will spend most of your MATLAB time using the Editor, so you should keep this visible as well. Multiple files can be

kept open at the same time, and these will be stored in different tabs, similar to the tabs of an Internet browser.

There is a small dark-gray horizontal bar on top of the scroll bar in the Editor window. If you click-and-drag that bar downward, it will split the Editor in two, allowing you to view the same script in two different locations. This is useful if you have a long script and want to compare the code on lines 200–220 with the code on lines 10–30. Double-clicking the splitting horizontal bar will make it disappear.

3. *Current Folder*. This window is similar to a File Explorer you might see in Windows or Mac. It shows your current directory, all the files in that directory, and the subdirectories.

4. *Workspace*. This window shows all of the variables that are stored in MATLAB's memory buffer. It also provides some information about each variable, such as the size, dimensions, and type (cell, double, structure, etc.). If you double-click a variable, a new window will open that shows the contents of that variable as a spreadsheet, similar to Excel. This works only on some types of variables.

5. *Command History*. This shows the previous commands that were typed into the Command window. If you keep the important lines of code in the Editor, you might not need to use this Command History window.

6. *Figures*. This is where your data are visually represented. The Figures window is not visible unless you have figures open. When there are multiple figures open, you can choose to have them be grouped together into one window, in which case you can view different figures by selecting their tabs, similar to selecting different tabs in an Internet browser, or you can choose to have each figure be its own separate window. The former is useful if you have many figures and want to keep them organized; the latter is useful when you want to inspect several different figures at the same time.

Each of these windows can be "docked" (stuck) inside the main window or "undocked" (free to be moved around on your monitor, outside the main MATLAB GUI). If you have multiple monitors or have a big monitor, you might prefer to have some windows undocked. If you like to keep your MATLAB session neat and tidy, you might prefer to have all windows docked (I like having everything docked).

Windows can be moved around the MATLAB GUI by clicking-and-dragging the colored bar on the top of each window. As you move your mouse around with the left button kept down, MATLAB will show an outlined frame so you can see where the window will be once you release the mouse button.

At any given time, only one window is considered the "active" window. The active window is the one to which keyboard presses are directed. You can select which window is active by clicking on it, and you can scroll through active windows by pressing Ctrl-Tab (Apple-Tilde or Ctrl-Tab in Mac, depending on your settings). You can see which window is active by looking at the bar on top of the window—if it is colored, it's active; if gray, it's inactive.

You can see that there are many ways to custom-tailor your MATLAB experience. My preference for layout is to have the Command window, the Editor, and the Figures visible and docked, and all other windows hidden, as shown in figure 4.2. Spend some time trying out different layouts; you want your MATLAB experience to be comfortable and efficient. As you become more comfortable with MATLAB, your preferences will likely change.

Tip to students: You might think that changing your layout preferences counts as "working in MATLAB," but your supervisor is likely to disagree; therefore, try to keep the monitor facing away from your supervisor.

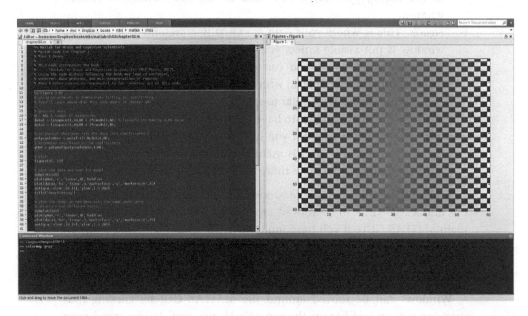

Figure 4.2
This shows my preference for organizing the MATLAB window.

4.3 Color-Coordinating MATLAB

If you like black font on a white background, then you will be happy with MATLAB's default color scheme. But if you feel that the MATLAB visual environment should reflect your colorful personality, then you might want to take a few minutes to change MATLAB's color preferences.

This can be done by selecting the *Home* tab on the top left, and then the *Preferences* button. The MATLAB *Preferences* window allows you to specify many options for how MATLAB interacts with you. Click on the option *Colors* in the left tab, and uncheck "Use default colors." Then modify the colors under *Programming Tools* (under the main *Colors* option). Pick some nice colors. Just make sure to pick colors that facilitate your experience—dark-gray text on a black background may look great on the cover of a 1980s metal band album, but it might not be the best color scheme for spending hours staring at MATLAB code.

You can also change other visual features here, including the font size and type (menu option *Fonts*). Don't use a font size that is too small, though—doctors say it's bad for your eyes. It is possible to change the font type; however, in programming it is best practice to use a monospace font such as Courier. This helps to keep the code visually clean and easy to read.

4.4 Where Does the Code Go?

All MATLAB code is run in the Command window. You can type in code manually or copy and paste code. This is useful when you have only a few short lines of code to run. In practice, however, you will want to save the code you are writing to keep it organized for the future. This is what scripts are for.

MATLAB scripts are just plain text (ASCII) files. They have an extension of .m, but they are normal text files and can be opened in any non-MATLAB text editor, even Microsoft Word. Having code in a script is nothing more than a convenient way to store a lot of code so you don't need to memorize and retype hundreds of lines of code each day. Just having code in the script, however, does not mean the code will run. The code will run only if it is evaluated in the MATLAB Command window.

Let's try an example. Open a new script (click on the *Home* tab, then *New Script*). Type figure, text(0.5,.5,'hello world!') in the script. Did anything happen? No, nothing happened. You wrote some code in the Editor but that does not run the code. You need to execute the code (we often say "evaluate" or "run" the code, perhaps because it's a bit macabre to go around constantly thinking that you should be executing things). Evaluating the code can be done using any of the following methods; try them all to gain familiarity.

1. Manually type the code into the Command window.
2. Copy the code from the Editor and paste it into the Command window (using Ctrl-C, Ctrl-V, or by selecting with the mouse and right-clicking).
3. Select the code in the Editor using the mouse and then right-click and select *Evaluate selection*.
4. Same as no. 3 but press the F9 key (depending on your keyboard and operating system, you might need to press an additional function key).
5. With the Editor as the active window, press Ctrl-Enter (Apple-Enter on a Mac).
6. Save the script in the current directory (call it "myFirstScript.m") and then in the command line, type `myFirstScript` and then Enter.
7. Same as no. 6 but press the F5 key (perhaps with an additional function key) with the Editor as the active window.

Taking any of the above actions will cause something to happen (if you take actions 1 or 2 and nothing happens, try pressing the Enter key to run the code). The code you wrote tells MATLAB to open a new figure and then draw the text *hello world!* at locations [0.5 0.5] on a standard Cartesian axis.

You can edit MATLAB code from any text editor you want. However, it's best practice to use the MATLAB Editor, for three reasons: First, you can run code directly without copy-paste annoyances (via actions 3, 4, and 5 above); second, you won't have to worry about other text editors putting hidden formatting characters into the script that will confuse MATLAB; third, you can take advantage of MATLAB's text coloring to help you read the code.

Normally, this would be a good time to save the script so you don't lose your work. But this one line of code is fairly easy to reproduce and is not very useful for real data analysis, so saving is unnecessary (although if you want to print it out and hang it on your mother's refrigerator, I guess that's probably okay). But you should always keep in mind that saving scripts each time you make a few changes is a good idea.

4.5 MATLAB Files and Formats

There are several native MATLAB file formats. Three of them you will encounter very often, and several others you should be aware of but will infrequently use, particularly as a novice-to-moderate MATLAB user.

1. filename.m: As mentioned above, .m files are plain ASCII text files that contain code. They can be scripts or functions (a subtle but important distinction that is explained later). They are very important because they contain a detailed record of exactly what code gets run and in what order.

   Because these are just plain text files, they do not necessarily need to be labeled ".m." You could also name them .txt, .dat, .code, or .anything. However, keeping MATLAB scripts and functions with a .m extension helps your computer and other users anticipate the contents of the file.

2. filename.mat: These files contain MATLAB-readable data. They are binarized and compressed, so you cannot see the contents of the file with a text editor. You can import .mat files into MATLAB through the Command window by typing `load filename.mat`, or by clicking on *Home* and then *Import data*, or by double-clicking the file in the Current Folder window. In practice, using the `load` command will be the most frequent method to import data. To create MATLAB data files, use the function `save`. You'll learn more about how to import and export data in chapter 8.

3. filename.fig: .fig is short for figure. Figures contain graphical representations of your data. They can also contain raw data and other information, although this is less efficient than saving your data in .mat files. .fig files are also used for creating and saving your own GUIs (see chapter 32).

4. .mex*: These are library files for compiled code. "Compiled" means that the code is binarized rather than printed in ASCII letters. Compiled code can run faster than noncompiled code, and it is also not viewable with text editors, thus making the algorithms hidden (this is often used in proprietary functions). Compiled code is operating system specific, meaning that a .mex file from a Mac will not work on a Windows computer, and a .mex file for a 64-bit machine is unlikely to work on a 32-bit machine.

Compiling your own MATLAB code is not discussed in this book, because in practice it is often more hassle than it's worth. This is because a lot of the time costs associated with neuroscience data analyses come from importing and exporting data and from using functions that are already compiled such as the fast Fourier transform. These processes are not significantly improved by compiling code. Compiling code is most often helpful in computational neuroscience, in which a few basic functions need to be called trillions of times.

4.6   Changing Directories inside MATLAB

MATLAB is always working "in" a directory (the terms *directory* and *folder* are used interchangeably). By default, MATLAB starts in a home directory in Windows and Mac, and in the directory from which it was called in Linux. Being able to change directories is important because your data and scripts are probably kept in different locations for different experiments. The address bar at the top of the main MATLAB GUI tells you the current directory. You can also type `pwd` or `cd` in the command window, and MATLAB will print its current directory.

There are three ways to change directories. You can use the mouse in the Current Folder window; you can use the "Browse for folder" button, which is the small folder icon immediately to the left of the address bar; and you can change directories in the command line using the Unix function `cd`. Many Unix commands also work in the MATLAB command, so there is some overlap between learning Unix and learning MATLAB.

You need to be comfortable using the `cd` function from the Command window, because you want your analysis scripts to be able to change directories. To move MATLAB into a subfolder of the current directory, type `cd subfoldername`. Typing `cd ..` will take you up one folder in the folder structure. You can also use absolute directory names: `cd C:/Users/mxc/folder`. You will learn more advanced ways to use `cd` in chapter 8, but it's good to gain some familiarity with the command already.

To create a new subfolder in the current folder, use the function `mkdir` (make directory). For example, type `mkdir tempfolder`. Then check in a File Explorer that a new folder appeared. You can also type `cd tempfolder` in the MATLAB Command window. The command `ls` will list all of the files in the current directory.

4.7   The MATLAB Path

A little experiment will help teach you about the MATLAB path and how MATLAB knows where to find files. To begin this experiment, you need to cd MATLAB into a folder that is not the default starting location. For example, you can create a new folder on the desktop and cd MATLAB into that folder.

Open a new script. You already know how to do this via the menu. You can also type edit in the Command window or you can press Ctrl-n. We will write a script to generate some random numbers, compute their average, and then display the average. Type the following code in the script:

```
numbers = randn(10,1);
average = mean(numbers);
disp([ 'The average is ' num2str(average) ])
```

If you are really new to programming and have no idea what you just typed, then don't worry, it will all become clear over the next few chapters. Briefly: The first line produces 10 numbers drawn at random from a normal distribution and stores those numbers in a variable called numbers; the second line computes the average of those 10 numbers and stores that average in the variable called average; the third line prints a message to the Command window that includes that average. You can run the code and see that it prints the average of 10 randomly generated numbers. To convince yourself that these are randomly generated, run the code multiple times. (The function randn generates random numbers around a mean of zero, so don't be surprised when the average is close to zero each time you re-run the code.)

Save the script in the current directory using the name "showmean.m." Now you can call that script from the command line by typing showmean and then press the Enter key. Notice that you are not running this script from the code in the Editor window. In fact, you can close that file in the Editor window and run it again from the Command window. When you type showmean in the MATLAB Command window, MATLAB reads that text file and evaluates it line by line. It doesn't matter if the file is currently open in the Editor window.

Now create a new folder called "temp" in the current directory and cd into that folder. Type showmean again. What happened? You got an error about Undefined function or variable 'showmean'. That doesn't make sense! You just created and evaluated the file, so why did it crash now?

The answer is that MATLAB is no longer able to "see" that file. MATLAB cannot see all of the files on your computer. It can only know about files that are in its *path*. A path is a list of folders that MATLAB searches through to find files. The current directory is always in the MATLAB path, and there is also a saved path that you can edit.

Now we will change the MATLAB path to include the folder in which the file showmean.m lives. Click on the menu *Home* then *Set Path*. You can see a long list of folders; this list is MATLAB's path. **Warning! Do not remove folders from the MATLAB path unless you are certain you know what you are doing.** Instead, you want to add a folder. Add the folder that contains showmean.m and click *Close* at the bottom of the window. You should not click the *Save* button. This is a bit confusing. The *Close* button really means "save the path for this session but do not permanently save this new path," and the *Save* button really means "save the modified path forever." If you click *Close*, your changes will take effect in this MATLAB session, but those changes will be lost when you close MATLAB and open a new MATLAB session.

Now that the folder is in MATLAB's path, try again to run `showmean`. Now it works! The reason why it works now is because the script showmean.m is in a folder in MATLAB's path, and so MATLAB can find that file even though you are not in the folder where the script lives.

You can modify the path from the command line or in a script using the functions `addpath` and `genpath`. Similar to clicking *Close* in the menu, adding folders to the MATLAB path from the command line will not permanently change the MATLAB path. It is generally a good idea not to modify the MATLAB path permanently, and instead to add folders to the path from scripts. This is particularly the case when using MATLAB on a shared computer or server. On server computers that allow different users, you can store your own specific modifications to the path. You'll read about this in the MATLAB etiquette section later in this chapter.

4.8 Comments

Comments are the second most important aspect of programming (after the code itself). A comment is a piece of text that is embedded in the script but is not evaluated by MATLAB. In MATLAB, comments are indicated by a percent sign (%). Comments can be placed after code, but code cannot be placed after comments. Always use comments. But keep comments brief. Examples:

```
% The next line of code computes the average
average = mean(data); % data is a variable
```

If you have many lines of code to comment, you can use block comment formatting:

```
%{
    These are comments but have no percent sign.
    Instead, they are encased in curly brackets.
%}
```

4.9 Cells

A cell is a grouping of related lines of code. For example, you might have a cell for finding and importing the data, another cell for filtering the data, another cell for separating the data according to different experiment conditions, and so on. Cells are demarcated by a double-percent sign (%% at the start of the line; %%% will not create a cell). In the *Colors* option of the *Preferences* window, you can define active cells to be in a different color. Cells are a great way to separate blocks of code, so make sure to use them. If you think about programming as a written language, then cells are the paragraphs.

One useful feature of cells is that you can run all of the code in a cell—but none of the code outside the cell—by pressing Ctrl-Enter (Apple-Enter on a Mac). That saves a lot of time and is really convenient, particularly when you want to test some code after changing parameters or input files.

4.10 Keyboard Shortcuts

MATLAB keyboard shortcuts can save you time and reduce the hassle of having to switch between the keyboard and the mouse (but don't keep your hands on the keyboard too long—your supervisor and I don't want you to get repetitive stress–related wrist injuries). There are several built-in keyboard shortcuts, and you can define new keyboard shortcuts in the *Home*, *Preferences* menu window. Below is an explanation of some of the most commonly used keyboard shortcuts.

• Ctrl-Enter: When pressed in the Editor, this will evaluate all of the code within the highlighted cell.
• Tab: The Tab key can be used to move four spaces over, but it can also be used to complete a function, variable, or file name. In the Command window, type showme and then hit the Tab key. MATLAB autocompletes the script name showmean, because this is the only variable or function or file

that starts with showme (assuming you still have this script and it is still in your path; if not, you can follow the earlier section to create this file and put it in MATLAB's path). If there are multiple possible options, MATLAB will give you a list. For example, type sh and then Tab. You will see an alphabetical list of all functions, variables, and files that start with sh. Tab-complete also works with function inputs, including cd (change directory) and ls (list files in the current directory).

• Ctrl-r: When pressed in the Editor window, this will comment out the entire line. You can comment multiple lines at a time by selecting several lines and then pressing Ctrl-r.

This and the next few keyboard shortcuts are considered the "Windows defaults"; on Mac and Linux operating systems, the keyboard shortcuts are a bit different. In the *Home, Preferences, Keyboard, Shortcuts* menu option, you can select whether to use the Windows Default or the Emacs Default (Mac and Linux). I prefer the Windows default shortcuts, even when using Linux and Mac.

• Ctrl-t: Uncomment a commented line or multiple selected lines of code.
• Ctrl-i: "Smart indent." This automatically puts spaces into the selected lines of a script in order to improve readability. For example, smart indent would turn the code

```
for i=1:3
for j=i:10
disp(i*j)
end
end
```

into a more readable version of the same code:

```
for i=1:3
    for j=i:10
        disp(i*j)
    end
end
```

In general, you should try to make your code look visually clean and organized. It will help you and others read and understand the code. This is particularly important when you have multiple nested loops and other command statements.

• Ctrl-c: The same as in many other programs, Ctrl-c will copy selected text. However, if no text is selected and if you press this combination in the

Command window, Ctrl-c will break whatever MATLAB is doing and give you back control of the command. This is useful if you run a long script and realized you made a mistake or if you are stuck in an infinite loop and want to break out of it. (Wouldn't it be nice if life had a simple keystroke for this situation?)

• F9: If you highlight some code in the Editor window and press F9, the highlighted code will be evaluated in the Command window. If you are using a laptop or a Mac, you might need to press some other key combinations (e.g., function keys) to get the computer to recognize that you want to use the F9 key instead of whatever else is also mapped onto that key, such as "play-next-song."

• F5: Pressing F5 will run whichever script is currently active in the Editor window. Note that F5 runs the entire script, equivalent to calling the script from the command line as you did with the showmean example earlier. Pressing F5 automatically saves the script first, so it is redundant to save and then press F5.

• Up-arrow: In the command line, press the up-arrow key to see a list of previous commands, in reverse order from when you typed them. Then keep pressing the up-arrow key or down-arrow key to scroll through them. Once you've found a command you want to run again, you can press the Enter key to run the code immediately or you can press the Tab key to print it in the command line without executing it.

You can also filter through previous commands by typing the first few letters of the command. For example, type sh in the Command window, and then press the up-arrow key. Now you see only the previous commands that start with sh. If you are using MATLAB for the first time and followed this chapter, the only previous command that matches this filter is showmean. If you have a tendency to type out your frustrations, you might see some other previous attempted commands.

• Standard keyboard shortcuts: Many other system-wide shortcuts also work in MATLAB, such as Ctrl-s for save, Ctrl-arrow to skip forward/backward by one word, Home/End keys, and so forth.

4.11 Help Box and Reporting Variable Content

MATLAB provides two features for quick help while you are programming. One is called "function hints" and is a small window that tells you how to complete functions in terms of the number and types of inputs (figure 4.3). You can close the little window by hitting the Esc key, and you can disable

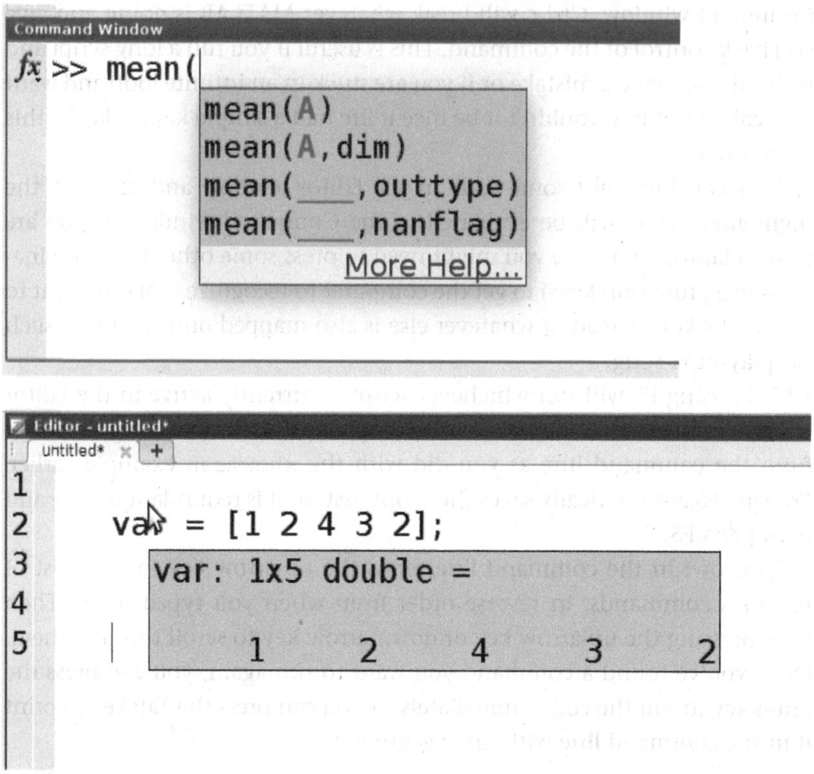

Command Window

fx >> mean (

```
mean(A)
mean(A,dim)
mean(____,outtype)
mean(____,nanflag)
            More Help...
```

Editor - untitled*

untitled* × +

```
1
2        var = [1 2 4 3 2];
3        var: 1x5 double =
4
5                1       2       4       3       2
```

Figure 4.3
MATLAB can provide hints while you are typing. Function hints (top panel) remind you of the inputs to a function you are currently writing. Variable content hints (bottom panel) tell you the size, type, and contents of a variable in the current workspace. These hints can be disabled if you find them annoying.

all function hints in *Home, Preferences, Keyboard*. (I find this option distracting and leave it disabled.)

A second useful tool is used to show the contents of a variable. If you hover the mouse on top of a variable name in the Editor window, a small window will show you information about that variable (similar to the information you get from typing whos <variable>). This is useful because you can see the contents and sizes of variables without having to type whos or evaluate the variable. It works only after you create the variable; that is, MATLAB will not guess the content of a variable based on the code, it will only print information that is contained in its buffer. If you don't see these little windows, check the "Enable datatips in edit mode" option in *Home, Preferences, Edit/Debugger, Display*.

4.12 The Code Analyzer

MATLAB has a built-in program that analyzes the code in the Editor window. It will give various warnings and error messages. A warning is given when the code analyzer detects that some piece of code will most likely work, but could be problematic in some circumstances or could be optimized to improve efficiency. For example, write the following code in the Editor:

```
for i=1:3, asdf(i)=i; end
```

You will see a red squiggly line under the variable asdf, and if you hover the mouse over this variable, you will see a warning message about the variable asdf growing with each iteration, along with a suggestion to preallocate for speed. This is MATLAB's way of telling you that you should initialize variables to improve speed and prevent possible unwanted effects. You will learn in chapter 5 what this warning means and why MATLAB is right to warn you.

To tell MATLAB to ignore this warning, you can right-click on the underlined piece of code and select for the warning to be ignored, either on this line or in the whole script. Warnings in the code have no consequence for whether MATLAB will run the code, so forcing MATLAB to ignore the warning has only aesthetic value, not programming value. Other warnings are issued if the function you are using will change in future versions of MATLAB.

If the code analyzer gives an error message, it means that MATLAB will crash on that line. Sometimes, when you are in the middle of programming, you will get error messages because lines or loops are incomplete. That means MATLAB is analyzing your code faster than you can type. You can ignore these error messages while you are still programming (or type faster).

The locations of warnings and errors are indicated in the scroll bar as orange (warnings) and red (errors) horizontal lines. By hovering the mouse on top of these horizontal lines, you can see what the message is, and by clicking on the horizontal bar, MATLAB will take the cursor to the offending line.

It is good practice to inspect all of the code analyzer's warnings. Most of these warnings contains useful suggestions for improving code and avoiding mistakes. But do not feel obligated to correct all of them to MATLAB's satisfaction (MATLAB is satisfied when the colored square on top of the scroll bar is green). The code analyzer is not perfect, and its suggestions are not demands. There are times when the warnings can be ignored and when the offending code is actually the best or the only way the code can be written.

4.13 Back Up Your Scripts, and Use Only One Version

Needless to say, you should back up your analysis scripts. I will go a bit further and state that it's really stupid not to back up your analysis scripts. They are very small and therefore very easy to back up.

However, it is dangerous and sloppy to keep multiple different versions of the same script on different computers. You don't want to be in a situation where you change different parts of the script on your laptop at home, on your desktop computer in your office, and on your USB stick on a different computer. You'll have three slightly different scripts that do three slightly different things, and it can be quite a headache to figure out what was most recently modified in which version.

The best solution to this problem is to use a cloud-based sharing/ archiving system such as Dropbox, Copy, Box, Google Drive, Github, or any other similar service. This solves two problems at the same time: It allows you to use one version of the script on all of your computers (or any computer with Internet access), and it automatically backs up your files every time it updates (which, for a continually Internet-connected computer, is each time you save the file) so you don't have to worry about remembering to back up your files manually. Better still is to have a shared folder for the entire research group, so that everyone can keep their analysis scripts in a centralized location. This also facilitates sharing code with your colleagues.

While we're on the topic of backing up, don't forget to back up the experimental data as well. Usually, experimental data are too big for cloud-storage solutions, unless you have paid service subscriptions or an unusually small amount of data. The data should be stored in at least two physically separate locations, and ideally one of these locations is a RAID storage or other similar backup device. Losing analysis scripts is annoying and time-consuming; losing data is terrible and potentially suspicious if the data are lost during or shortly after publication.

4.14 MATLAB Etiquette

If you are the sole user of MATLAB on your computer, then go ahead and personalize anything and everything. But if you use the same copy of MATLAB that other people share (this happens often in computer clusters, network servers, and small research labs), use some etiquette and be nice to the other MATLAB users.

Mainly, MATLAB etiquette means not changing the path, colors, or other settings. If you want to have your own personalized default settings, put all of your settings into a script called startup.m, and put this file in your local MATLAB home directory. When MATLAB starts, the last thing it will do before letting you use it is to search for and run a file called startup.m. You can also add motivational encouragement to the startup.m so it prints out a friendly message each time MATLAB starts. Try adding the following code to the end of your startup.m, then open a new MATLAB session:

```
disp('Stop facebooking and get back to work!!');
```

The MATLAB local home directory is something like C:/Users/<username>/ Documents/MATLAB/ for Windows, /home/<username>/.MATLAB for Unix systems, and is something like /Users/<username>/Documents/ MATLAB for Mac.

Finally, if you are using MATLAB on the same computer that other people are using, for example a shared server computer in your department or research group, then be mindful of when to run the big memory-intensive analysis scripts. If you need a lot of computer resources for an analysis that takes 7 hours to run, your colleagues will appreciate it if you start the analysis at 8 p.m. rather than at 8 a.m. That's called MATLAB Karma, and ancient superstitions suggest that building up MATLAB Karma reduces the number of programming errors you make. It's probably just legend, but to my knowledge this has not been definitely scientifically disproved.

5 Variables

Unless you have superhuman memory abilities and superhuman typing speed, you are going to need variables. Variables are basically just convenient placeholders for information (mostly numbers, sometimes letters). There are several types of variables and ways to combine variables. This chapter will introduce you to working with variables of different types. Learning how to work with different types of variables is the first hoop to jump through on your way to becoming a good programmer.

5.1 Creating and Destroying Variables

Creating variables in MATLAB is simple: You assign a value to them. For example, to create a variable called `asdf` that is a placeholder for the number 7, simply type the following (not the `>>`, just the text thereafter):

```
>> asdf=7
asdf = 7
```

MATLAB, being a polite program, has confirmed what you just wrote: The variable `asdf` is a placeholder for the number 7. Each time you type in some code, MATLAB will confirm by printing out its interpretation of what you typed in. This can get annoying after a while, and it can also considerably slow down your analyses, because it takes time to print out all of those confirmations. Thus, in practice, it is best to ask MATLAB not to confirm each line. The semicolon is MATLAB's polite way of letting you say "shut up":

```
>> asdf=7;
>>
```

In general in this book, MATLAB code will be printed without the command-line prompt (">>"). The prompt is used only when necessary to illustrate output formatting.

You can reassign any value to an existing variable in the same way that you first created the variable. Keep in mind that the value previously assigned to the variable will be lost.

```
asdf=7.6
```

All variables take up space in MATLAB's memory buffer. Often, you will want to clear variables out of the memory buffer. You might want to clear variables to create space for other variables, you might want to clear variables to avoid confusing the contents of the variables across different data sets, or you might want to clear variables just because you don't like them anymore (MATLAB is sometimes better than real life).

To clear a variable from MATLAB's buffer, type `clear asdf`. You can also type `clear as*` to clear all variables that start with `as`. Type `clear` or `clear all` to clear all of the variables in MATLAB's buffer.

5.2 Whos Are My Variables?

To see a list of the variables in your workspace, use the function `whos`:

```
>> whos
>>
```

If you've followed this chapter so far, you won't see anything printed after `whos`, because we cleared out all of the variables from MATLAB's buffer.

```
>> asdf = 10;
>> whos
Name   Size   Bytes   Class    Attributes
asdf   1x1    8       double
```

The function `whos` also shows you important information about the type and sizes of the variables. As you learn more about MATLAB, these categories of information will make more sense. In general—during learning and when writing analysis code—it's a good idea to type `whos` often to make sure that variables have the sizes and properties that you expect them to have.

5.3 Variable Naming Conventions and Tips

There are a few rules, and a few guidelines, for naming variables. Here are the important rules.

1. Variables cannot contains spaces. This is sensible; otherwise, it would be impossible to know how to parse the code. Instead of spaces you can use the underscore character: `name_of_variable`.

2. Variables cannot start with numbers. The exception to this rule is the imaginary operator, which is the square root of –1. For the imaginary operator, use `1i` or `1j`. You'll learn more about imaginary numbers and how they are used in data analysis in chapter 11. Variable names may contain numbers, though, and this can be useful. For example, you can have variables named `frequency4plotting` or `data2export`.

3. Variables cannot contain most non-alphanumeric characters. Non-alphanumeric characters include !, @, #, $, %, ^, &, *, (,), -, and +. Most non-alphanumeric characters have other uses in MATLAB, such as power (^), function handle (@), or comment (%). The main exception to this rule is the underscore (_), as mentioned above.

Those are the rules, and MATLAB will give an error if you try to break them. In addition, there are a few guidelines to improve your programming experience and make your code easier to read.

1. Don't use the same name as existing functions. In the previous chapter you were introduced to the function `mean`, which computes the average of the input numbers. Technically, you can also use `mean` as a variable name (such as `mean = 7;`), but this will take precedence over the *function* `mean`, which is likely to cause errors or at least confusion. If you would like to use a variable name and are unsure whether it is also a function name, type `which variablename`. The `which` command tells you where a function is located on your computer. If MATLAB tells you that the function or file name is not found, you are safe to use it.

2. Use capital letters carefully. MATLAB variables are case-sensitive. The variables `asdf` and `aSdf` are completely different. For computers, this distinction is easy because the ASCII codes for `asdf` and `aSdf` are different. But for humans, the distinction can be confusing. Use capital letters in variable names only when it improves readability; for example, `dataToExport` instead of `datatoexport`.

3. Use meaningful variable names. Avoid using variable names like `m` or `x`, particularly if you are going to use the same variables throughout a long script. The best variable names are those that provide information about the content of the variable. Examples of good variable names include `averageSpikes`, `frequencies2analyze`, `times2save`, and `conditionLabels`. (In this book I sometimes use variable names that

are shorter than optimal, in the interest of preventing code from over-
flowing into the following line.)

4. Don't use variable names that are really long. MATLAB has a maximum
    number of characters for variable names, which is 63 in version R2015a.
    But you should try to keep the length of variable names shorter
    than that. Really long variable names make really long lines of code,
    and really long lines of code are difficult to read and difficult to debug.
    Try to make variable names as short as possible while still being
    meaningful.

There are a few other variable naming tips that will be presented throughout this book in the context of other material.

5.4 Variables for Numbers

This is the easiest and most commonly used type of variable. You've
been using variables for numbers since you first learned algebra. The example earlier in this chapter with the variable asdf was a numeric variable.
Variables for numbers can contain single numbers (also called scalars),
arrays, matrices, and high-dimensional matrices (greater than three dimensions). Chapter 10 provides an introduction to matrix algebra and will
describe matrices in more depth, but you are already familiar with the idea:
A matrix of numbers is a list of numbers that have some geometric organization, like a line (one dimension), box (two dimensions), or a cube (three
dimensions).

Arrays are one-dimensional matrices and can be created using square
brackets:

```
anArray = [1 3 2 5 3 6 7 9];
```

Two-dimensional matrices can be created by using the semicolon to
indicate where the next row of numbers begins:

```
aMatrx = [1 3 2; 5 3 6; 7 8 9];
```

What is the size of aMatrix? It has three numbers and then a semicolon,
then another three numbers and a semicolon, then another three numbers.
So it is a 3 × 3 matrix. You can confirm this by typing whos or whos
aMatrix.

Try to create aMatrix, but put the first semicolon between the 3 and 2
instead of between the 2 and 5. MATLAB gives an error because each row of
a matrix must have the same number of elements; matrices cannot be
jagged.

Once you have some variables, you can use them in mathematical expressions exactly like you would in paper-and-pencil algebra. Most mathematical operations use the keys you would expect:

```
a = 4;
b = 3;
b^2
a+b
a*b
(a*4) + (b/2)
a*4+b/2
```

What is the difference between the last two lines of code? For MATLAB, there is no difference; those two lines produce exactly the same result. But which line is easier for a human to read? Clearly, it's the penultimate line, because the parentheses and spaces provide visual grouping that helps you understand which variables go together. The lesson here is that spaces and parentheses—even when technically unnecessary—should be embraced as a programming technique to help make your code easier to read.

Numbers can be represented in MATLAB in different formats, including integers or floating-point decimals. Here I'll discuss "doubles" and "singles," which are both floating-point types. MATLAB uses 64 bits of detail to represent double-precision numbers and 32 bits of detail to represent single-precision numbers. Most numeric data such as time series data are represented as floating-point numbers, which means they include decimal points. Images are often integer-based (no decimal points). Some MATLAB functions work better with or can only accept floating-point numeric input. Thus, in practice it is often useful and occasionally necessary to convert your data to double-precision floating-point, which can be done using the function `double`:

```
data = double(data);
```

If you have very large data sets, it is often useful to convert the data to single precision before saving. This can reduce file size by up to a factor of 2. "But space is cheap" people often retort. Yes, hard drive storage space can be cheap, but reading files into and out of MATLAB costs time. If your analysis scripts involve a lot of importing and exporting of data, then reducing file sizes is not only about reducing data storage but also about reducing analysis time.

However, converting from double- to single-precision format involves some data loss. The question then is whether any actual information is lost. Extreme precision (greater than 32 bits) is often unnecessary in neuroscience data, considering the level of noise that is present. An example of significant information loss by changing from double to single precision is with magnetoencephalography data that are stored in teslas (T). In this case, the data are in the range 10^{-14} to 10^{-16} T, and converting to single precision may involve information loss. However, in this case, you could multiply the original, double-precision data by, for example, 10^{15}, and then convert to single precision. Most other neuroscience measurement ranges are not affected by reducing the precision to single. You should check carefully whether you will lose information, but if not, then converting to single precision is something to consider when saving data to disk.

As mentioned earlier, several MATLAB functions require double-precision data, including many filtering functions. If you convert your data to single precision for saving, remember to convert it back to double precision for analyses.

5.5 Variables for Truth

MATLAB has a special type of variable for "logical" (sometimes also called "Boolean") values. Logical variables contain only two states: true or false. Logical variables are often used in data analysis; for example, as a toggle to decide whether to run a piece of code or to mask a large data set to extract information out of a subset of the data. Logical variables contain entries that look like normal numbers, but MATLAB interprets the number zero to mean "false" and any other number to mean "true."

To create a logical variable, you can use a double-equals sign, which is MATLAB's way of letting you ask a question. Try this:

```
isEverythingOK = 1==2
```

This looks confusing at first. We are using 1==2 to *ask the question* "Does 1 equal 2?" The answer can only be true or false. MATLAB then stores that answer in the variable called isEverythingOK (the single equals sign). You can type whos to see that isEverythingOK is a logical variable (i.e., its class is type logical).

The MATLAB treatment of logical variables can be confusing. Sometimes it is possible to use a logical variable as a normal number. For example, if you type a = isEverythingOK + 3; and type whos, you will see that the variable a is a double, which means that MATLAB automatically converted isEverythingOK to the *number* zero and then added 3. You can also confirm this by setting isEverythingOK to be the result of 1==1, and then test that "true" plus 3 is 4.

Logical variables can also be created using the functions true and false:

```
var = true; % or var=false;
var + 4
```

In general, avoid the temptation to rely on MATLAB's ability to convert logical variables to numbers on the fly. It can make your code more flexible when used correctly, but it can also produce confusion and errors. You will see a few examples in later chapters where this trick is used to simplify some code, and other examples where it will produce errors.

5.6 Variables for Strings

Variables can also store characters. This becomes useful for storing file names, experiment labels, and so on. Use single quotes to let MATLAB know you want to use strings:

```
nameVariable = 'Mike X Cohen@#$%';
```

Note that spaces and non-alphanumeric characters are fine here; unlike in variable names, these are just normal letters. Numbers inside quotes are not recognized as numbers, but rather as strings. And MATLAB in turn stores strings according to their ASCII codes.

```
string4 = '4';
number4 = 4;
number4 + string4
```

MATLAB doesn't give an error here, but it gives an answer that you might not expect: 4 + '4' is 56. You are probably now wondering what crazy universe MATLAB comes from to produce 4 + 4 = 56. The ASCII code for the *string* '4' is the number 52, and so 52 + 4 = 56. Try this in MATLAB: 4 + '1234'. Until a few seconds ago, you would have expected to see the number 1238, but instead you get four numbers (53, 54, 55, 56). Now you know that MATLAB thinks of '1234' as containing four separate characters (strings, not numbers) that are represented using four separate ASCII codes.

If you feel confused now, then don't worry; that's normal. The situation of 4+'4'=56 is an example of one of the worst mistakes you could make when programming in MATLAB, because MATLAB does not do what you expect it to do, but it also does not give an error because you did nothing illegal. The lessons here are (1) always do sanity checks, and (2) always be mindful of variable types. Needless to say, you should avoid at all costs performing mathematical operations between strings and numbers.

5.7 Variables for Cells

What if you want to store numbers and strings in the same variable? You can use square brackets to concatenate strings, so you might initially try something like this:

```
a = [52 'hello' 52]
```

And again you get something that causes confusion. In the previous section, you learned that the ASCII code for the string '4' is the number 52. This means that to MATLAB, [52 'hello' 52] is the same thing as ['4hello4']. Assuming that isn't some offensive Internet meme (my apologies if it is), 4hello4 is not a meaningful statement.

Instead, to have different types of data in the same variable, we use cells. You can think of a cell like a cage. It isolates and stores something, and different cages can have different sizes and store different types of beasts. Cells are indicated in MATLAB using the curly brackets, {}. Below is an example of a cell array:

```
celery{1} = 52;
celery{2} = 'hello';
celery{3} = 52;
```

Now if you type whos, you will see that variable celery is a 1 × 3 cell array. You can type celery in the Command window to see the contents of the cells.

Cells, like number variables, can be multidimensional matrices. You can see this by initializing an empty two-dimensional cell array:

```
celery2d = cell(2,4)
```

Each element (cage) in the cell array can be accessed by indexing, which you will learn about soon.

5.8 Variables for Structures

Structures are perhaps the most useful variable type for storing a lot of different kinds of information. You can think of a structure as if it were a small database. Each property is called a "field" and is indicated by a period. Let's call our structure mouse.

```
mouse.name = 'mickey';
mouse.surgeryDate = '24 March';
mouse.geneType = 'wildtype';
mouse.age = 65; % in days since birth
mouse.numElectrodes = 3;
```

Typing whos reveals that mouse is a 1 × 1 structure. You may be tempted to disagree with this size, because we typed in five pieces of information. But all of these fields are part of one instance of mouse. Soon we'll create additional instances. You can access the data inside each field as if they were normal variables. For example: mouse.age + 10.

You can already see that structures are useful ways to store different kinds of information. Structures can also come in arrays and matrices, which makes them powerful tools for storing a lot of information.

```
mouse(2).name = 'minney';
mouse(2).surgeryDate = '2 April';
mouse(2).geneType = 'GAD2Cre';
mouse(2).age = 82; % in days since birth
mouse(2).numElectrodes = 6;
```

Now typing mouse will not show the contents, but rather all of the field names. You can type mouse(1) to see the data from the first element in the structure mouse and mouse(2) to see the contents of the second element. In addition to listing all properties of one element in the structure, you can also extract information from one field over the different elements of the structure array.

```
mouse.age
mouse.geneType
```

Note that the results are printed out as separate answers, one for each element of the structure. You can use square brackets, [], or curly brackets, {}, to concatenate them. Try the above two lines again, encasing each in square brackets or in curly brackets. What are the differences and when would you want to use square brackets versus curly brackets?

5.9 The Colon Operator

The colon operator is not a variable per se, but it is important for creating and accessing parts of variables. The color operator, very simply, is a way to count numbers automatically. Type the following into the MATLAB command:

```
1:5
1:2:5
```

What is the difference between these two lines? I'm sure you noticed that the first command counts by 1 while the second command counts by 2. Implicitly, the command `1:5` is the same thing as `1:1:5`. In English, the command `x:y:z` translates to "count from x to z in steps of y." The stepping part need not be integers. Try evaluating `1:.5:5`. You can also try `1:.000000001:5`, but this requires some patience. If you run out of patience, press Ctrl-c (or Apple-c) in the Command window. Counting will never go higher than the number specified; `0:2:5` will stop at 4 rather than at 6. Counting backward involves stepping by `-1.5:-1:-5`.

You can use square brackets to concatenate numbers defined by the colon operator and individual numbers. For example, if you want the numbers 1 to 3, 6, and 8 to 13:

```
[1:3 6 8:13]
```

5.10   Accessing Parts of Variables via Indexing

Big variables are great, but sometimes you want to access only part of a variable. Let's say you have a variable that ranges from 1 to 200 in steps of 2:

```
numbers = 1:2:200;
```

If you want to know the number in the fifth position of that array, use parentheses: `numbers(5)`. This is called *indexing*: You are indexing the fifth element of the variable `numbers`.

Now combine your new colon operator skills with indexing. How would you access the third, fourth, and fifth elements in the variable `numbers`?

Of course you know the answer: `numbers(3:5)`.

If you want to access elements 1–3, 6, and 8–13, use square brackets inside the parentheses: `numbers([1:3 6 8:13])`.

You can also combine logical arrays with indexing to produce something called logical indexing. Let's say we want to get all of the elements of `numbers` where the index is larger than 50. That means we want to ignore the first 50 elements of `numbers`, and extract only elements 51 through the last (there are 100 elements in total). Of course you could do this the easy way: `numbers(51:100)`. But the hard way is much more fun (and will be much more useful for many applications in the future).

Before writing the solution in code, let's think about how this should work. To use logical indexing, we want MATLAB to associate a "true" or a "false" with each element in the array. For example:

```
c = 1:3;
c([ true false true ])
```

The second statement returns [1 3], because we used logical indexing to extract the first and third elements while ignoring the second element. Going back to the original problem, what we want is an array of Booleans such that the first 50 are false and the next 50 are true. One solution to this problem is

```
logicalIdx = 1:100>50
```

The abbreviation idx at the end of logicalIdx is often used for indexing variables. Now use the variable logicalIdx inside the variable numbers:

```
numbers(logicalIdx)
```

And voilà, we have only the elements of the variable numbers that are contained in positions 51 to 100.

This works because although logicalIdx might look like it contains zeros and ones, those are really falses and trues. There are two ways that this can produce an error. First, you can actually use zeros and ones instead of falses and trues:

```
numbers(double(logicalIdx))
```

Now you get an error because MATLAB cannot access the zeroth entry of a matrix (MATLAB starts counting at one, not at zero, just like normal human beings).

Next let's just take all ones. Re-create the logical array to be logicalIdx=1:100>0. This variable is now all trues, because all of the numbers between 1 and 100 are greater than zero:

```
numbers(double(logicalIdx))
```

Now what happens? You get all ones. A bit weird. To see if this can make more sense, try evaluating numbers(double(logicalIdx)+1), and then using +2 instead of +1, and so on.

What happened here is that you are accessing only the first element (or the second or third), but you are re-accessing it 100 times. It's equivalent to numbers([1 1 1 1 1 1 1 1]) but with 100 ones. It can get confusing, so you should always be careful with this. Don't worry, you'll have plenty of opportunities to learn about indexing throughout this book.

5.11 Initializing Variables

Initializing variables means reserving a space in MATLAB's memory buffer for the variable before you actually create it.

Think of MATLAB's memory buffer as a restaurant. It's a big restaurant, but it's also very popular. Initializing variables is like calling ahead of time and making a reservation for a group of 20. The restaurant will set a table with empty chairs, and whenever you and your friends arrive (they don't all need to arrive at the same time) there are reserved places for them. Now imagine that you don't make a reservation but just show up by yourself. No problem, the restaurant will make a little table just for you. Then one of your friends shows up, and the waiter brings out another small table and puts it next to yours. Then three more of your friends show up, and the waiter has to bring more tables and chairs. And so on until you get to 20 people. In the end, it's possible to end up with the same table of 20 people, but it could take a lot longer (20 is a cute number; in practice you might have 20 billion friends showing up one at a time, each demanding a table and chair). And if the restaurant is nearly full, your friends who show up late might not be allowed in. Or maybe the entire restaurant will crash and you'll have to reboot.

There are other reasons why initializing variables is important, but appreciating these reasons requires some MATLAB skills that are beyond the scope of this chapter (and they don't really fit into the restaurant analogy). Among the many reasons to initialize is that it forces you to think carefully about what you expect to be contained inside variables. And the more you think carefully about your code, the more likely it is to be efficient and devoid of errors.

Initializing variables is very easy. You simply create the variable you want before you start populating it. You can initialize variables to be all zeros (generally the most common way to initialize), or all ones, or NaNs (NaN is "not a number"; this is the result of division by zero), or logical trues, or many other things. The examples below illustrate the varied ways to initialize variables:

```
mat1 = zeros(3);
mat2 = zeros(3,3);
```

What is the difference between mat1 and mat2? What does this tell you about how this MATLAB function works? Below are a few additional ways to initialize variables:

```
mat3 = ones(2,4);
mat4 = 7.43*ones(2,4) + 10;
mat5 = true(1,3);
mat6 = nan(8,2,3,4,5,2);
```

In practice, it is usually best to initialize variables to zero or NaN, which makes it easy to identify when part of a matrix was not filled in.

5.12   Soft-coding versus Hard-coding

If you are thinking that this section will be about softcore and hardcore [insert impolite term here], then you are about to be disappointed. Soft-coding and hard-coding refer to using variables to define parameters (soft-coding) versus typing in specific parameter values as numbers (hard-coding). Here is an example of soft-coding:

```
nbins = 10;
<200 lines of code...>
results_mat = nan(nbins,1);
<300 more lines of code...>
for bini=1:nbins
    <more code...>
end
```

The parameter nbins is soft-coded, meaning that you assigned the parameter value (in this case: 10) to a variable at the top of your script, and then the rest of the script uses this variable, rather than using the number 10. Here is the same script hard-coded:

```
<200 lines of code...>
results_mat = nan(10,1);
<300 more lines of code...>
for bini=1:10
    <more code...>
end
```

This second set of code would produce exactly the same result as the first set. But what if you wanted to change the analysis to 12 bins instead of 10? You would have to search through the entire code for every place where the number 10 was used. And it could get confusing because there might be other parameters or variables set to 10 and you would need to make sure to change only the relevant ones. And if you missed one of those instances, MATLAB might crash or—much worse—it won't crash but your analysis will be flawed.

In the soft-coding example, changing the number of bins is very easy. You just change the one parameter at the top of the script, and voilà, the rest of the script is adapted with no further adjustments.

You should soft-code as much as possible, even if it seems unnecessary because your script is small. Analysis scripts have a tendency to grow, and what may seem like one little innocent hard-coded parameter in the early development of an analysis script may turn into a huge headache in the end.

5.13 Keep It Simple

MATLAB has more options for variable types than discussed in this chapter. Here you learned about the variable types most commonly used in neuroscience analyses (floating point, string, logical, cells, and structures). There will be situations in advanced applications in which other variable types are needed. But whenever possible, try to stick to the basic variable types.

Be mindful of which variables have which types. You saw how mixing variable types in mathematical equations can be legal but confusing (e.g., 4+'4'). Unless you have a strong reason otherwise, keep numbers as double precision (unless saving in single precision after checking that information is not lost), text as strings, and Booleans as logical.

Finally, always keep in mind that programming is just a tool to facilitate science, not an end in itself. Try to keep your code as simple as possible. Brain science is already challenging enough as it is.

5.14 Exercises

1. Initialize a 30 × 4 matrix. Then add a value into the (31,10) position. Type whos after each step. What happened, and what does this tell you about how MATLAB deals with matrix sizes and adding new content to existing variables?

2. One of the important concepts in programming is to *use the output of one command as the input into another command*. The code below uses

square brackets to put the ages of the mice into one array called ages, and then to convert that variable to logical by testing whether the ages are less than 70 days. Perform the same operation using one line of code.

```
ages = [mouse.age];
ages = ages<70;
```

3. Which of the following lines of MATLAB code are legal, and which will produce an error? Of the legal lines, which are likely to produce unintended results, and how could you fix them? First think of answers and then test them in MATLAB.

```
datamat(1:.5:end)
1:4:3
[ [1 2:3 4:5] (6:10)' ]
[ [1 2:3 4:5]; (6:10) ]
cvar = 4;-5
var = ones(1,10)';
var = [1:10'];
```

4. Determine whether the following three lines produce different results, and explain why or why not. Again, first think of your answer, then confirm in MATLAB.

```
0:.1:(2>1)
(0:.1:2)>1
0:.1:2>1
```

5. Using the colon operator and square brackets, create an array that counts from 1 to 10 and then back down to 1. Make sure 10 is not repeated. Then, add 4 to all elements in the array. Then test whether each element is equal to 8.

6. What is the difference between the following two lines of code?

```
clear a
clear a*
```

7. Using the mouse structure created earlier, write one line of code that will produce the total number of electrodes in both animals (the MATLAB function for summing is sum).

8. Initialize a variable to be a 50 × 40 matrix of zeros. Then set the first 10 and the last 5 elements in the first dimension to have the values of 10 and 5, respectively, for the first 20 elements in the second dimension.

9. Convert the following mathematical expressions into MATLAB code.

(a) $4 + \dfrac{5}{4}$

(b) 19×48^{-4}

(c) $\dfrac{4+3}{8}$

(d) $4 + \dfrac{3}{8}$

(e) $\dfrac{-(4+5^3-.39)}{2^4 \times 17.26}$

10. What are the differences among the following lines of code?

```
n = zeros(3,5)/0;
n = 3+ones(3,5)/0;
n = nan(3,5);
```

11. Which of the following variable names are legal versus illegal? Of the
    illegal ones, how would you fix the names? Of the legal ones, are these
    good names to use, and why or why not?

```
data4analysis
7mac11
iHeartData!
j
data-set
data set
filter
thisOldVariable
variablesRgr8
this_is_a_variable_i_will_use_for_analysis_results
polyfit
```

12. Which of the following lines of code will produce valid matrices?
    What's wrong with the ones that produce an error?

```
aMatrx = [1 3 2 3; 6 7 8 9];
aMatrx = [1 3 2 3; 5 3 6; 7 8 9];
aMatrx = [1 3; 5 6; 7 8];
aMatrx = [1 3 2; 5 6; 7 8 9];
```

13. For the matrix or matrices in the previous exercise that are valid, perform the following basic arithmetic operations on them: add the number 4 to all elements; multiply each element by 6.4; subtract 100. Then, use indexing to apply these three operations: only to the first element of the matrices; only to the first row of the matrices; only to the second column of the matrices.

6 Functions

6.1 Introduction to Functions

Now you are ready to learn about functions, and about how to write, modify, and debug functions. A function is basically the same thing as a script, except that it is typically not modified as often as a script is, and it operates in its own miniverse (more on this later). All functions have a function name, which is generally the same thing as the name of the file that contains the function.

You've already seen a few functions in the previous chapters. The function mean, for example, computes the average of the input numbers. The function disp displays some text in the Command window.

Most functions take inputs and give outputs. The function mean, for example, takes numbers as the input and gives the average of those numbers as the output. More generally, functions have the following syntax:

```
output = functionname(input);
```

Let's try a few functions to give you hands-on experience. First, we will create an array of normally distributed random numbers:

```
x = randn(15,1);
```

This example looks a little different from the canonical function syntax shown above, because there are two inputs instead of one. The function randn generates a matrix of random numbers drawn from a normal distribution (mean of 0 and standard deviation of 1), up to however many dimensions you want. To generate a $5 \times 25 \times 17$ three-dimensional cube of random numbers, you would type randn(5,25,17). If you provide only one input (randn(10)), it will produce a 10×10 matrix.

Now let's use the size function to see how big the variable x is. The size function takes a matrix as input and gives the number of elements across each dimension as the output. This is an important function for debugging and finding programming errors.

```
sizeX = size(x);
```

The variable `sizeX` is a 2-element array that tells you the variable x has 15 elements in the first dimension and 1 element in the second dimension. Notice how I used an informative variable name; if this were part of a really long script containing many variables, there would be little ambiguity about the meaning and contents of variable `sizeX`. Let's test a few more functions before moving on.

```
minX = min(x);
maxX = max(x);
medX = median(x);
```

You're starting to get the idea.

6.2 Outputs as Inputs

An important concept in programming is using the output of one function as the input to another function. This saves time and space while programming. Let's say you want to initialize a matrix of zeros to be the same size as our random number matrix. This could be done as follows:

```
sizeX = size(x);
zmat = zeros(sizeX);
```

Or it could be done more efficiently:

```
zmat = zeros( size(x) );
```

The output of the function `size` is being used as the input to the function `zeros`. Of course, it is possible that you want to create the variable `sizeX` for later use in the script. But if you need to get `size(x)` only once, there's no need to create a new variable and take up a new line of code. Let's try something else:

```
zmat = zeros( max(x) );
```

It is possible, though incredibly unlikely, that you did not get a MATLAB error from running that line. Why did MATLAB give an error? The function `zeros` expects the input to be the number of elements along each dimension of the matrix of zeros. What is `max(x)`? It's different each time you generate a new variable x, of course, but it might be somewhere around 2 or 3. I got 2.511... It is impossible to generate a matrix that has

2.511 elements in the first dimension (MATLAB deals only with integer dimensions). We can fix this by applying the concept of using the output of a function as the input to another function, but this time doing it twice:

```
zmat = zeros( round( max(x) ) );
```

Now the output of max is the input to the function round, and the output of round is the input to the function zeros. There is no limit to how many functions you can nest like this, but code starts getting confusing after too many nested functions. I recommend using three to four nestings at most. Note also that I added spaces to make the code more readable, although the spaces are not necessary.

MATLAB usage tip: If you click on or move the cursor on top of one of the parentheses, it and the complementary parenthesis are underlined. This helps you keep track of where function calls start and end. In this particular example there is little ambiguity, but it becomes useful when using multiple embedded functions with multiple inputs.

6.3   Multiple Inputs, Multiple Outputs

You've already seen that some functions take more than one input. In many functions, the later inputs are optional. Consider the difference between these two lines:

```
size(x)
size(x,2)
```

If you run this code in MATLAB, then you can probably figure out on your own what the "2" means: The second input in the size function returns the number of elements in that dimension. The variable x is a 15×1 matrix, and so the second dimension has 1 element. Try replacing the "2" with a "1."

Some functions also have multiple outputs. Multiple outputs can be separated by commas and need to be enclosed by square brackets, like this:

```
[out1, out2, out3] = function(inputs);
```

Try this one, for example:

```
[val,idx] = max(x);
```

Both outputs from the function max are used very often in data analysis, so it's worth spending some time to understand this one. The first (and default, if you request only one) output of the function max returns the largest value in the input matrix. But where is that maximum value located in the variable x? Is it in the first position or the last position or the fifth position? That's where the second output becomes useful. The second output tells you the index (position) in the matrix where the maximum value can be found.

You can confirm this by checking the value of idx, outputting x into the Command window (just type x in the Command window to display the contents), and confirming that idx correctly identifies the element that contains the maximum value. This is an example of sanity-checking the code, and you should get in the habit of sanity-checking your code whenever possible.

If the maximum value is repeated several times, the second output will tell you only the first position where that maximum is found.

```
[maxval,maxidx] = max([ 1 2 1 1 5 5 1 5 ])
```

The variable maxidx tells you the maximum value (5) can be found in element 5, although that maximum value also appears in other elements. Exercise 4 in section 6.14 will ask you to program a solution to finding repeated maximum values.

6.4 Help

Nearly all functions in MATLAB have a help text that helps you understand what the function does and what its inputs and outputs are. You will soon learn that these help texts are just comments in the beginning of the function file.

To see a function's help text, just type help <functionname>. Try it:

```
help mean
help max
help randn
```

MATLAB function help texts are usually fairly helpful. Sometimes they can be difficult to understand at first, but the more experience you have with MATLAB, the better you will be able to interpret the help texts.

Many functions have longer help files that can be viewed in a separate window by typing doc <functionname>. For example, type help fft. You'll see some information about the fast Fourier transform. Don't worry now about reading it or trying to understanding it—you'll learn all about the Fourier transform in chapter 11. Now type doc fft. Notice how the pop-up window gives you more detail and also has some examples with plots.

Whenever you see examples in function help or doc files, don't hesitate to try them by pasting the example commands into the Command window. They usually produce plots that help you understand what the function does.

And of course, you can also look up MATLAB functions on the Internet. If you search for "MATLAB <function> examples," you are likely to find examples and explanations of the function on MATLAB's website and on other websites.

More generally, there are thousands of MATLAB functions. It would be a waste of space to list them all here and describe their basic functions. You will learn the most important functions as you go through this book and get more experience with MATLAB. If you know what you want a function to do, but you don't know whether there is a function to do it, try searching the Internet. Memorizing a list of functions is as useful for learning to program as memorizing the dictionary is for learning to speak a language.

6.5 Functions Are Files

In the Command window, type `edit  mean`. A file called mean.m will appear in the Editor window. That is the function `mean` right there. It's just a text file, really not much different from the script you wrote in the previous chapter.

You can scroll through the `mean` function. Rather long just to sum and divide by N, wouldn't you think? MATLAB functions are designed for maximum flexibility and robustness. The function `mean` can handle various kinds of inputs, including matrices and different variable types, and it deals with them appropriately. That's why the additional overhead is necessary.

At the top of the function file are lots of lines of comments. These are the comments that print out in the Command window when you type `help mean`.

Being able to open functions as text files is great because if you want to know how a certain function works, you can just look through the code. (You can open these files in any text editor, not only in MATLAB.) You can open function files by typing `edit <function>`, `open <function>`, by selecting the typed name of the function (in the Command window or in the Editor window) and pressing Ctrl-d, or by highlighting the typed function name, right-clicking, and selecting Open Function.

Not all functions are ASCII text files. Some functions are compiled (binarized), which means you cannot view their contents. Compiling functions is done either for speed or for proprietary reasons (e.g., to maintain secrecy of algorithms). When a function is compiled, then you can see only the help text. Try, for example, edit sum.

6.6 Writing Your Own Function

Wouldn't it be fun to have a function that you wrote? Let's do it. We'll create a function that takes a vector of numbers as input, displays some basic descriptive statistical information about those numbers, and then gives some outputs.

Open a new script and convert it to a function simply by stating at the top of the script that it's a function. On the first line of the file, type:

```
function stats = mikeIsANiceGuy(data)
```

Yes, we're naming the function mikeIsANiceGuy. When you write a book, you can name functions after yourself. The input will be called "data" and the output will be called "stats."

Typically, right underneath the function declaration is the commented help text, as you saw with the mean and sum functions. Although you might be tempted to jump right into the fun part of programming and get back to the help section later (or never), this is not good programming skills. Remember from chapter 1 that step 1 of programming is to think and plan your code before writing any actual code. So we will start by writing the help text, and that will provide the plan of action for the rest of the function.

```
% mikeIsANiceGuy Compute descriptive statistics on input data
% stats = mikeIsANiceGuy(data) displays and returns descriptive
statistics:
% mean, median, minimum, maximum, count
% INPUTS:
% data: a 1D vector of numbers in single or double precision
%
% OUTPUTS:
% stats: a vector containing mean, median, minimum, maximum, count
%
% code written by <your name here> <your email address here>
```

Save the file. Function files should always have the same name as the function name. Test the function by typing help mikeIsANiceGuy in the Command window. You will see all of the text that you wrote in the comments.

Before starting to write code, you should make an outline of the entire function using comments. There are three parts to our function, and we will therefore use three cells for (1) checking the input data, (2) computing the statistics, and (3) displaying the outputs.

Checking the input data requires if-then statements, which will be the topic of the next chapter. In this chapter, we'll assume that the input data are always valid. Leave this cell blank for now.

Next, compute the statistics.

```
%% compute the descriptive statistics
meanval    = mean(data);
medianval  = median(data);
minval     = min(data);
maxval     = max(data);
N          = length(data);

% group outputs into one vector
stats = [ meanval medianval minval maxval N ];
```

Notice how I spaced the equals signs to be vertically aligned. This improves readability and aesthetics.

Finally, we want to display the statistics in the Command window. To do this, you can use the function disp in combination with the function num-2str (this converts a number into its character representation, like turning 4 into '4') or you can use the function fprintf. The function fprintf is more complicated but has increased flexibility for formatting the text and is the function of choice for writing out data to a file. You'll learn more about fprintf in chapter 8.

```
%% display output stats in order of output
disp([ 'Mean value is ' num2str(meanval) '.' ])
disp([ 'Median value is ' num2str(medianval) '.' ])
fprintf([ 'Minimum value is ' num2str(minval) '.' ])
fprintf('Maximum value is %g.\n',maxval);
fprintf([ 'Number of numbers: ' num2str(N) '.\n' ])
disp('Done.')
```

The second use of fprintf may look a bit confusing; this syntax will be explained in chapter 8. For now you should just appreciate that there are several ways to print information to the Command window. Notice that the square brackets were not necessary in the final disp command, because no variables were concatenated. For displaying simple messages, I generally prefer the disp function.

It is time to use the function and make sure it works. Remember that you should always start with sanity checks. We'll do that here by simulating data where we know the ground truth of what the output should be. Let's set the input data to be the numbers 1:5. We know even before calling the function what the descriptive statistics will be. Now verify.

```
data = 1:5;
stats = mikeIsANiceGuy(data);
```

Congratulations! You've just written your first function (with only a little help from me). This is, admittedly, not a very useful function, but it's good to start simple and then build up.

6.7 Functions in Functions

Functions can be embedded inside other functions. You'll see this more in chapter 32 (graphical user interfaces), but let's do a simple introduction here. Rather than using the MATLAB mean function, we'll create our own function that sums and divides by N.

Add the following lines of code to the end of the mikeIsANiceGuy script:

```
function y=myMean(x)
y = sum(x) / length(x);
```

And then replace the meanval line with meanval = myMean(data);. Now you have a subfunction. Subfunctions are private in the sense that they cannot be called from the Command window (try it) or from any other function; they can be called only from within the function file. If you click on the *Editor, Go To* buttons, you will see a list of functions inside that file. Selecting one will take the cursor to that function.

6.8 Arguments In

There is an important lesson in myMean about how MATLAB deals with naming inputs and outputs: The input into myMean is called data, but inside the function myMean the input is called x. MATLAB pays no attention to the names of inputs and outputs when they are passed through; it cares only about their order.

In fact, you don't even need to specify all of the inputs. If you define the function to have one input called varargin (variable arguments in), all of the inputs arrive as cells in the cell array called varargin. Try replacing data with `varargin` in the function definition of `myMean`, and then add this line before the descriptive statistics:

```
data = varargin{1};
```

Re-run the script to make sure it still works. Then undo this change so data is listed as the input to the function.

6.9 Think Global, Act Local

You probably thought you were finished learning about variables after the previous chapter. But there is one more thing you need to understand about variables before really understanding how functions work. Variables can be *local* or *global*.

And before you understand the difference between local and global variables, you need to understand that MATLAB can create and destroy miniverses at will (a miniverse is a miniature universe). Variables that are created and live inside one miniverse do not know about variables that live in other miniverses, just like we do not know about our doppelgangers in parallel universes. And when the miniverse is destroyed, so are all of its variables (don't worry, I'm sure our universe will not be destroyed anytime soon). These variables are called "local" variables, because they are local to a miniverse. Some variables can be endowed with powers that allow them to be seen, accessed, and manipulated across miniverses; these variables are called "global" variables. (I guess this is where the analogy of our real universe breaks down, sci-fi movies notwithstanding.) All variables are local by default, and any variable can be anointed as a global variable by declaring it to be `global`.

```
global variablename
```

In the first line of the `mikeIsANiceGuy` file where the function is defined, remove the input by deleting `(data)` (including the parentheses). Save the file and try running the function again from the command line. What happened? It crashed, and MATLAB wrote that there is no variable called data. But this makes no sense, because you *did* create a variable called data. You can even type `whos` to confirm that the variable data still exists.

Put the `(data)` back and run the function again. You can see inside the function that several variables were created, including `meanval`, `minval`, `N`, and so forth. But where are those variables? They don't appear when you type `whos`. But they must exist, because they are used and printed, and MATLAB did not produce any errors or warnings.

You can guess where this is going: These variables are local variables. They exist only in one miniverse. The variable data exists in the default miniverse. MATLAB calls this miniverse "base" and it is the one you work in when MATLAB starts. When you called the function mikeIsANiceGuy, MATLAB created a new miniverse just for that function. The variable data does not exist in that miniverse, and the variables meanval and so forth do not exist in the base miniverse. And miniverses are stacked, because the variables x and y exist only in the myMean function's miniverse, which itself is a miniverse inside the other miniverse. Passing an input into a function is one way to pass information into that function's miniverse.

Let's make data a global variable. In the Command window, type global data. MATLAB doesn't like to reassign existing local variables to be global variables, so you might get a warning or error (depending on your MATLAB version) when you run the above declaration. You can first clear the variable data and then define a global variable called data. Then you need to reassign data=1:5.

From inspecting the output of whos, you can see in the "Attributes" column that the variable data is indeed global. So, now in theory, this global variable should be accessible from any miniverse, including from inside the function, right? Let's try. Delete the (data) from the function declaration at the top of the function file, and then save. Run the function without passing through any inputs; just type mikeIsANiceGuy.

Oops, the function still crashes. That's actually fairly sensible—it would be inconvenient to have *all* global variables *always* accessible from *every* function *all* of the time. Instead, we need to make this particular function share this particular global variable. Inside the function and before any calls to variable data appear, type global data. Save the function and run the function again. It works!

The important lesson here is that defining variables to be global allows them to be accessible from any MATLAB miniverse, but only if they are declared.

I suppose now would be a good time to tell you that the official word for miniverse is "stack." Sometimes the term "workspace" is also used. Maybe I've watched too many movies, but I find that thinking of them as miniverses is more intuitive and more fun.

6.10 Stepping into Functions

We will now step into the miniverse that MATLAB creates when running the function mikeIsANiceGuy. In the file of that function in the Editor window, click on the little horizontal line to the left of the final disp function, just to the right of the line number (figure 6.1). You'll see that line turn into a red circle. If you see a gray circle instead, it means there are unsaved changes in the file. (You can also put the cursor on that line and press F12; or select *Editor, Breakpoints, Set/Clear* from the menu.) This is called a "breakpoint," and it will allow you to visit the function's miniverse. If you don't see line numbers on the left side of the Editor window, you can change the settings in the *Home, Preferences, Editor/Debugger, Display* menu.

Before running the function again, create some new variables in the workspace. These can be called anything you want and can contain anything you want. You just want a few variables to help illustrate the miniverse concept.

Now run the function mikeIsANiceGuy using the same code you used before. It should bring you inside the function's miniverse on the line where you placed the breakpoint. MATLAB lets you know you are inside a nondefault miniverse by changing the command prompt from >> to K>>. Now type whos. What happened? All of the variables you just created are "gone," and all of the variables created inside the function are listed. Don't worry—the variables in the base miniverse are still there, they are just not immediately accessible.

You can shift between miniverses (this is starting to sound like the plot of a bad sci-fi movie) in the *Editor, Function Call Stack* menu. Select "base" and then type whos again. Select the "mikeIsANiceGuy" stack again to go back to the function miniverse.

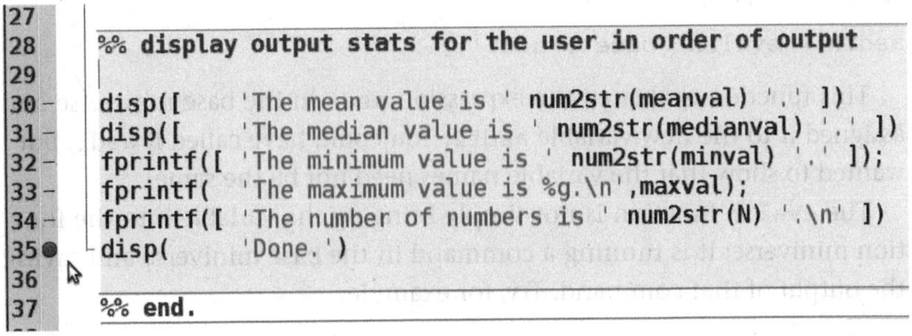

```
27
28      %% display output stats for the user in order of output
29
30 -    disp([      'The mean value is ' num2str(meanval) '.' ])
31 -    disp([      'The median value is ' num2str(medianval) '.' ])
32 -    fprintf([ 'The minimum value is ' num2str(minval) '.' ]);
33 -    fprintf( 'The maximum value is %g.\n',maxval);
34 -    fprintf([ 'The number of numbers is ' num2str(N) '.\n' ])
35 ●    disp(      'Done.')
36
37      %% end.
```

Figure 6.1
To step into a function, place a breakpoint in the file where you want the function to freeze. Breakpoints can be created by clicking on the horizontal line between the line number and the code.

Now place another breakpoint inside the subfunction myMean. Then run the line that calls myMean and you'll be in the myMean miniverse. Going back to the *Call Stack* menu will allow you to move across the three miniverses. Try typing whos in each stack.

Now destroy the miniverses by quitting the function. You can do this by typing dbquit in the Command window or by clicking the big red button labeled *Quit Debugging* in the *Editor* menu. The prompt in the Command window should return to >>.

You know from the previous section that assigning variables to be global is one way to share information across miniverses. There are also other ways to transmit information across miniverses. Step back into the function and type the following:

```
assignin('base','meanvalFromFunction',meanval)
```

The function assignin assigns a new variable in another stack (miniverse). This line will create a new variable in "base," will name that variable meanvalFromFunction, and will set the value of that variable to be the value of meanval. Quit the function and type whos.

The tricky part of the function assignin is that the second input is the *name* of the new variable in the base stack, while the third input is the *content* that you assign to that new variable. Try running this line again from inside the function, but now putting single quotes around 'meanval'. What is the new value of the function meanvalFromFunction?

If you are inside a function and want to access a local variable from the base stack, you can use the function evalin. Let's try it. Create a variable in the base workspace: asdf = 10:10:100;. Then step into the function again and type:

```
asdf2 = evalin('base','asdf');
```

This function evaluated the expression asdf in the base miniverse and assigned it to the new variable asdf2. You could have called it asdf, but I wanted to show that the variable names need not be the same.

The evalin function is not simply bringing the variable into the function miniverse; it is running a command in the base miniverse and saving the output of that command. Try, for example:

```
asdf2 = evalin('base','asdf>50');
```

So far we've been stepping into functions where we want to step into them. Another way to step into functions is to have MATLAB automatically step into them when things go wrong. In the *Editor, Debug* menu, select *Stop on Errors*. This means MATLAB will automatically step into a function when the function crashes. It will take you to the line that crashed and print out, in lovely red font, the error message. This is a useful method for debugging scripts and functions.

Let's try this by intentionally making an error in the script. In the line that gets the number of elements ("N = ..."), change `data` to `data1`. Clear all existing breakpoints in the file, save the function, and run it again.

There is more to know about stepping into functions, but you now know enough to continue exploring on your own. Using this little script, try the different operations under the menu options for *Editor, Breakpoints* and *Editor, Debug*. Try modifying and saving the code while you've stepped-in to see what happens. Being comfortable with the debug option is an important part of learning how to program in MATLAB.

One note about the `clear` function: Typing `clear` clears all variables in the local workspace, not in other stacks. Typing `clear all` will also clear—in addition to all of the variables—breakpoints, pointers, javascript references, and other things. In general, it is best just to use `clear` and not `clear all`.

6.11 When to Use Your Own Functions

With great power comes great responsibility. Okay, let's be honest here: Being able to create your own functions is a pretty small power, but it does come with some responsibility. Before you start writing functions for everything, there are two things to keep in mind.

First, writing too many functions promotes bad programming and increases the likelihood of errors. With many functions, it is easy to get confused about what different functions do, what parameters are hard-coded in different functions, whether different functions overwrite or append data, and so on. Generally speaking: more functions, more problems.

Second, functions can make code-sharing annoying, because you will need to share all of the additional functions as well (also called dependencies). If you have just a few functions, it might be only a minor hassle. But if you write many functions and store them in various places on your computer, keeping track of all these dependencies can become a headache. And you might forget which functions are called in which scripts, particularly if some functions call other functions. When I teach classes about programming or data analysis, students lose points if they turn in assignments that require functions that are downloaded or otherwise not bundled with MATLAB.

The conclusion is that you should create your own functions sparingly and only when it will significantly improve your analysis pipeline. I very rarely create my own functions.

6.12 When to Modify Existing Functions

You should be extremely hesitant to modify existing functions, particularly if they are functions that come with MATLAB. In fact, let's just agree that you should never modify existing MATLAB functions. If you want to change the behavior of an existing function, the best way to do this is to make a copy of the function file and change its name (also change the function name in the top of the file to avoid confusion). You could add your initials to the end of the function to personalize it, such as `meanMXC` instead of `mean`.

Modifying existing functions in toolboxes or other third-party code will also make it annoying to share your code with others, because you will also need to share the functions that you modified, and this could cause other scripts that rely on those functions to crash or produce invalid results.

6.13 Timing Functions Using the Profiler

One of the best tools for improving the efficiency of your code is to use MATLAB's profiler. It's a tool that, when turned on, will track how much time is spent on each line of code. Try it using the following code.

```
profile on
mikeIsANiceGuy(data)
profile viewer
```

The viewer will open and will highlight how long each line of code took (figure 6.2). For a puny function like this one, the profiler is not terribly insightful. But for long scripts and functions, the profiler can help you identify inefficient code, unnecessary loops, and other bottlenecks to computation time. It is fun to use, helps you hone your programming skills, and can shave minutes to hours off your analysis time.

6.14 Exercises

1. The code below will not produce an error, but it is a bad line of code. Why? (Hint: Check out the size of zmat when you run it repeatedly.)

```
zmat = zeros(round(max(randn(30,1))));
```

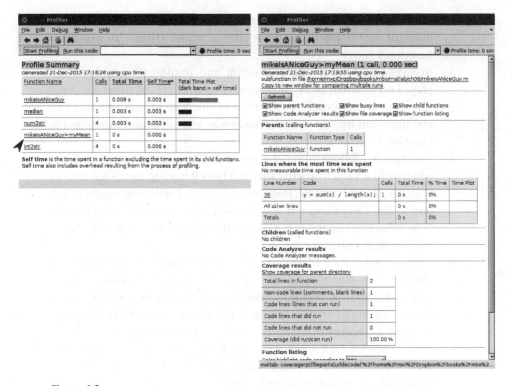

Figure 6.2
The profiler (left panel) reports how long MATLAB took to run each line of code and each function. By clicking on one of the functions, another window will appear (right panel) to provide details about the time-consuming parts of that function. The profiler helps you find inefficiencies in your code.

2. Your friend Suzie needs help. Her code successfully creates the variable zeromat, but there are too many lines of code. She offers you one piece of chocolate for each line you can eliminate, while obtaining the same final output. Can you earn three pieces of chocolate?

```
N = 14;
randnums = rand(N,2);
rn_size = size(randnums);
zeromat = zeros(rn_size(1),1);
```

3. Using the function sort instead of max, write a function that finds the maximum value of an input vector.
4. The second output of the function max tells you only the location of the first instance of the maximum value. Write code to find all

instances of the maximum value. Test your code using a variety of different vector inputs to make sure it works. (Hint: Try using the function find.)

5. In mikeIsANiceGuy, the maximum and minimum values are displayed on different lines. Although we haven't yet discussed the function fprintf in detail, on the basis of the code in that function can you figure out how to print the maximum and minimum values on the same line?

6. Generate a random number from a uniform distribution, and test whether that value is greater than 0.5 (you'll have to figure out how to write an if-end construction; see help if). Report the results of the test in the Command window. For example, the code should print "The value is 0.123 and is not bigger than .5." Put this code into a separate function that you can call.

7. Generate a $4 \times 3 \times 2$ matrix of normally distributed random numbers. This is a three-dimensional cube of numbers. How does MATLAB represent such a matrix, and how does it show the matrix contents in the Command window? Then compute the average and maximum values along each dimension. What are the sizes of these results?

8. NaN stands for "not a number" and is the result of dividing zero by zero. Find two ways to create NaN in MATLAB. What happens when you divide a non-zero number by zero?

9. NaNs can create problems in many functions. Some functions also have built-in methods to ignore NaNs. From the following vector, test whether NaNs disrupt the max, mean, and sum functions. If they do, use the help files or perhaps search the Internet for methods to apply these functions without disruption from NaNs (come up with at least two solutions to this problem).

```
v = [1 0 0 1 NaN 1 NaN 0];
```

10. The function randperm generates a random reordering of integers from 1 to the first input. For example, randperm(2) could return [1 2] or [2 1]. How can you generate a random sequence of even numbers between 10 and 20 (for bonus points, use only one line of code)? Solve this problem using randperm, and then find another solution without using randperm.

11. Evaluating the code below produces the output [8 13]. Without reading the help file, describe what the function strfind does. When you think you understand it, check the help file, and try changing the code to confirm your expected outputs.

```
strfind('Hello, my name is Mike.','m')
```

12. Another function for searching strings is called strncmp. Evaluate the following line of code. Then reevaluate it, changing the third input to 2, 3, 4, and 5. Can you figure out what this function does based on the output?

```
strncmp('Hello, my name is Mike.','Hell',1)
```

13. Create a five-dimensional matrix of random integers. You can define the number of elements in each dimension. How many elements does the matrix contain in total? You can answer this question by using the functions size and prod or by using the function numel.

14. What defines "a line of code" in MATLAB? Is it the Enter key or a semi-colon? Try the following lines of code in MATLAB to find out.

```
asdf=
4;
asdf=4; asdf2=6;
asdf = ...
4;
```

7   Control Statements

You are on your way to becoming a good beginner programmer. The next step is to learn how to control MATLAB in flexible ways. This is where *control statements* (also sometimes called command statements, flow statements, or flow-control statements) become necessary. There are only a few control statements that you will use in programming, but they are important and you will use them often.

7.1   The Anatomy of a Control Statement

All control statements evolved from the same primordial prototype. Understanding this prototype will help you learn the specific control statements in the same way that understanding Latin will help you learn specific Romance languages such as French, Italian, and Spanish (not to be confused with romantic languages like German). Figure 7.1 illustrates the anatomy.

```
<control type> <conditional>
    <content>
    <¿alternatives?>
end
```

Figure 7.1
Gross anatomy of MATLAB control statements. Try to keep this template in mind as you learn about each specific control statement.

7.2   If-then

If-then statements (sometimes simply called *if* statements) are perhaps the easiest control statements to understand and to implement, because you use this logic every day of your life. "*If* it is raining, *then* I'll bring an umbrella." All programming languages have if-then constructions, though the syntax differs slightly across languages. In MATLAB, the "then" is unnecessary.

```
if 2>1
    disp('I like big numbers.')
end
```

You can clearly see the anatomy of this simple if-then construction. In if-then statements, the conditional part is always a logical—true or false. Whatever code you write in the conditional must return a single logical answer. If the conditional statement is not a logical, MATLAB will either try to interpret it as a logical or give an error. Here is a good and simple sanity check to make sure the condition is appropriate: Evaluate the code on the "if" line (but not including the if) and make sure it returns a single value that is either 0 (false) or 1 (true).

If the answer to the statement after if is true, then MATLAB runs all the code between if and end, line by line. It is common practice and good programming technique to indent these lines of code. This helps you see which code is inside the control statement. If you type the code above into the MATLAB Editor, try deleting the spaces before the disp function, then highlighting the three lines, and then press Ctrl-i or right-click and select Smart Indent.

But wait, there's more! If-then statements can be expanded to include elseif and else. These are alternate routes that you can have MATLAB take. In the next example, we will test whether a randomly generated number is bigger than or smaller than certain cutoffs. (The important parts of this code are lettered for reference.)

```
r = randn; % with no inputs, a single number
if r>0% A
    disp([ num2str(r) ' is positive.' ])
    elseif r>2% B
        disp('Snuffleupagus is real!')
    elseif r<0 && r>-1% C
        disp('r is small but I''ve seen smaller');
    else % D
        disp('r is really small')
end
```

Let's start with the code section labeled A. Nothing new here, this is the same format as the previous example. The code tests whether the random number in variable r is positive.

Now for code B. Here's a question: If you run this code 1 billion times, how often will it print messages about the existence of Snuffleupagus? The answer is exactly zero. And not because Snuffleupagus isn't real—it's because as soon as one of the statements in an if-elseif-else-end construction is true, the rest of the statements don't matter; they are ignored. If r is greater than zero, the statement in A is true, its associated content is run, and the if-then control statement is finished. Nothing else gets evaluated. It can never happen that the logical result of A is false and the logical result of B is true. If you were to swap the order of A and B, this would work, because it is possible for a number not to be greater than two, but to be greater than zero.

Now let's discuss C. The lesson here is that the conditional statement can have multiple elements, as long as the entire conditional statement can be true or false. Here we are saying that this conditional is true only when r<0 *and* when r>-1. In human language, you'd say when the number r is between –1 and 0.

Also notice in the disp function that there are two single quotation marks (not a double quotation mark) in I've. That's not a typo—it's MAT-LAB's way of distinguishing between a single quotation mark to indicate a boundary of a string and an apostrophe in text.

MATLAB allows AND (&&) and OR (||; this is not the letter l or the number 1, but the vertical line character that is usually above the Enter key on a keyboard; in Unix it's called the "pipe" operator). "And" means both statements must be true; "or" means MATLAB will consider the entire line to be true if at least one of the statements is true. For example:

```
True:  5>3 && 9>3
False: 5>3 & 9<3
True:  (5>3) | (9<3)
False: 5<3 | 9<3
```

As usual, the parentheses in the third line above are not necessary, but probably you agree that they make that line easier to interpret.

"And" and "or" operators are not limited to pairs; you can have as many as you want. You will see this often in real data analysis scripts. For example, you might want to extract the data from condition "4" AND cells that have a firing rate greater than 10 AND that show differential responses in the localizer task AND only during the anesthetized state AND so on.

Finally, we get to part D in the `if-elseif-else-end` construction. An `else` statement just before the `end` captures anything that gets left over when none of the previous statements is agreeable (your German colleagues might call this the *resteficken* option). In the above example, anything that doesn't fit into the previous `if` or `elseif` statements is a number smaller than −1. Actually, there is one possibility (exceedingly rare, but it could happen) for that `else` statement to run but without the number being negative. See if you can figure out what it is.

If-then constructions can be nested inside each other.

```
r = rand;
if r>0
    disp('r is definitely positive')
    if r>2, disp('>2'); end
    if r>1
        disp('>1'); end % don't do this!
else
    disp('r is most likely negative')
end
```

You can see above that an entire `if-then` statement can be on a single line if the conditional and the content are separated by a comma or a semicolon. This is okay if the statement is very short, like in this example, but you should avoid putting too much code in one line.

You can also see that it is technically legal to have the `end` appear on the same line as the content while both are on a different line from the `if` statement. But don't do it. It becomes difficult to see where each `if` statement ends. It will also confuse MATLAB and elicit a warning about confusing end statements.

Finally, with this code in the Editor window, click on or put the cursor position over an `if`, `else`, `elseif`, or `end`. MATLAB will automatically underline the associated `if` or `end`. In the toy examples you've seen here with only a few lines of code, this may seem like an unnecessary feature. But in real analysis scripts, when there are dozens or hundreds of lines of code between an `if` and its `end`, and when there are multiple nested command statements, these visual programming tools can be really useful.

Let's get back to the function in the previous chapter that computes, displays, and exports some descriptive statistics on input data. (In case you forgot, that function was called `mikeIsANiceGuy`.) In the previous chapter, we skipped the first part of the function, which was to check the input data to ensure our function will work. Now that you know how if-then statements work, we can finish this function.

What are the conditions on the input variable? It must (1) contain numbers, (2) be a one-dimensional vector, and (3) be more than a scalar (a single number). Let's start with the first test.

```
%% check inputs
% check that data are numeric
if ~isa(data,'numeric')
    help mikeIsANiceGuy
    error(Input must contain numbers!');
end
```

The function isa checks whether the data type of the first input matches the second input. The output of this function is a logical variable. The tilde (~) before it is the MATLAB expression for negation—it turns trues into falses and vice versa. Try it by typing ~(7==7). The answer is false (0). That may be confusing at first, but it is accurate: 7 equals 7 (true), and negating this makes it false. The code above checks whether the data are numeric. If the result of that test is true, the conditional is false, and the content isn't run. If the result of that test is false (this would happen, e.g., if the input data were strings), the conditional is true, and then the content is run. The content prints out the help file for this function and then gives an informative error message—that dreadful red text that you see in the Command window. When the error function is called from inside a function, MATLAB immediately quits the function without running any further code. Before moving forward, sanity-check that this input test works by inputting correctly formatted data (and receive no errors) and incorrectly formatted data (and receive an error).

Now for the second condition, that the input must be an array. (Technically, this function won't crash if the input data is a matrix, but that complicates matters and so we'll stick to vectors here for simplicity.) We can test for this condition using the size function.

```
% check that the input is a vector
if sum(size(data)>1)>1
    help mikeIsANiceGuy
    error('Data must be a vector!');
end
```

I admit, that's a weird piece of code. sum(size(data)>1)>1 is actually four separate pieces of code that are nested within each other. When you are trying to decipher nested code, always start by finding the innermost piece of code, and then work outwards. The innermost code is usually inside the most number of parentheses. As I explain below how this conditional statement works, you should follow along in MATLAB and run each piece of code. That's a good exercise to learn how to break down and understand these kinds of code lines.

The innermost code here is data, which is the input data. The next piece of code going outwards is size(data). This you know—it returns the number of elements along each dimension. Using the example from the previous chapter, the output of this function will be the two-element vector [1 15].

Next, that vector is tested against the number 1 ([1 15]>1). This is a logical test that will return another two-element vector, except that this vector will contains zeros and ones. In this case, the result will be [0 1] for [false true].

Next we apply the sum function to this vector. That converts the vector from logical to numeric, which means the *logicals* [false true] become the *numbers* [0 1] (I think I may have written in a previous chapter to avoid doing this, but let's pretend I have a terrible memory). Their sum is one. Finally, we test whether this number is greater than one.

Think about what will happen to this conditional statement if the data had more than one dimension of numbers, for example if it were a 2×15 matrix. The first logical test would be [1 1], and the sum would be two, which is greater than one, and then the conditional would be true and the script would give an error.

That's quite a mouthful of code to have in one line. Don't worry if you're not at the level of being able to produce code like that—it's more important to know how to piece apart and understand that kind of code. Never be intimidated by complicated-looking code. Just take a deep breath and work your way through it one function or variable at a time. And use your trusty code-inspecting tools: evaluate, plot, and check sizes. Keep in mind that all code—even the hairiest, ugliest code—is just composed of small and simple pieces.

Finally, the third input check. This is the easiest. The data should contain more than a single number. I'll leave that one to you to solve. You can use the functions size, length, or numel. The function numel stands for "number of elements" and returns the total number of elements in the entire matrix, regardless of how many dimensions it contains. The function length returns the number of elements in whatever happens to be the longest dimension of the matrix. And size, as you know, returns the number of elements along each dimension.

Now you are finished with your first MATLAB function. Before using any new function, you should always sanity-check it. I know, you already did some sanity checks, but we made modifications since the last sanity check, and anything new must be checked. Try calling this function again, inputting data that you expect to work and data that you expect to get caught by the input checks. Are there any other input checks that we forgot to include?

7.3 For-loop

For-loops are just as important as if-then statements. They are initially slightly less intuitive because you don't often use for-loops in language or in thinking the way you use if-then statements. But you will quickly see the logic.

```
for i=1:10
    disp(num2str(i))
end
```

You can see the prototype control statement in there. The main difference between for-loops and if-end constructions is that the content is not run only once; it is run as many times as there are elements in the looping variable. In the code above, the looping variable is i, and it goes from 1 to 10 in integer steps. The variable i is not explicitly changed inside the loop—that is, there is no piece of code that changes the value of i—but MATLAB changes its value on each iteration (an "iteration" is one of several runs through the content between for and end). When you run the code, you'll see that the value of i starts at 1 and goes up to 10 in integers.

The variable i is more generally called the "looping variable." The looping variable does not need to start at 1, nor does it need to be integers.

```
for i=-5:1.1:6
    disp([ 'My favorite number is ' num2str(i) '.' ])
end
```

However, it is often convenient to keep looping variables as integers, because looping variables are often used as indices into matrices.

```
for i=1:5
    sq(i,1) = i;
    sq(i,2) = i^2;
end
```

In this loop, we are creating a matrix called sq, which will be a 5×2 matrix. The first element of the second dimension is the value of the looping variable at each iteration, and the second element of the second dimension is the squared value of that looping variable. Try running that loop again using 1:.5:5 as the looping range. Which part of the code causes MATLAB to produce an error and why?

For-loops are used very often in programming. You will have loops over conditions, loops over frequencies, loops over time points, over electrodes, over trials, over subjects, and so forth. Similar to if-end constructions, loops can also be nested inside other loops.

In real MATLAB scripts, there are many nested loops, if-then statements, and other control statements. You can imagine that this much nesting can become confusing. So here are three MATLAB programming tips about writing loops and if-end statements.

1. Most important: Avoid loops and if-statements whenever possible. Loops bring confusion and errors, and they will slow down your code. They are often necessary but sometimes can be avoided. You will see many examples throughout this book of places where loops can be eliminated.

2. Give your looping variables meaningful names, and end them with the letter i or idx. For example, a loop over conditions might have a looping variable called condi. Other loops go over trials (triali), frequencies (freqi), time points (timei), channels (chani), and so on. This naming convention makes it easy to recognize and remember the looping variables. Equally sensible alternatives include trialno or trialnum. The important things are interpretability and consistency.

3. Put a comment after each end statement to indicate which loop it is ending. Sometimes it can be useful to write additional information. You might initially think of including the line number at which the loop started, but unfortunately this is usually not a good idea: Each time you add or remove a line of code, this number will change.

As with many other programming tips, these may seem unnecessary in the toy examples in this chapter. But when you have four for-loops spanning hundreds of lines of code, calling your looping variables a, s, d, and f is just asking for trouble. That said, if you have small and simple loops comprising just a few lines, then it's fine to use a short meaningless variable name.

Below is an example of a well-written nested series of for-loops.

```
for chani=1:nChannels
    for condi=1:nConditions
        for freqi=1:nFrequencies
            tfd(chani,condi,freqi) = analysis_res;
        end % end frequency loop
        moreMatrix(chani,condi) = otherstuff;
    end % end condition loop
end; % end channel loop
```

This may look intimidating, but don't worry, after a few more chapters you'll be able to write code that looks like this. Also, a small note: MATLAB does not require semicolons after the end statement, although it's fine to put one there if you like the way the end appears to be winking at you.

7.4 Skipping Forward

You may want to skip iterations in a for-loop. Imagine, for example, that you want to run an analysis on hundreds of data files, and each data file takes a few hours to process. Some of the data files have already been analyzed, and you don't want to waste time reanalyzing those files. In the next chapter you'll learn more about how to determine if files exist on the computer. Now you'll learn how to skip iterations.

Let's start with something simple. We will use the first for-loop introduced in this section, but now we want to print a message only when the looping variable is odd (I don't mean if the number is weird or has erratic behavior; I mean if the number is indivisible by two). How would you program this?

There are several ways to do this—in programming there are often multiple ways to solve a problem. One way is to use the control statement continue. This is a useful statement, but it is a bit counterintuitively named, because, in fact, when this statement is called, it does the opposite of continue—it breaks out of the loop and goes immediately to the next iteration of the loop. (I suppose the reasoning is that it continues to the next iteration of the loop.)

```
for i=1:10
    if mod(i,2)==0, continue; end
    disp([ 'Running iteration ' num2str(i) '.' ])
end
```

The function mod might be new to you; it computes the modulus, or remainder, of a division between the first input and the second input. The remainder of integers after dividing by two is zero for even numbers and one for odd numbers. Here's a quick sanity check:

```
[ 1:10; mod(1:10,2) ]
```

The code above displays two row vectors, one containing the numbers 1:10 and the second containing the remainder after dividing by two. (How could you modify this code to produce two column vectors? I can see three ways to do it.)

So, the for-loop above first tests whether the value of the counting variable at each iteration is even or odd. If it is even, it runs the continue command, which tells MATLAB to bypass all of the code after continue and before end, and skip to the next iteration.

Now that you know about the mechanics of writing for-loops, it is time to learn the most important thing there is to know about for-loops: Avoid them whenever possible! (I realize I wrote this a few pages ago, but a bit of repetition facilitates learning.) For-loops require extra code, increase computation time (thus slowing down your analyses), and open up new opportunities for confusion and for MATLAB errors.

Many examples will be presented throughout this book where loops can be avoided by using computations and operations on matrices. Below is one example:

```
p3 = zeros(1,10); % always initialize!
for i=1:10
    p3(i) = i^3;
end
```

The four lines above can be accomplished faster, more elegantly, and with less possibility for errors using the one line below.

```
p3 = (1:10).^3;
```

7.5 While-loop

While-loops are very similar to for-loops. With a for-loop, you specify how many iterations to have; with a while-loop, the iterations keep going until you tell the loop to stop. Observe:

```
i=0;
while true
    i=i+1;
    disp([ 'Give me ' num2str(i) ...
        ' pieces of chocolate!' ])
end
```

Run that code. Don't hold your breath waiting for it to stop, though. That loop keeps running and running without stopping (better than the Energizer bunny). You can break it by pressing Ctrl-c in the Command window or by unplugging your computer (if it's a laptop, you'll have to wait for the battery to drain). Why does it never stop? Because `true` (the conditional statement) is always true. Try running it again, changing the conditional statement from `true` to `false`. Now it displays nothing, and the value of the variable `i` is never updated, because the content was never evaluated.

There is another new piece of code, which is the ellipsis (...), the three dots at the end of the line. Those are used to wrap long lines of code to the following line. You can try writing something after the ellipsis, and you'll notice that MATLAB treats what you write as comments (green text). Ellipses can sometimes be useful for organizing really long lines of code. Personally, I find them annoying and use them infrequently. You'll see the ellipsis several places in this book for page-formatting purposes.

Anyway, let's modify that while-loop so it eventually finishes.

```
i=0;
while i<50
    i=i+1;
    disp([ 'Give me ' num2str(i) ...
        ' pieces of chocolate!' ])
end
```

Now the loop will eventually stop, because `i<50` is true for the first 50 iterations (when `i` is 0 through 49). When `i` is equal to 50, `i<50` is false, which stops the while-loop. The conditional statement is reevaluated each time the code gets to the last content line—the line just before `end`. If the conditional is still true, it runs all the content again.

Unlike with for-loops, counting variables in while-loops are not automatically updated or changed. That's why we needed `i=i+1;` in the while-loop, whereas we did not need such a statement in the for-loops.

While-loops are often used during optimization algorithms, in which a parameter is continually modified until it is below a certain threshold. In these situations, the while-loop quits either via a "logical toggle" or via the `break` command. Below is an example using a logical toggle.

```
r = 50+rand*50;
toggle = true;
while toggle
```

```
    r = r/1.1;
    if r<1, toggle=false; end
    disp([ 'Current value is ' num2str(r) ])
end
```

The first line creates a random number between 50 and 100. How do I know it will be between 50 and 100? Because the function rand (not to be confused with the related function randn) generates uniformly distributed random numbers between zero and one. When those numbers are multiplied by 50, it scales the range from zero to 50. And then adding 50 brings the range of possible numbers to between 50 and 100. Here is another good opportunity for a sanity check: Type hist(50+rand(1000,1)*50,100). You'll learn more about how to use the hist function in chapter 9, but briefly, that line creates a histogram with 100 bins.

After creating this random number, the while-loop keeps dividing the variable r by 1.1 and printing its current value. In between these lines of code, it checks whether the current value of r is less than one, which in this case is our stopping criteria. If r is less than one, the toggle is set to false, and then the while-loop will stop the next time it checks whether toggle is true. Because MATLAB only checks whether toggle is true after it completes the final line of the content, the disp command will display the final value of r, which will be less than one.

An alternative method to stop a while-loop is to use the function break. Observe:

```
r = 50+rand*50;
while true
    r = r/1.1;
    if r<1, break; end
    disp([ 'Current value is ' num2str(r) ])
end
```

In this example, when r is smaller than one, the disp command is not evaluated. As soon as MATLAB sees the break command, it immediately breaks out of the while-loop without evaluating any more content. This difference between logical toggles and break is subtle but important, because they can lead to different behaviors in the loop.

You can also use the break command in a for-loop:

```
for i=1:10
    if i>6, break; end
    disp([ 'You are number ' num2str(i) '.' ])
end
```

Many programming goals can be achieved equally well using if-then statements, for-loops, or while-loops. In general, when you know beforehand exactly how many times to run through the loop (we'd call this "the number of iterations"), you should prefer a for-loop; and when you don't know beforehand how many iterations there will be, you should prefer a while-loop. If you don't like that heuristic, then just use whichever code structure you feel most comfortable programming.

7.6 Try-catch

Try-catch statements are also fairly intuitive, and you definitely use them often in your life, particularly when applying for academic positions or trying to get research funding. "Try this, if it doesn't work, here's the plan B." We start with an example.

```
e = [1 4];
try
    if e(3)>6, disp('e(3) is big.'), end
catch me;
end
```

Run the code above. MATLAB *should have* produced an error, because you are trying to access the third element of a two-element vector. In try-catch statements, MATLAB tries to run the code between try and catch. If it works, then MATLAB ignores all the code between catch and end and proceeds with the rest of the script. If the code produces an error, however, MATLAB will not crash; instead, it will put some information about the error into the variable me (in case you thought I was being cutely poetic, me stands for *MATLAB exception* and is the standard variable name for these situations). Then the script will continue to run the code after end. If this were inside a function, me would also tell you which function ("stack") and on what line of that file the error occurred. This is useful for debugging.

MATLAB always wants a variable to be the first piece of code after the catch statement. After that variable, you can include additional lines of code if you want to have a backup plan B.

```
try
    if e(3)>6, disp('e(3) is big.'), end
catch me;
    if e(2)>6, disp('e(2) is big.')
```

```
        else disp('e(1) is my friend.')
    end
end
```

Perhaps you are thinking that try-catch statements are the best way to write scripts that produce no errors. But be careful here—you should *want* your script to produce errors, because errors tell you that something went wrong. If you have too many try-catch statements, your script could be full of horrendous and dangerous errors without you knowing about it. It's like blocking the pain receptors in your arm when your hand is on the stove, and only 15 minutes later do you start wondering why the kitchen smells like something is burning.

7.7 Switch-case

You may have heard the expressions "save the best for last" or "last but not least." Those don't apply here. The switch-case control statement is the one I see least often in data analysis scripts. It provides the same kind of control as if-elseif-else-end statements. You might go the rest of your life and never program a switch-case statement, and you'd have lived a full and rich life with no regrets. But it is good for your general MATLAB knowledge to be able to interpret and possibly use switch-case constructions.

```
species = 'rabbit';
switch species
    case 'rabbit'
        disp('rabbit analyses')
    case 'rat'
        disp('rat analysis')
    otherwise
        disp('alien analyses')
end
```

The switch command works by holding the conditional statement "in mind" and then testing whether that matches the various cases. The first case that is satisfied is run and then the switch-case command finishes. The otherwise statement at the end works the same as else in if-else constructions, and is optional.

7.8 Pause

Let's learn one last function here that is not technically a control statement, but is used to control the timing of processes in MATLAB. The function pause is used to ask MATLAB to hold all processes until some time has passed or until the user provides some input. For example, run the following code.

```
pause(1); disp('time''s up!')
```

Now run it again, changing the "1" to "2" or "5." I'm sure you've already figured out that that input instructs MATLAB to freeze for that number of seconds. The input need not be integers; you can also have MATLAB wait for 100 milliseconds (pause(0.1)) or 7 milliseconds (pause(0.007)).

Typing pause with no input will cause MATLAB to freeze until a key is pressed. One situation where you might use the pause command is when looking through many results in sequence (e.g., the results from each of 20 channels). In this case, you can write a loop that shows each result in the figure and waits until you press a key to show the next result.

7.9   Exercises

1. Using the mod function, can you turn the vector 1:15 into 1–2–3–1–2–3 and so on? You have to do it in one line of code, and the result needs to start at 1 (not at 2 or 0). It should be scalable such that changing only one number in your code will move from counting in threes to counting in fours, sevens, or any other arbitrary number. This is a pretty handy trick to know.

2. What's wrong with this code, and how can you fix it?

```
if randn>1
    disp('big.')
elseif randn>-1 && randn<0
    disp('small')
else disp('super-small.')
end
```

3. This follows up on exercise 6 in the previous chapter. Create a 4 × 8 matrix of uniformly distributed random numbers. Loop through all rows and columns and test whether each element is greater than 0.5. Report the results of the test, but this time the printout should be more specific. For example, the code should print "The value in the 3rd row and 7th column is 0.123 and is not bigger than .5." Make sure the code prints out "1st," "2nd," "3rd," and so on, not "1th," "2th," and "3th."

4. Your friend Tim comes to you with this code and with tears in his eyes. He knows it's awful and it doesn't work, but no one—not even poor Tim—should ever be embarrassed to ask for help. Figure out what Tim is trying to do and fix his code to get this for-loop to work (but you have to keep the for-loop).

```
for i=-5:1.1:6
    sqxy(i,1) = i;
    sqxy(i,2) = i^2;
end
```

5. Typically, matrices are initialized before the loop. But sometimes it is necessary to initialize a matrix *inside* the loop. Modify the previous exercise to initialize the matrix sqxy inside the for-loop. Make sure it gets initialized only once!

6. Remember that the best kind of for-loop is no for-loop. Write new code that will replace the for-loop in the previous exercise. Make sure the results are identical. One way to determine whether two results are identical is to compute the mean squared error between them. Compute the point-wise difference between your new variable and sqxy, square the differences, and take the average of the squared differences. If the average is zero, the results are identical.

7. Find and fix the errors in the following code. Next, write new code to map asdf to data without the if-then statement. Any time you see an if-then statement, you should think about whether you can solve the problem without using a control statement.

```
asdf = 1;
if asdf=1
    data = 10;
elseif asdf=0
    data=5;
end
```

8. Rewrite the try-catch codes in section 7.6 using only if-then statements. Your code should not produce any errors.

9. When using AND or logical statements, MATLAB makes a distinction between && and &, and between || and |. Read the MATLAB help files for these operators to learn when to use which. Write two commands, one using && (or ||) and one using & (or |).

10. What's wrong with the following code?

```
for i=1:6
    i=3;
    varA(i) = i*43;
end
```

11. Can you make an infinite loop using a for-loop? How about using an if-then construction?

12. Determine whether the following lines of code will produce an error. First come up with your answer, and then type the code into MATLAB to confirm.

```
i=3; i=rand(i);
i=[3 2]; r=rand(i); r(3,3)
i=[3 2]; i(1), i(2)
i=3; i(1), i(2)
```

13. What are the differences between loops A and B below? Which will print out more messages in the Command window? Or will their outputs be the same?

```
%% loop A
i=0;
while i<5
    i=i+1;
    fprintf('%g ',i)
end
fprintf('\n')
%% loop B
i=0;
while i<5
    fprintf('%g ',i)
    i=i+1;
end
fprintf('\n')
```

8 Input-Output

8.1 Copy-Paste

The easiest way to get data into and out of MATLAB is simply by copying and pasting. If you have a programmer's mentality, such a brutish method may be offensive to your delicate dignity. To be sure, when you have a large amount of data, many separate files to import, or multidimensional matrices, the copy-paste method will not scale. But for small amounts of data, particularly if you need to get the data into or out of MATLAB only a few times, don't feel ashamed to open Excel or some text editor and copy-paste the numbers.

8.2 Loading .mat Files

If you already have data stored in a .mat file, the task is easy. You can load the data directly into the MATLAB workspace or you can load the data into a structure. The data file filename.mat that comes with the online MATLAB code for this chapter contains three variables: `frex`, `timevec`, and `tf_data`.

```
load ch08filename.mat
```

The first line will put the three variables directly into MATLAB's workspace. (Recall from chapter 6 that if MATLAB is not in the folder where that file is located or if the file is not in MATLAB's path, you'll get an error about the file not existing.) Typing `whos` will confirm the presence of these three variables. If you have any other variables in MATLAB's workspace, you might want to clear them for simplification. A clean MATLAB workspace is a clean mental workspace. Now try running these lines:

```
clear
datafile = load('ch08filename.mat');
```

The second command will load the file but will *not* put the variables directly into MATLAB's workspace. Instead, the variables will become fields of a structure called `datafile`. If you type `whos`, you will not see the variables from filename.mat, but if you type `datafile`, you will see them:

```
>> datafile =
    tf_data: [30x193 double]
    timevec: [1x193 double]
    frex: [1x30 double]
```

One advantage of loading the contents of a file into a variable is that it prevents overwriting data. For example, if you have a variable `timevec` in MATLAB, loading filename.mat will overwrite the existing `timevec`, whereas loading the file contents into the variable `datafile` preserves a separation between `timevec` and `datafile.timevec`. However, this can also be a source of confusion, so be mindful (it's good programming practice to give different variables different names).

The second method is also useful for loading in data from multiple subjects. Imagine you have filename_rat1.mat, filename_rat2.mat, and so on for 15 rats who volunteered to participate in your experiment. The following code will load in the data for all rats without any overwriting.

```
data = cell(15,1);
for fi=1:15
    data{fi} = load([ 'data_rat' num2str(fi) '.mat' ]);
end
```

The contents of `data{3}` will contain the data only from rat number 3, and so on.

You may not know *a priori* how many files there are or even what the entire file name is. In this case, you can use the function `dir` to locate files. This function is adopted from the DOS command and returns directory information into a variable.

```
allFiles = dir;
someFiles = dir('*rat*.mat');
```

Let's first examine the variable `allFiles`. It is a structure array with one element per file or folder in the current directory. The structure contains several fields, the most important one being `allFiles.name`, which stores the name of each folder or file in the current directory. In addition to the files and subfolders, you will also see two elements for the entries "." and ".."; these are standard Unix pointers for the current directory (".") and one

directory higher (".."). They are always the first two entries from a nonselective dir call. You can remove them by typing

```
all = all(3:end);
```

Let's say you want to have a list of all files in the current folder, and
you want to ignore all of the subfolders (including "." and ".."). You can
use the structure field .isdir. This field tells you whether each entry is a
directory (1) or not (0). To solve this problem, you could write a for-loop
with an embedded if-then statement to check whether the isdir field in
each entry of allFiles is 0 or 1. Or you could help reduce the number
of control statements in the world and solve this problem using one line
of code.

```
allFiles = allFiles([allFiles.isdir]==0);
```

This line of code says to find the entries in allFiles that have all-
Files.isdir false (in other words, that they are files and not folders), and
then reassign those to be the new allFiles variable. This is another example of a dense line of code with multiple embedded pieces. Try running it
piece by piece, from the inside out: allFiles.isdir, [allFiles.isdir],
and so forth. However, even when removing the directories, there are still
items in the allFiles variable that are not related to the rat data files. You
can see this by typing

```
{allFiles.name}'
```

Now that you know about the nonselective use of dir, it's time to discuss the selective use of dir. The structure variable someFiles (bottom of
page 104) contains only the files in the current folder for which the file
name has the letter string "rat" somewhere in the middle and ".mat" at the
end. The asterisks are wildcards indicating that any other letters can be
before the "r" or in between the "t" and the ".mat" but not after ".mat."

How would you find out information about the third file in the variable
someFiles? Of course you already know the answer: someFiles(3). If
you want to know the names of all files, you need to concatenate the field
"name" across all elements in the structure. These file names are most sen-
sibly placed in a cell array, so curly brackets are used:

```
{someFiles.name}
```

It is, admittedly, confusing to know when to use (), [], and {} for grouping
different variable types. The rules are that parentheses are used for functions, numerical indices, and structure indices; square brackets are used for
concatenation; and curly brackets are used for strings and cells. Sometimes,

multiple types of brackets are legal but produce different behavior. I think the best way to learn this aspect of MATLAB programming is not to try to memorize the rules, but rather to try them all and learn which ones produce the results you want.

The function `dir` returns only the files in MATLAB's current directory. To access files in another directory, write the full path into the input of the `dir` function. Using full paths when reading in files is always a good idea because it allows the code to produce the same result regardless of the directory MATLAB happens to be in.

```
files = dir('/path/mbcs/ch08/*rat*.mat');
```

Somewhat annoyingly, the field `files.name` will NOT contain the full path; only the file names. You will need to rewrite the path when loading in data. Soft-coding is a great solution here, as it often is.

```
filedir = '/home/mxc/mbcs/ch08/';
files = dir([ filedir '*rat*.mat' ]);
data = cell(size(files));
for fi=1:length(files)
    data{fi} = load([ filedir files(fi).name ]);
end
```

If the file path changes, or if you use the same script on different computers, you need only change the variable `filedir` and the rest of the script should work as expected.

The most common source of errors when reading in files in this way is that the file names or paths are incorrectly specified. If you see an error `Unable to read file <filename>. No such file or directory.`, the first thing to check is whether the folder path is correct. Keep in mind that Windows interprets both "/" and "\" as indicating folder separations, but Mac and Unix systems treat these two characters differently. (Hint: Check out the function `filesep`.)

Another useful function for seeing MATLAB files is `what`. Typing `what` reveals only the MATLAB files (.m, .mat, .fig, etc.) in the current directory, and not any files with non-MATLAB extensions. It should not be confused with the function `why`, which is something slightly different.

Another option for selecting files is to use a graphical user interface (GUI). MATLAB has a built-in GUI for selecting files or folders. It is easy to program and intuitive to use but is not a scalable solution for importing dozens or hundreds of files, nor is it useful if you want the analysis to run without any user input or intervention (figure 8.1).

```
[filename,filepath] = uigetfile('*rat*.mat');
```

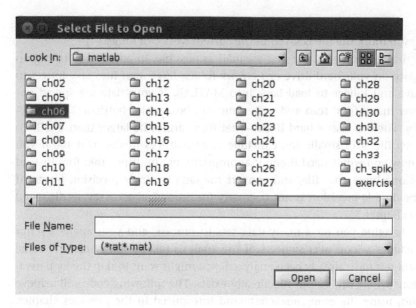

Figure 8.1
Graphical user interface for selecting files and folders.

The input string acts as the file filter to help select the right file names. Try running this line of code again using different filter strings.

One final note about importing files: Reading MATLAB files from a network server is slower than reading files that are stored on a local drive. If your data are kept on a remote server, depending on how many files there are and how often you need to import them, it might be worth copying those files to the local hard drive while doing the analyses.

8.3 Saving .mat Files

Now let's talk about writing data from the MATLAB buffer to a hard drive using the native MATLAB file format .mat. To save data in a .mat file, simply type save. This will save all of the data in the entire workspace into a file in the current folder called matlab.mat. This is generally not a good idea: The file name is uninformative, and you might have many gigabytes of data in MATLAB's workspace while you need to save only a few megabytes. Thus, you should specify the file name, and also specify exactly which variables should be saved in the file.

```
save('output_filename.mat','var1','var2','var3');
```

The variable names must be in single quotes. Without specifying a folder destination, the file is saved in the current MATLAB folder.

It's best to keep .mat files as small as possible. In addition to taking up space on your hard drive or network folder, large .mat files take longer to save and longer to load back into MATLAB. If your data are spread out over many files, read and write time can be a major bottleneck to analyses. There is also a hard limit—.mat files cannot be larger than 4 GB. To keep file sizes small: don't include large matrices in the .mat file if you know you won't need them; if appropriate, use multiple .mat files instead of one huge .mat file; and convert the data to single-precision format if the double precision is not necessary for numerical accuracy, as discussed in chapter 5.

Imagine you have many data sets to process, and your analysis script includes a loop over subjects. If the analysis takes a long time and some data sets have already been analyzed, you might want to skip the loop iterations where the output file already exists. The following code will achieve this, using the continue command introduced in the previous chapter. This code also shows you how to automate the creation of output file names so that they are easily associated with input file names.

```
ddir = '/path/mbcs/ch08/';
filz = dir([ ddir '*_file2analyze.mat' ])
for fi=1:length(filz)
    outfname = [ddir filz(fi).name(1:end-16) 'res.mat' ];
    if exist(outfilename,'file'), continue; end
    load([ ddir filz(fi).name ])
    <lots of analysis code...>
    save(outfname,'var1','etc')
end
```

There are a few new concepts in this code. First is the function exist, which tests whether the first input string exists. You can test whether a variable, function, or file exists. If the file does not exist, the output of the exist function is false, which means the continue statement is not evaluated, and then the analysis script continues. If the file does exist, the result is a non-zero number that gets interpreted as true, which means the continue command gets evaluated, and this loop skips to the next iteration in the loop (the next data set). This construction is useful when you are adding new data files into the folder, because you won't need to change the code to process the new data without re-running the existing data sets.

This code also shows how to soft-code the name of the output file. This is good programming because it guarantees that all file names have the same format, differing only by the data set–specific identifier, such as a subject number or date in which the data were collected. The "(1:end-16)" gets rid of the string "file2analyze.mat" in order to replace it with "res. mat." There is nothing special about the number 16; that just happens to be the number of characters in "file2analyze.mat."

More generally, it is good programming practice to have a consistent file-naming convention for all data sets in your research. This facilitates reading in and writing out data files without manually selecting and specifying each individual data set name.

A final minor but important point: There is nothing special about the extension .mat. MATLAB-formatted data files can be given different extensions in the save command, including .set or .dat or .MATLAB (e.g., `save('test.asdf')`). However, unless you have a specific reason, it is best to use .mat so everyone else knows it's a MATLAB-format data file.

8.4 Importing Text Files

Importing ASCII text files can be simple if those files contain only numbers and if every row of data contains the same number of columns (i.e., a full matrix). Whenever possible, you should strive to make your ASCII data files follow this organization.

```
data = load('textdata_easy.txt');
```

Life isn't always so easy, though. You might have an ASCII data file that has different numbers of columns in different rows or that has strings and numbers mixed together. Before moving forward, use a text editor to inspect the file "headache_data.txt." It starts with text that has nothing to do with the data we want to import. The data we care about are stored as one row per trial, but the different stimulus and outcome parameters are stored in different columns on different trials. Even describing the (lack of) organization in this file gives me a headache. Please, never organize your scientific data in this way. The only redeeming part about this nightmare of a data file is that it gives us an opportunity to show how flexible MATLAB can be when importing data.

For these kinds of files, you will need the `fread` function. This allows maximal control over importing the data, although it also requires some additional overhead code. The first step is to create a pointer in MATLAB to the file you want to open.

```
fid = fopen('headache_data.txt','r');
```

The variable `fid` stands for file identifier (it is common practice to use this variable name, although you could of course use any variable name you want). The function `fopen` creates a pointer to the file name specified in the first input. The second input tells MATLAB that we are going to read from this file (as opposed to write to this file, which is the topic of the next section). The variable `fid` may look like a normal variable (it might be "3" for example), but MATLAB actually uses this variable as a pointer to a location on the hard drive where that file is stored.

The next step is to read from the file. There are a few different ways of reading in text files, but it's often easiest to read them in line by line. You might not know in advance how long the data file is, so a while-loop is preferred over a for-loop. The while-loop should stop when the pointer comes to the end of the data file. This is done using the function `feof`, which stands for "file-end-of-file," and which returns "true" when the pointer is at the end of the file and "false" otherwise.

```
fid = fopen('headache_data.txt','r');
datavar = [];
row = 1;
while ~feof(fid)
    aline = fgetl(fid);
    aline = regexp(aline,'\t','split');
    datavar(row,:) = cellfun(@str2double,aline);
    row = row+1;
end
fclose(fid);
```

There are several things to explain about this code. First, we initialize the variable `datavar` to be empty. In previous chapters, I wrote that you should always initialize matrices to be the size of the final matrix. Here, however, the full size is not known in advance. At least with this empty initialization you can be confident that any previous `datavar` variable will be cleared out rather than being appended. An alternative is to initialize the variable `datavar` to be, say, 100,000 rows long, and then cut out the all-zero rows after the while-loop finishes.

Next we get to the `aline` line. The variable `aline` contains a line of data from the file, using the function `fgetl`. Notice that each time we use an f* function, we specify that we want to get data from the file associated with the pointer `fid`; this gives you the freedom to read data from multiple files in the same loop. A "line" in a text file is defined as all characters between two next-line characters (a next-line character is what you insert into a text file when you press the Enter key). The variable aline is a string. We want to extract the entries in that string that are separated by tabs. For this, the function regexp is used, which splits aline according to tabs. You could change the second input to split the data according to spaces, commas, or any other delimiter.

The function regexp can be used to convert one long string of characters into a cell array. Some practice will give you familiarity with it.

```
stext = 'Hello my name is Mike and I ride a bike.';
s1 = regexp(stext,' ','split') % parse by spaces
s2 = regexp(stext,'M','split') % parse by M
```

By typing whos, you can see that stext is a character array while s1 and s2 are cell arrays. Each cell is defined by breaking the character string each time the delimiter appears (space or M for s1 and s2). Change Mike to mike and re-run the code. You will see that MATLAB treats delimiters as case-specific.

Now that you have the data in a cell array of strings, you need to convert the strings into numbers. There are two ways to accomplish this. One is to use the function str2double, and another is to use the function sscanf. Let's see how these functions work:

```
aNumber = '42';
str2double(aNumber)
sscanf(aNumber,'%g')
```

The function sscanf is faster than str2double (demonstrated later in the exercises) and therefore might be preferred for long files. Neither function will work on cell arrays, however, so you would need another loop over each element of the cell. This is cumbersome.

Instead, you can use the cellfun function. The idea of cellfun is to perform some function on each element of a cell array ("fun" here means "function," although it's also fun to use). It works by specifying the function to be performed using the @ symbol, and then specifying the cell array to which that function should be applied.

```
cellfun(@str2double,s1);
```

The cell array s1 contains one cell with a number; the rest of the cells are converted to NaN. You can also use cellfun to extract the number of elements in each cell by typing cellfun(@length,s1).

There's more to importing this headache file to produce a usable data matrix. The exercises will guide you along the process to complete the import.

After you've finished reading in a file, you should always ask MATLAB to close that file, using the command `fclose(fid)`. If you leave the file open, your computer operating system might block access of other programs trying to access that file. Without closing the file, you might get a warning from your operating system like "This action cannot be completed because the folder or file is open in another program."

8.5　Exporting Text Files

Similar to importing text files, there is an easy way and a difficult way to write out text files from MATLAB. The easy way, which you should use whenever possible, involves the function `dlmwrite`. You specify the output file name, the name of the variable to write, and the delimiter. This function works only when the data are numeric and only when you can write out a one-dimensional or two-dimensional matrix.

```
dlmwrite('filename',data,'\t');
```

The more complicated but also more flexible method involves `fprintf`. This is the method to use when you have rows of different lengths or if you want to write strings and numbers in the same file. The code below will print some text before each number, which is not possible when using `dlmwrite`.

```
fid = fopen('data2write.txt','w');
fprintf(fid,'accuracy\treactionTime\n');
for r=1:size(data,1)
    fprintf(fid,'v1: %g\tv2: %g\n',dat(r,1),dat(r,2));
end
fclose(fid);
```

Let's go through this code. First, notice that we defined `fid` using a second input of `'w'` instead of `'r'`. The `'w'` indicates that we want to write to a file. The `fprintf` line inside the loop is a bit complicated. The first input specifies where the data should be written to. In this case, the pointer to the file (variable `fid`) tells MATLAB to write to that specific file. If you omit the pointer, the information will print to the Command window, which you saw in chapter 6. The second input specifies the information to be written into the file. The `\t` writes the tab character, and the `\n` writes a new-line character (same as the Enter key). The `%g` is a placeholder for a number that should be written out as a string. There are several other placeholders for different types of information (`%s`, `%i`, etc.); you can search the MATLAB help files or the Internet for `fprintf` to learn more of the details. These different printing options don't change the fundamental organization of the exporting shown above. Finally, the inputs after the second input are the data corresponding to each placeholder.

Now that you know how to parse the `fprintf` line inside the for-loop, you should be able to parse the `fprintf` line before the loop. At first glance you probably read it as "accuracy SLASH treactionTime SLASH n." But now you know better: it is actually "accuracy TAB reactionTime NEWLINE."

8.6 Importing and Exporting Microsoft Excel Files

To import Excel-formatted files, use the function `xlsread`. This function can separate the numeric data from the text data and also optionally imports the data in all cells as strings in a two-dimensional matrix.

```
[numericdata,textdata,alldata] = ...
            xlsread('exampleExcel.xls');
```

The variables `numericdata` and `textdata` might have missing rows or columns if those rows or columns contain no numeric or string data in the Excel sheet. That is, the second row of `numericdata` might correspond to the third row in the Excel file if the first row contains no numbers. This behavior is a potential source of confusion or errors. The safest way to deal with Excel-formatted data is to use the third output (`alldata`). You will need to convert the strings to numbers, as described earlier. The complement to `xlsread` is `xlswrite`.

```
xlswrite('outputdata.xlsx',data);
```

R and SPSS software programs are often used in the behavioral and social sciences for data analyses involving analyses of variance and other mixed-effects models. As of 2015, MATLAB does not provide standard input and output methods compatible with SPSS's or R's file formats. You will need to export the data from those programs to a text file or to an Excel-formatted file.

8.7 Importing and Exporting Hardware-Specific Data Files

Most or perhaps all data acquisition hardware systems export data files in a specialized, and usually compressed, format. These file formats vary greatly, and it is not feasible for MATLAB to provide cookbook code to import every type of file. However, unless it is secretive proprietary data, manufacturers are generally happy to provide formatting details and often will provide MATLAB functions to import their files. Some MATLAB toolboxes such as eeglab, fieldtrip, and SPM provide functions to import file types that are often used in neuroscience data acquisition.

Although manufacturers sometimes provide code to export data from MATLAB to their native format, you should avoid writing out data from MATLAB to specialized formats, for two reasons. First, some manufacturers require licenses or additional software to be installed in order to import the data into MATLAB, and thus keeping the data in the manufacturer's format may limit the number of computers that you can use to re-import the data. Second, keeping data in the MATLAB file format facilitates sharing data.

8.8 Interacting with Your Operating System via MATLAB

Many basic file manipulations can be done from the MATLAB command. For example, you can copy, move, and delete files by following these examples:

```
copyfile('file1.txt','newfolder/file2.txt')
movefile('file1.txt','newfolder/file2.txt')
delete('newfolder/file2.txt')
```

Be careful when using the delete function—unlike when you move a file to the trashcan, the delete function actually deletes the file (irretrievably!) from your computer.

MATLAB users on Mac and Unix operating systems can use the function unix('commands here') to have MATLAB send the command to a Unix terminal. MATLAB users on Windows can use the function dos('commands here'). These functions are useful, for example, if you want to have MATLAB start non-MATLAB programs.

8.9 Exercises

1. Is there a difference between these two lines of code? Why or why not?

```
all([all.isdir])=[];
all = all([all.isdir]==0);
```

2. In section 8.4, the variable datavar was initialized as an empty variable. Adjust the code to initialize the variable to be 100,000 rows long, and then cut out the all-zero rows after the while-loop finishes.

3. There is a MATLAB function called `csvread`, which reads in comma separated values (CSV)-formatted files. The online code includes a file called somedata.csv. Based on the result of `help csvread`, figure out how to import the data. Next, open the file using a non-MATLAB text editor such as Excel or Notepad, and confirm that MATLAB imported the data correctly (e.g., what happens to missing entries in the CSV file?). Finally, generate a two-dimensional matrix of uniformly distributed random numbers between 10 and 35, and export this matrix to a file using `csvwrite`.

4. Section 8.2 had the following line of code. Rewrite this functionality using a for-loop and an embedded if-then statement. Not that I encourage using loops instead of single lines of code; it's just for practice.

```
allFiles = allFiles([allFiles.isdir]==0);
```

5. I claimed that the function `sscanf` is faster than the function `str2double`. Now it's time for you to prove it. Write code that times how long it takes to convert a string to a number using each of these functions. To measure computation time, use the functions `tic` and `toc`. You might not see much of a difference with one function call, so you should re-run the test by looping over 1 million numbers.

6. Each of the following lines of code contains an error. Find and fix them.

```
data = dlmwrite("datafile.txt");
data = load('rdata' num2str(fi) '.mat');
```

7. Inspect the following two lines of code (variable `var4fname` is a string variable that contains the name of the output file). Neither contains an error, but one will do something other than the intended action. Which one is it and why?

```
save(var4fname,'var1','varB')
save('var4fname','var1','varB')
```

8. Are there differences among the following lines of code?

```
save(var4fname,'var1','varB')
save(var4fname;'var1';'varB')
save(var4fname,'var1';'varB')
```

9 Plotting

Visual inspection of data is very important, particularly in neuroscience. Visualization allows you to inspect your data for quality and to identify artifacts, it allows you to do sanity checks, and it allows you to interpret the results in more meaningful and insightful ways. For exploratory analyses and for multidimensional results, interpretation might be nearly impossible without visual inspection. Have I convinced you that it's important to plot your data? (Did you even need to be convinced?)

9.1 What You Need to Know Before You Know Anything Else

Before starting to learn how to visualize data, you need to learn a few basic MATLAB commands that will help you work with figures. A figure is a window in which data are drawn, and you can open a new figure simply by typing `figure`. When you issue a plotting command, MATLAB will draw the result in whatever is the most recently used figure. If there are no figure windows, MATLAB will automatically create one.

The figure opens as a blank gray window. The top left of the window says "Figure 1" (figure 9.1). You can have many figures open at the same time. Try it:

```
figure, figure, figure
```

You can also specify the number of the figure.

```
figure(1)
figure(100)
fignum=400; figure(fignum)
```

Run that first line (`figure(1)`) before running the other two. There were three figures before running that line, and there are still three figures after. This happened because you called figure no. 1, and figure no. 1 already

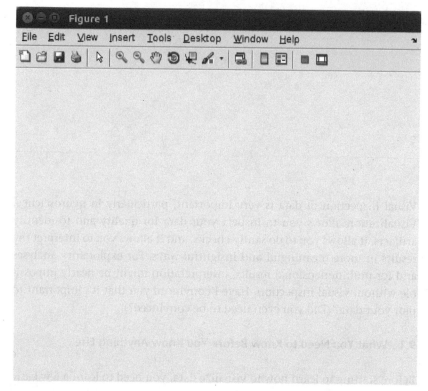

Figure 9.1
A typical MATLAB figure window.

existed. Instead of creating a new figure, that command made figure no. 1 the active figure. The next two lines of code demonstrate that figures need not be consecutively numbered. If you want to make a good impression, start counting figures at a large number like 846; your supervisor might think that you've been meticulously (or perhaps obsessively) creating hundreds and hundreds of data plots.

Although there is no strict limit to the number of figures you can have open, in the interest of maintaining your sanity (and your computer's graphics card capacity), try to avoid having too many figures open at once. I like to have all of my figures docked inside the main MATLAB GUI, but you should arrange them however you find most comfortable.

All plotting commands draw data into the "active" figure. This is the figure you last opened, most recently mouse-clicked, or called directly by number (as in, `figure(4)`). When you issue a plot command, MATLAB does not draw the data directly in the figure per se; instead, it creates an "axis" inside the figure, and then draws the data in that axis. The axis is the white area inside the gray figure. Having data drawn in axes is useful for creating many axes inside the same figure, which you will see later. Type plot(rand(4)) and you will see that a new axis was created in the active figure.

To close figures, you can click on the "x" in the top of the figure window. You can close figures from the command line:

```
close % closes only active figure
close(10) % closes figure #10
close([2 3 5:8])
close all
```

Finally, if you want to remove the drawing from the figure, type clf, which stands for clear figure. That will bring you back to a fresh and empty figure. Typing cla will clear the axis. As you go through this chapter, you might find it useful to clear the figure or axis before each section or to open new figures.

9.2 Plotting Lines

The MATLAB function to plot a line is called plot. It can take several inputs, the most important being what to plot on the y axis.

```
plot(1:10)
```

If only one input is provided, MATLAB assumes that the x axis should be indices (1 to N in integer steps). You can also specify the x-axis values.

```
x = 1:2:20;
y = x.^2;
plot(x,y)
```

Notice that the new plot overwrote the previous plot. That is, when you asked MATLAB to plot new data, the old data were first removed from the plot. You can plot multiple lines on top of each other using the hold command.

```
plot(x,y)
hold on
plot(x,log(y))
hold off
plot(x,y.^(1/3))
```

If you have MATLAB version 2014a or earlier, all of the lines on your plot are blue. If you have MATLAB 2014b or later, each line is automatically colored differently while the `hold` command is toggled. Even if you have the latest MATLAB version, it's better to maintain control of the line colors yourself rather than let MATLAB pick the colors, in part because this behavior is version-specific, and in part because losing control over the colors of the lines can lead to confusion about which line corresponds to which data vectors. An optional third input allows you to specify the line color (see figure 9.2 for a grayscale version of this figure).

```
cla
plot(x,y/50,'r')
hold on
plot(x,log(y),'k') % log is the natural log
plot(x,y.^(1/3),'m')
```

After running this code to see three lines, run the code again but delete the `hold on` command. Are you surprised to see the three lines still plotted? The `hold` property of an axis is not cleared when you type `cla`. It is, however, when you type `clf`. Now replace `cla` with `clf` and run the code again (still without `hold on`). It is also possible to plot multiple sets of lines without using the `hold` command, if those lines are in matrix form or if you use the function `plotyy`. You'll see examples of this later and throughout the book.

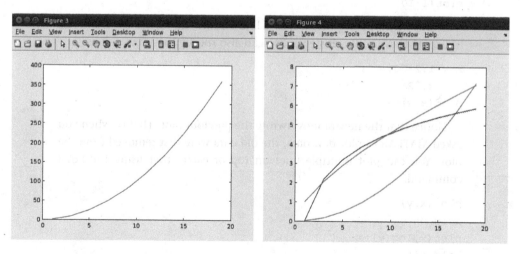

Figure 9.2
Two MATLAB figures illustrating a line plot of y^2 (left plot) and y^2, $\ln(y)$, and $y^{1/3}$ (right plot).

In addition to line color, you can also specify the marks used at each point, the line thickness, and several other line properties.

```
hold off
plot(x,y,'ro-','linewidth',9) % default width is 1
```

You can type help plot to see more plotting options. I won't list them all here, but throughout the rest of the book you will see many different plotting options and styles.

If you have labels for the lines, you can use the legend command.

```
plot(x,y,'bp')
hold on
plot(x,log(y),'r*--')
legend({'y=x^2' ; 'y=log(x^2)'})
```

There are two things to notice about the legend function. First, it takes a cell array as input, and each element of the cell corresponds to each plot item in the order in which it was called. Second, when plotting text in a figure, MATLAB interprets some characters differently. For example, the text x^2 is printed as x^2, because MATLAB interprets the caret symbol (^) to indicate that the next character should be superscripted. We'll come back to this issue in section 9.10.

9.3 Bars

Creating bar plots is just as simple as creating line plots.

```
bar(x,y)
```

An optional third input specifies the width of the bars.

```
bar(x,y,.2) % try other numbers
```

Bar plots often have associated error bars. The MATLAB function errorbar is helpful here. In this function, you specify the x and y data as above, and then also the size of the error bars (figure 9.3). By default, MATLAB uses this input to create symmetric bars, such that when you specific error e, the error line is $e/2$ on top of the bar and $e/2$ below the bar. You can create asymmetric errors by inputting separately the upper and lower error sizes.

```
e = 100*rand(size(x));
errorbar(x,y,e) % symmetric
errorbar(x,y,e/2,e/8) % asymmetric
```

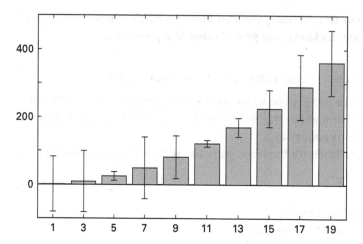

Figure 9.3
Bar plots with different-sized error bars on top.

Professional-looking error bar plots typically have both bars and error bars without any lines connecting them.

```
bar(x,y), hold on
errorbar(x,y,e,'.')
```

9.4 Scatter Plots

Basic scatter plots can be created with the plot function, but the scatter function provides additional options that can be particularly useful when you need to show an extra dimension in your visualization.

```
scatter(x,y,'o')
```

It is the additional optional inputs that make the function scatter useful. Let's imagine you want to plot the relationship between firing rate and firing variability, and you simultaneously want to indicate the cortical depth from which the neurons were acquired (thus, three dimensions of data in a two-dimensional plot). Firing rate and variability can be shown on the x and y axes, and the depth can be shown by the color of each dot (figure 9.4).

```
n = 100;
frate = linspace(10,40,n) + 10*rand(1,n);
fvar = frate + 5*randn(1,n);
ndepth = linspace(100,1000,n);
scatter(frate,fvar,100,ndepth,'filled')
```

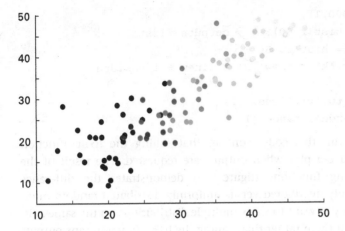

Figure 9.4
One use of the MATLAB `scatter` function is to add a color to each dot, which can be used to illustrate an additional dimension of information.

The input `100` above specifies the size of the markers, `ndepth` defines the color intensity of each circle (note that the size of `ndepth` is the same as the size of the to-be-plotted vector), and `'filled'` specifies that the color should fill the circles rather than having empty circles with only a colored outline (try that line again without that final input). Obviously, these are fake data, but the interpretation of such a result would be that neurons with higher firing rates (*x* axis) have more variable responses (*y* axis), and that neurons found deeper in the cortex ("hotter" colors toward dark red) have higher firing rates.

9.5 Histograms

Data distributions can be illustrated in histograms. MATLAB will cut the data into *N* bins and create a bar plot showing the number of elements in each bin.

```
r = randn(1000,1);
hist(r,50) % 50 bins (default is 10)
```

Rather than immediately showing the histogram, the `hist` function can return the *x* and *y* values. This is useful if you want to show the histogram in a line plot rather than a bar plot, and is useful for certain data analyses, such as computing entropy and discretizing continuous variables. The outputs of the `hist` function will become important, for example, in model fitting (see chapters 28 and 29).

```
ru = rand(1000,1);
[y_r,x_r] = hist(r,50); % y outputs first
[y_ru,x_ru] = hist(ru,50);
plot(x_r,y_r,'k','linew',2) % linew = linewidth
hold on
plot(x_ru,y_ru,'r','linew',2)
legend({ 'randn';'rand' })
```

When you run this code, confirm that calling the hist function does not produce a plot when outputs are requested. The result of the separate plotting functions (figure 9.5) demonstrates the difference between normally distributed versus uniformly distributed random numbers. Whenever you want to show multiple distributions on the same axis, it's better to plot them rather than calling the hist function sans outputs multiple times.

MATLAB recently introduced a new function to create histograms, called histogram, which provides some additional functionality beyond hist. You'll learn more about this in chapter 29. If you are feeling adventurous, you can try to reproduce figure 9.5 using the histogram function.

9.6 Subplots

Subplots allow you to create multiple windows (axes) inside a figure. To create subplots, define the spatial layout of the figure in terms of the number of rows and the number of columns, and then specify which cell in that

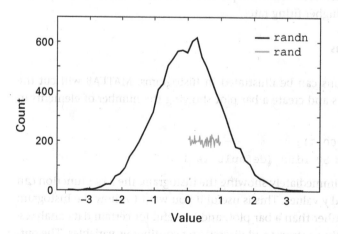

Figure 9.5
Histograms (distribution plots) of normally distributed and uniformly distributed random numbers.

grid you want to be the active plot (the one to which the next plotting command will be issued). Figure 9.6 illustrates this concept, and the code below shows an example of a 2 × 2 figure arrangement.

```
subplot(221) % commas unnecessary for <10 subplots
plot(r), title('normal random numbers')
subplot(2,2,2) % but commas help differentiate
plot(ru), title('uniform random numbers')
subplot(223)
plot(x_r,y_r), title('distribution of normal')
subplot(224)
plot(x_ru,y_ru), title('distribution of uniform')
```

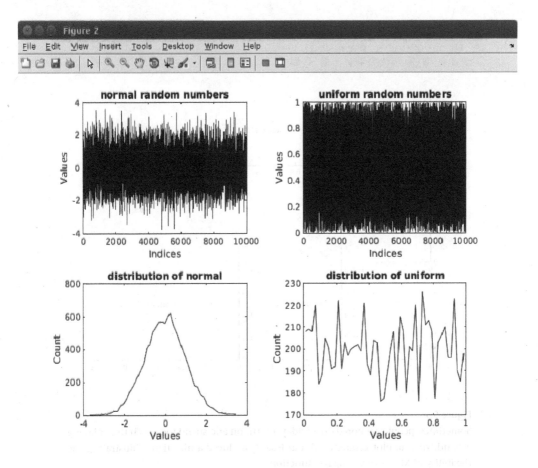

Figure 9.6
Four subplots in one figure. Subplots are useful to show several related plots.

You are not confined to having subplots in perfect grids. Subplot geometry can be mixed.

```
subplot(221), plot(r)
subplot(222), plot(ru)
subplot(212), plot(x_r,y_r)
hold on, plot(x_ru,y_ru,'r')
```

Notice that I changed the 2×2 geometry to a 2×1 geometry. Figure 9.7 shows an even more flexible use of the subplot function. In fact, you can create any number of axes with any arbitrary sizes and locations. This is done using the set command, which you'll learn about later in this chapter and again in chapter 33.

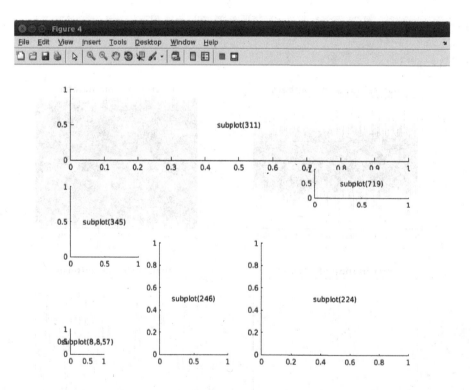

Figure 9.7
Sometimes, people get bored on a Friday afternoon and spend too much time playing around with subplot geometry. But at least I produced a nice figure illustrating the flexibility of MATLAB's subplot function.

9.7 Patch

The patch command draws flat "tiles" of any shape you define. Patches are often used in neuroscience illustrations; for example, to indicate regions of statistical significance in line plots or to draw continuous error regions around time courses. Other MATLAB drawing functions use patches (such as contourf, which you will learn about below), particularly when exporting figures to vector-format files.

To create a patch, specify the x and y coordinates of a polygon, and MATLAB will take care of the rest. You must also define the color of the patch as a third input.

```
x = [1 2 3 4 3 2 1];
y = [9 9 7 4 1 3 2];
patch(x,y,'r')
```

The xy coordinates must be specified going around the polygon, as shown in figure 9.8. This often means listing x-axis coordinates forward and then backward, as is done in the code above. Watch what happens if the points are specified in ascending order (figure 9.8, lower panel).

```
[~,idx] = sort(x);
patch(x(idx),y(idx),'r')
```

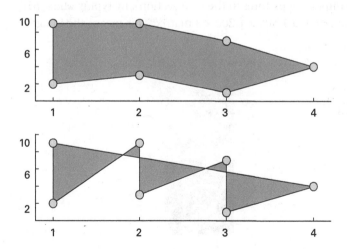

Figure 9.8
The patch function can be used to make a colored region around specified boundary coordinates. The order of the inputs is important—these two different patches share the same seven boundary points.

This may seem like a simple point in this toy example, but it can get tricky in real analyses. For example, if you have a patch that is defined by a region in time, you will need to specify time (the *x* axis) going forward and then going backward.

Using handles to patch objects will allow you to change their qualities such as transparency, edge color, and thickness. You will learn about handles later in this chapter, and the exercises will help develop your patch-making skills.

9.8 Images

Image data are shown very often in neuroscience. Not only pictures of imaging (fMRI and optical imaging) but also many complex results in, for example, electrophysiology are often best viewed as images. An image is a matrix of numbers, and the value at each pixel can be mapped onto a color and shown in a picture (figure 9.9).

There are several functions you can use to show images in MATLAB, one of which is called imagesc.

```
pic = imread('saturn.png');
imagesc(pic)
```

The function imread can be used to read data from many picture formats (more on importing pictures in the next section). By typing whos pic you can see that pic is a 1,500 × 1,200 × 3 matrix. This means the image is

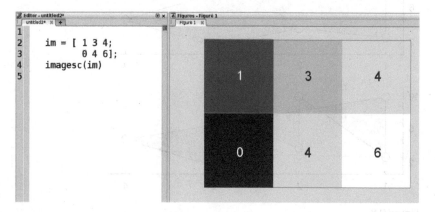

Figure 9.9
Matrices containing data or images can be visualized by mapping the numerical values in the matrix onto color intensities in the image.

1,500 pixels wide and 1,200 pixels high. The third dimension contains the red, green, and blue channels (figure 9.10). The function imagesc can take a three-dimensional matrix as input, but the third dimension must have exactly three elements. Watch what happens when we try to make an image from a matrix with an oversized third dimension.

```
pic2 = pic;
pic2(:,:,4) = pic(:,:,1);
imagesc(pic2)
```

The code above creates a new matrix in which the fourth element of the third dimension is the same as the first element of the third dimension (saying that out loud makes you feel like the narrator of a sci-fi story, but you'll get used it to). MATLAB gives an error because it understands how to plot RGB images but does not know how to interpret a fourth element in that matrix.

The function imagesc comes from the function image, but imagesc automatically optimizes the range of the color scale in order to span the range of the input matrix; image, in contrast, does not. You can try using both, but in practice there is really never a situation in which image is advantageous over imagesc.

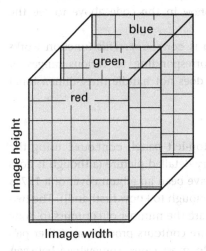

Figure 9.10
A color picture is represented in MATLAB (and many other computer programs) as a three-layer matrix in which the value at each pixel encodes the amount of that color at that pixel.

The function `imagesc` also handles two-dimensional matrices; in this case, the matrix element values signify pixel intensity, and the figure's color legend maps those intensity values onto color or grayscale values. Let's look at each of Saturn's color channels, while simultaneously integrating and applying concepts you learned in previous chapters.

```
colorchans = { 'red';'green';'blue' };
for chani=1:3
    subplot(2,2,chani)
    imagesc(pic(:,:,chani))
    axis off
    set(gca,'clim',[0 255])
    title([ colorchans{chani} ' channel' ])
end
```

The code `axis off` turns off the axis in each plot. You can try running this code again with `axis on` (or deleting that line) to see the difference. You'll learn more about the function `set` below; briefly, this changes the color limit to be between 0 and 255.

You can also use the function `imshow` to display image data: `imshow(pic)`. This function does some additional axis and image processing to show the picture. Single-layer images are shown in grayscale, for example. Replace `imagesc` with `imshow` in the code above to see the difference.

Another function for showing data is `contourf`. This function works by creating a number of patches corresponding to various ranges of color values. The function `contourf` does not handle three-dimensional inputs.

```
contourf(pic(:,:,1))
```

As you can see from figure 9.11, top-left panel, `contourf` using the default settings does not produce a very detailed picture (although it does look neat ... the color version could have been an album cover of a 1970s rock band [for the record, I'm not old enough to know firsthand]). The two most commonly used optional inputs are the number of contours to draw and the color of the contour lines. More contours produce smoother pictures but take more time to render. In most cases, somewhere between 20 and 50 contours is sufficient. The color of the contour lines, which is black in the top-left panel of figure 9.11, can be set to `'none'` for smooth-looking plots.

Figure 9.11
The image of Saturn using `contourf` with different inputs, `imagesc`, and `surf`.
Don't try to clean the page—those white dots are moons, not printing errors or dust
specks.

```
contourf(pic(:,:,1),40) % 10 is the default number
contourf(pic(:,:,1),40,'linecolor','m')
contourf(pic(:,:,1),40,'linecolor','none')
```

`contourf` is an extension of the function `contour`. The function contour only draws contour lines without filling any colors between them. We
will use it later in the book to outline regions of statistical significance on
top of plots created by `contourf`. For now, try a few different options to get
a feel for how `contour` works.

```
contour(pic(:,:,1))
contour(pic(:,:,1),'linecolor','m'), axis off
contour(pic(:,:,1),1,'linecolor','k'), axis off
```

A related function is called `surf` (readers from the west coast of the
United States or France may be disappointed to learn that this function
stands for surf*ace*, not for surf*ing*). The `surf` function produces something
like a `contourf` plot, except that it's in three dimensions.

```
surf(pic(:,:,1))
```

When you run this code on your computer, your figure will look all black and will be difficult to interpret. Try changing the shading option, which is done in figure 9.11, bottom-left panel.

```
shading interp
```

Maybe you can sort-of make out the image. The prominent features are the ring and the moons off to the left. Type the command `rotate3d` and then left-click-and-drag the figure until the three-dimensional surface makes more sense. You will see that in a surface plot, the color and height (z axis) provide the same information. For comparison, run one of the `contourf` lines above and keep the `rotate3d` command on—you'll see that `contourf` and `surf` produce identical images when looking "directly down" onto the image, and rotating reveals that `contourf` is a colored flat surface.

There are several noticeable differences between `contourf`/`surf` and `imagesc` (figure 9.11). These are extremely important to keep in mind, because getting confused about the differences may lead to misinterpretations of your results. You can see one of the key differences in figure 9.11, but running the code below will create a separate figure that highlights the distinction.

```
subplot(121)
contourf(pic(:,:,1),40,'linecolor','none')
title('contourf')
subplot(122)
imagesc(pic(:,:,1))
title('imagesc')
```

Notice that the x axis is the same for both plotting functions, but the y axis is upside-down! By default, `imagesc` draws in so-called matrix form (also sometimes called *ij*-mode), which means the first row of the matrix is plotted at the top of the figure; `contourf` draws in *xy*-mode, which means the first row is plotted at the bottom of the figure. Both formats are "correct," although you will find the row-1-at-the-bottom format to be more intuitive in most visualizations. There are a few more differences between `imagesc` and `contourf` that will come up later in the book.

9.9 Get, Set, and Handle

`get` and `set` is a pair of functions that allow you to access and modify many properties of figures and parts of figures. You will probably use `get` and `set` functions nearly every time you produce a plot or image in MATLAB, so it's good to be comfortable with using these functions.

Each axis, figure, and plot object has a variety of associated properties. These properties can be accessed with the `get` command, and they can be changed with the `set` command. Let's start with the `get` command. This function takes two inputs: the object to query and the property value to return.

```
plot(rand(3))
get(gca,'xlim')
yTik = get(gca,'ytick');
```

The first input in both `get` commands is `gca`, which stands for "get current axis." The "current" axis is the active one—the one most recently used, or most recently mouse-clicked. The first `get` command in the code above returns the x-axis limits. This is a two-element vector that indicates the starting and ending points of the *x* axis. Try replacing the `xlim` with `ylim`, `zlim`, or `clim` (color limit). The second `get` command returns the values of the ticks (the small horizontal black lines) on the *y* axis and stores those values in a variable called `yTik`. You can type `get(gca)` to see all of the properties of the current axis.

Remember the concept of using the output of one function as the input to another function? Let's try that here by plotting a horizontal line at $y = 0$. We want this line to go completely from the left to the right side of the plot, but we want to accomplish this without knowing *a priori* what the x-axis limits are.

```
hold on
plot(get(gca,'xlim'), [0 0],'k')
```

The `get` command is useful to *access* properties, but you need the `set` command to *change* them. The `set` command works using "property-value" pairs. This means you first specify the property that you want to change, and then you specify the new value of that property.

```
set(gca,'xlim',[0 2])
set(gca,'ytick',[0 .5 .8 .91])
```

Multiple property-value pairs can be specified in the same `set` command, as long as those properties refer to the same object (in this case, gca).

```
set(gca,'xlim',[0 2],'ytick',[.5 1],'xtick',0:.2:2)
```

More generally, `gca` and `gcf` are called "handles." Axes have handles, figures have handles, and—this is where the fun begins—plot objects have handles. Handles look like normal numeric variables, but MATLAB recognizes them as pointers to objects in a figure. They are similar to file pointers that you learned about in the previous chapter. It is good programming style to put an "h" at the end of handle variable names, so they are easily recognizable.

```
line_h = plot(1:10,(1:10).^2);
get(line_h)
get(gca)
```

Notice that the properties of the line plot and of the axis are different. That will become very useful, because in real analyses, you will have multiple plot objects (images, lines, patches, etc.) in the same axis, and you will want to change the properties of each item independently. The code below will change a few properties of the lines.

```
set(line_h,'linewidth',4,'marker','o',...
           'markeredgecolor','k')
```

Take a few minutes to try finding and changing other properties. Then close the figure and try re-running the set(line_h...) commands. Why does MATLAB give an error?

A nice feature of set is that it can take a vector of handles and change all of their properties simultaneously (assuming they have the same properties). For example:

```
plot_hs = zeros(1,100);
for i=1:100
    plot_hs(i)=plot(randn(max(1,round(rand*10)),1));
end
set(plot_hs(1:50),'color','k')
set(plot_hs(25:75),'marker','o')
set(plot_hs([1:10 20:5:100]),'linewi',4)
```

You can change properties of figures by referencing gcf (get current figure) rather than gca.

```
set(gcf,'color','m','name','Data, experiment 2b')
```

You can also create handles to specific axes or figures.

```
axes_h=axes;
figure_h=figure;
```

If you use the set function with a property but without specifying a value, MATLAB will return the possible values that can be specified for that property.

```
set(plot_hs(1),'marker')
```

9.10   Text in Plots

```
clf
text(.6,.4,'Yo!')
```

Already you get the idea of the text command. It plots the string in the third input at the x and y coordinates specified by the first two inputs. Unlike most other plotting functions (but similar to the patch function), repeated text commands will remain on the same plot even without typing hold on. Try running the text function a few more times, changing the three inputs (for now, keep the x and y coordinates between 0 and 1). If you are too lazy to think up numbers between 0 and 1, try this code:

```
for i=1:1000, text(rand,rand,'Yo!'), end
```

Now run text(1.05,.7,'outside'). Typically when you plot new data into a plot, MATLAB will automatically adjust the axis limits to show the new data. That doesn't happen with the text command (figure 9.12). Indeed, the word "outside" is partially cut off in the plot. Try running text(1.5,.7,'outside'). When text is plotted outside the current range of the axes, you would need to adjust the axes manually using the set function.

There are many properties of the text object that can be modified (including the text itself). As you might have guessed, changes to text objects require associating the text object with a handle and using set.

```
clf
txt_h = text(.5,.5,'Hello')
set(txt_h,'Position',[.2 .7])
set(txt_h,'color','m','String','Zoidberg')
set(txt_h,'HorizontalAlignment','Center')
```

As mentioned earlier, all of the above lines could have been put into one long line. That's possible because they all access the same plot object (the same handle). Type get(txt_h) to see a list of additional changeable options.

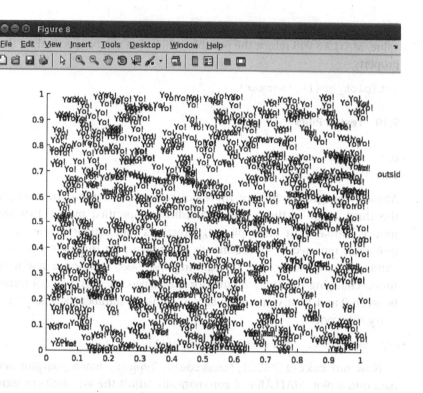

Figure 9.12
This figure shows that multiple calls to the text function do not replace existing plot items and that the plot axes are not automatically adjusted to new plot items.

In the text function and in other ways of displaying text in figures (titles, axis labels, etc.), some characters are reserved for modifying other characters. Two of them are underscore (_) and caret (^), which, respectively, make the subsequent character subscripted or superscripted. To show Greek characters, use the backslash and then the Latin word that corresponds to the Greek letter. Most of them are obvious (\mu for μ, \ alpha for α, etc.); the rest you can look up online. If you want to display one of the reserved characters, use a backslash in front of it. For example, to display an underscore, use _. Observe the effects of the following command:

```
set(txt_h,'String','Z_oidbe^r^g is an \alpha\_crab')
ylabel('Power (\muV^2)')
```

You can remove text objects (and, more generally, other plotting or axis objects for which you have handles) by passing their handles into the `delete` function.

```
delete(txt_h)
```

9.11   Interacting with MATLAB Plots

MATLAB allows you to interact with plots by zooming, panning, and extracting data. These features can be activated or deactivated by clicking on the appropriate button in the Figure *Menu* tab. Note that when the figures are docked, the Figure *Menu* tab is on the main MATLAB *Menu* bar; when the figures are undocked, the menu options appear on the figure window. Many of these menu options can also be activated in the command line or in a script by typing `zoom`, `rotate3d`, and `datacursormode` (this is called "Data Cursor" in the Figure *Menu* tab). Below is a description of these three interactive methods.

Zoom. Once the zoom is activated in a figure, there are three ways to zoom. One is to left-click somewhere on the plot, which will zoom to the area in the plot where the cursor is positioned by a zoom-factor that MAT-LAB guesses. Another is to click-and-drag a box in the figure, and then MATLAB will zoom into that section. Finally, you can use the mouse's scroll wheel or touchpad gesture to zoom in (scroll up) and zoom out (scroll down). Holding down the Shift key while zooming zooms out. You can also right-click and select to zoom out. In fact, you never really need to use the dedicated zoom-out button, because all the zooming out can be done while the zoom-in button is toggled.

Data Cursor is a useful utility. Run the following code to see how it works.

```
plot(randn(100,1)), datacursormode on
```

Now that the Data Cursor is activated, move the mouse close to a data point and left-click on the plot. The data point closest to the mouse will be selected, and a small window will open that displays the exact x and y coordinates of that point (figure 9.13). This is useful for picking peaks or troughs in the data and comes in handy during sanity checking.

Pan lets you click-and-drag the area of the plot (figure 9.13). This is useful if you zoom in to part of a plot and want to pan to another part while keeping the zoom factor the same.

Edit Plot lets you click on parts of the figure; for example, to delete them.

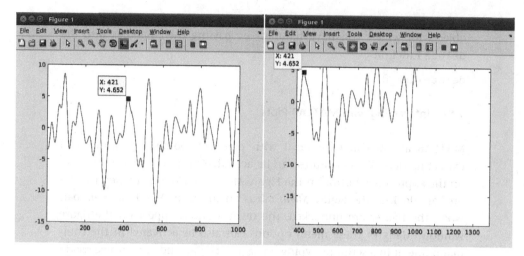

Figure 9.13

The Data Cursor tool (left panel) shows the *xy* values of a mouse-clicked plot coordinate. The Pan tool (right panel) is used to move the data sheet around in the axis.

Rotate 3D is used in combination with a three-dimensional plot created by, for example, `surf` or `plot3`, to rotate the axes using click-and-drag mouse operations. While rotating an axis, you can see some text on the bottom left of the figure, which says something like "Az: -1 El: 72." This indicates the azimuth and elevation to which you manually set the plot. You can manually set the three-dimensional view using the MATLAB `view` function. You'll see in chapter 33 how changing the view inside a loop can be used to create movies.

There are other ways to adjust the figure properties manually, which you can learn about on your own by randomly clicking different menu options. However, I recommend against adjusting figure properties manually. It might seem easier to do in the moment, but manual adjustments, though easy to do, are also easy to forget and difficult to reproduce. It takes a few extra seconds or minutes to set your plotting preferences using the `set` function, but once you have the figure set up how you like it, you or anyone else will be able to reproduce the exact same figure.

9.12 Creating a Color Axis

MATLAB has many built-in color schemes, with "jet" being the most popular in neuroscience and many other fields. It's easy to create your own color scheme, but before rushing off to invent new color palettes, check out the existing schemes to see if they suit your needs. Type `help graph3d` to see a list of color maps (you might think the color maps would be listed under `help colormap`, but that redirects you to `help graph3d`).

There are three categories of color maps. They are described below, and the online MATLAB code will illustrate each of these map types on the same data.

1. *Bipolar.* Bipolar maps start at some "cold" color (e.g., blue) that indicates negative values, and end at some "hot" color (e.g., red) that indicates positive values. Bipolar maps should be used when your data have negative and positive values. Except in rare situations, you should always use symmetric color maps, meaning that a color limit of –3 to +3 is preferable to –2 to +3.
2. *Monopolar.* These are color maps that are designed to range from zero to some non-zero positive or negative value. Avoid using bipolar color schemes for monopolar data to minimize confusion. For example, most people expect green to indicate zero and blue to indicate negative numbers, but if you use the jet map for monopolar data, dark blue indicates zero and green indicates a midrange positive number.
3. *Circular.* These are color maps in which the color "wraps around," such that the color on the negative end is the same as the color on the positive end. Circular color maps should be used only for circular or polar data. To understand a circular map, try to visualize a three-dimensional RGB space; most color maps are lines through this space, whereas a circular color map is a ring.

A color map in MATLAB (and in many other programming languages) is a look-up table (LUT) in which ranges of numerical values are mapped onto specific color values. A color map is defined as an $N \times 3$ matrix, where N defines the color resolution and 3 specifies the ranges for red, green, and blue colors. When creating an image, MATLAB will discretize the data into N bins and color each bin according to the color-value mappings in the look-up table. To get a feel for how these look-up tables work, consider figure 9.14. The three lines in each subplot correspond to red, blue, and green color intensities as a function of the data value bin. Shown here are three examples: jet (bipolar) and bone (monopolar) color maps (these maps come with MATLAB) and a circular color map that I created.

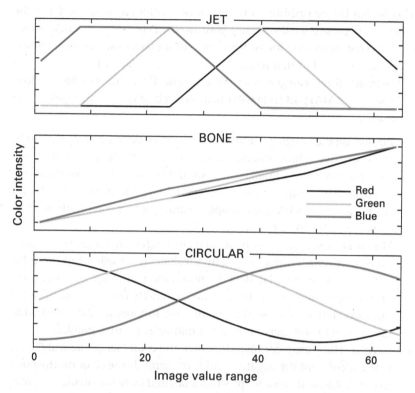

Figure 9.14
These lines are visual representations of look-up tables that MATLAB uses to map
image intensities to color intensities.

9.13 Saving Figures as Picture Files

Figures are great to look at in the MATLAB program, but eventually you will
want to save the figures to use in posters, publications, and so forth. To
export a figure in a format that can be used by other programs, click on the
Figure menu button, *File*, and *Save as*. There are many formats to save in.
The default figure format is a .fig file. This is a MATLAB-formatted file and
therefore easily accessible only in MATLAB. The main advantage of saving
figures as .fig files is that, when reimported into MATLAB, they retain zooming, panning, and other specifications.

However, pixel-based or vector-based files are generally preferable for
sharing and for publications. In pixel-based graphics (e.g., file formats
.bmp, .png, and .tif), the image is stored as a matrix of pixels; zooming-in

to the picture will reveal a block-like structure. In vector-based graphics (e.g., file formats .eps, .svg, or .pdf), image parts are stored as objects, and zooming in or stretching does not reduce the quality of the image. Programs for pixel-based images include Photoshop and Gimp; programs for vector-based graphics include Illustrator, Inkscape, and CorelDRAW (Inkscape and Gimp are free programs). In general, you should try to work with vector-based graphics because of the increased image quality. If you are unsure which format will work best, try exporting the figure using several different formats and import into your graphics software.

If you have a lot of rich graphics in your figures, they might get pixelated during export even though you selected .eps or .svg. If this happens, you need to force MATLAB to export as vector format. To do this, select *Figure, File, Export setup*. In the *Export setup* window, select *Rendering* from the left-hand side, and then choose *Custom renderer* to be "painters." From there, you can export directly, or select *Apply to Figure* and then follow the *Save as* instructions in the previous paragraph.

Manually exporting figures is fine if you want to export only a few figures. If you want to save many figures or if you want figures to be saved while a long analysis is running and you are off frolicking in dream land, you'll need to be able export figures automatically. This is done using the `print` function. It's easy to use: Input the handle to the figure you want to save (this can be a specified handle or `gcf`), the file type (with a "-d" in front of the type), and the file name. Below are a couple of examples.

```
print(gcf,'-dpng','test') % png format, test.png
print(gcf,'-dsvg','test') % svg
```

9.14 Exercises

1. Why does this code produce an empty plot? How would you fix it?

```
pic = imread('saturn.png');
contour(pic(:,:,1),10,'linecolor','none')
```

2. Open a new figure and run the first two lines of code below. Then, use the `set` function to set the x and y axes to go from –1 to +1. Next, open a new figure, and run the second line before the first line. What have you learned about the function `line`? Finally, change the code to make the lines from both functions magenta. Setting line colors in the `line` function is not the same as in the `plot` function.

```
plot([0 cos(pi/3)],[0 sin(pi/3)])
line([0 cos(pi/4)],[0 sin(pi/4)])
```

3. Patches can have handles as well. Modify the code that produced figure 9.8 to make the patch purple with a blue line around it. What other visual features of the patch can you change?

4. What is wrong with this code, and how can you fix it?

```
randImg = round(255*rand(800,600,4));
imagesc(randImg)
```

5. Another image plotting function is called pcolor. Run the code below and observe the differences between pcolor and imagesc. Then count the number of rows and columns in the two images. Is there a difference? The answer is yes, and you should search the Internet to figure out why.

```
ri = randn(10);
subplot(121), pcolor(ri)
subplot(122), imagesc(ri), axis xy
```

6. You can change the orientation of the result of imagesc by typing axis xy (the default is axis ij). Or use the set command. Create an image (e.g., Saturn, or whatever else) with a handle. Use the get function to see a list of properties of the image. Which corresponds to the direction of the y axis, and how do you flip the image so the result is consistent with the picture drawn by contourf?

7. Open a figure and plot some data. Then activate the zoom tool by typing zoom in the Command window. Confirm in the figure that the Zoom tool works. Then type zoom again in the Command window, and try to zoom again in the figure. Now type zoom on in the command line. Type zoom on again. What have you learned? The same rule applies for rotate3d and pan.

8. Convert the equation below into MATLAB code, and evaluate it from –8 to +8 in steps of 0.0132. The plot is probably difficult to view at the default settings; try adjusting the y-axis limits. Use the Zoom and Data Cursor tools to find the local minimum of the function at around $x = -1.78$. Somewhat confusingly, $\sin^{-1}$ indicates the inverse sine function (a.k.a. arcsine), unlike x^{-1}, which means $1/x$. What happens if you replace the 1/sin component with the mathematical function $\sin^{-1}$?

$$y = (x/5)^3 + 1/\sin(x^2) - 1$$

9. The logarithm has many uses, including transforming long-tailed data to a more manageable distribution. Search the help files to determine

the differences among the functions `log`, `log10`, and `log2`. Try taking the log of some functions. Plot x by `log(x)` (and `log10(x)` and `log2(x)`) for different x's. What happens when x is negative?

10. The Fibonacci series is simple to construct. Start with [0, 1] and then set each new value to be the sum of the previous two values.

```
fibseq = [0 1];
for i=3:1000
    <insert your code here>
end
```

Plot the Fibonacci sequence. It grows to galactic numbers very quickly. (I don't know if there is a specific minimum for a number to be "galactic," but I think we can all agree that 10^{208} is galactic.) In a different subplot in the same figure, plot the logarithm of the Fibonacci series. Is it more interpretable now?

11. The `pause` command with no inputs waits for the user to press a key before continuing to the next line of code. This can be useful when looking through results in sequence (e.g., data from each of 20 channels). Write a for-loop that, for each of 10 iterations, creates a 10×20 matrix of random numbers, displays the matrix as an image, and pauses between each image to allow you to inspect each result. It would be useful for the image to have a title indicating the image number and the number of remaining images to inspect. Where does the `title` function need to be relative to the `pause` function in the for-loop?

12. Adapt the code from the previous exercise so that on each iteration, the new matrix is the average of all previous matrices plus noise. Set the color limit to be the same for each image. What happens to each successive image?

13. The following two lines appear in the online MATLAB code. What's on top, the green dots or the red patch? Can you change the layer order by changing the order of the code?

```
plot(x,y,'o','markerfacecolor','g','markersize',15)
patch(x,y,'r')
```

14. What, if anything, is the difference between the following two lines of code?

```
scatter(x,y,'o')
plot(x,y,'o')
```

15. How would you use the set function to specify that the *y*-axis limits should be the same as the *x*-axis limits?

16. How's your trigonometry knowledge? You might remember that tangent is defined as the ratio of sine to cosine of an angle. As cosine approaches zero, the tangent blows up to infinity or down to negative infinity. Don't believe me? Try plotting it. Create a variable that goes from –5 to +5 in steps of 0.1 (these values are interpreted as radians) and plot the tangent of this vector. Does it blow up toward infinity? Not really. What happens if you go in steps of 0.01 or 0.0001?

II Foundations

10 Matrix Algebra

MATLAB was designed for working with matrices (MATrix LABoratory). The vast majority of neuroscience data are stored as matrices, which is one of the reasons why MATLAB is so useful for neuroscience analyses. You don't need to be an expert in matrix algebra or linear algebra to use MATLAB and successfully analyze data. But it is important to know the basics of matrix algebra in order to understand how to think about and use vectors and matrices, and as a prerequisite for understanding matrix-based computations such as principal components analysis (PCA) and linear least-squares modeling.

Matrix algebra is a big and fascinating topic that permeates many areas of science and engineering. This chapter provides a basic introduction to the concepts in matrix algebra that are most pertinent for neuroscience data analysis. The terms *matrix algebra* and *linear algebra* are often used interchangeably. Generally, linear algebra indicates an emphasis on the math and underlying proofs, while matrix algebra connotes the application of linear algebra to solving specific problems in engineering and science. Linear transformations and systems of linear equations can be easily represented and manipulated using matrices, which is why there is a large overlap between linear algebra and matrix algebra.

10.1 Vectors

Vectors are like simple matrices, so it's useful to start a foray into matrix algebra by learning about vectors. There are two ways to think about vectors: the algebraic interpretation and the geometric interpretation (figure 10.1). These two interpretations provide complementary insights into vector and matrix computations. The geometric interpretation is most often useful for gaining intuitive insights into problems in two dimensions or

Algebraic   Geometric

[3 −2]

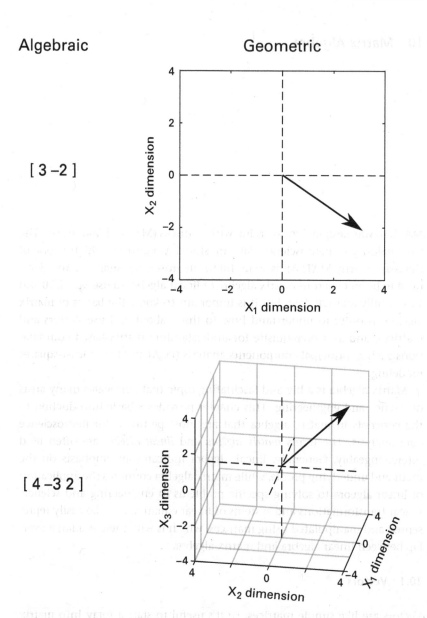

[4 −3 2]

Figure 10.1
A vector can be conceptualized as an ordered list of numbers (algebraic) or as a position in space (geometric). Illustrated here are vectors in two dimensions (top) and in three dimensions (bottom).

three dimensions, while the algebraic interpretation becomes useful for high-dimensional data.

The algebraic interpretation. A vector is an ordered list of numbers, such as [4 12] and [3 4 π 0.62]. The number of numbers is referred to as the *dimensionality* of the vector or the number of elements in the vector. For example, [4 12] is a two-dimensional (2D) or two-element vector, and the second element (the number in the second dimension) is the number 12.

The geometric interpretation. A vector is a line in an *N*-dimensional space (where *N* is defined by the number of elements in the vector) that starts at the origin of that space and goes to the coordinate specified by the numbers in the vector. For example, the vector [4 12] is a line that starts at the origin [0 0] and ends 4 units to the right and 12 units up. Technically, vectors do not need to start at the origin (the vector [4 12] could validly start at position [3,–6] and end at position [7,6]), but it is often more intuitive to think about vectors starting from the origin.

Vectors can be positioned either standing up or lying down (figure 10.2). In technical terms, a lying-down vector is called a *row vector* (because it has one row and many columns), and a standing-up vector is called a *column vector* (because it has one column and many rows). If you get a bit confused, think about a column vector like a tall but narrow Greek column and a row vector like a row of seats in a movie theater (or a live theater, depending on how cultured you are).

The sizes and dimensions of vectors and matrices are written as *M* by *N* (or *M* × *N*), where *M* and *N* refer, respectively, to the number of rows and columns. Rows are always listed first, then columns. That's really

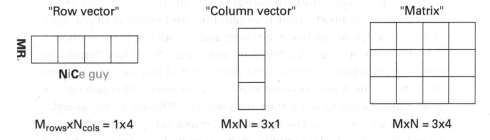

Figure 10.2
Illustration of vectors and matrices. A mnemonic to remember the order of describing the size of a matrix is to think of "MR. NiCe guy."

important. To remember the order, I think of "**MR. NiCe guy**" (**M** Rows by **N** Columns).

You will see later that vectors can also be thought of as matrices where one dimension has only one element: A row vector is a 1-by-*N* matrix, and a column vector is an *M*-by-1 matrix. You've already seen this convention in MATLAB when using the functions zeros, ones, and nan. To create a 1-by-5 row vector of zeros, for example, type: zeros(1,5);

A vector can be transformed from column to row or from row to column using the "transpose" operation, which involves swapping rows and columns. In mathematical formulation, the transpose is indicated by a capital T in the superscript: "**v** transpose" is $\mathbf{v}^T$. The transpose function is a single apostrophe after the vector. There is also the function transpose, which does almost the same thing (you'll learn about this in exercise 13, chapter 11).

```
v = rand(10,1); % 10D column vector
v
size(v)
v' % now it's a row vector
size(v')
transpose(v)
```

The transpose of a transpose is the original vector. It may seem a bit strange to take the transpose of a transpose, but the equivalence of **v** and $\mathbf{v}^{TT}$ (in MATLAB: v==v'') turns out to be important for some proofs in linear algebra. You'll see an example of this in the discussion of symmetric matrices (section 10.10).

10.2 Vector Addition and Multiplication

There are three types of mathematical operations that can be applied to vectors: vector addition, scalar multiplication, and vector multiplication. Each of these operations has an algebraic and a geometric interpretation.

Vector addition is accomplished by summing two vectors element by element (subtraction is defined as the addition of –1 times a vector). Vector addition is defined only between two vectors of equal dimensionality—it doesn't make sense to add a three-dimensional (3D) vector and an eight-dimensional (8D) vector. The algebraic interpretation of vector addition is to create a third vector by summing each corresponding element of the two vectors, so if **v** = [3 2] and **w** = [8 –4], then **v** + **w** = [3+8 2–4] = [11 –2]. The geometric interpretation of vector addition is to put the second vector at the head of the first vector, and then the sum is the new vector that goes from the origin to the head of the second vector (figure 10.3).

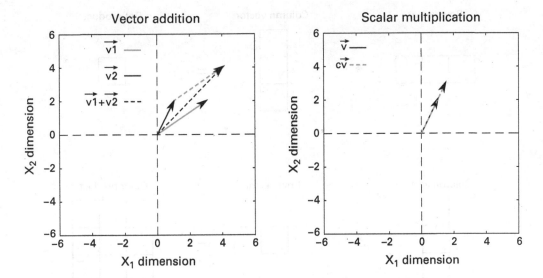

Figure 10.3
Geometric representations of vector addition and scalar multiplication. In vector addition, one vector is placed at the head of the other vector, and their sum is the new vector from the origin to the head of the added vector. In scalar multiplication, the vector is scaled by a factor defined by the number (that's why they call it a scalar).

A *scalar* is the linear algebra term for a single number. The algebraic interpretation of scalar multiplication is to multiply each element in the vector by a single number. If $v = [2\ 5]$ and $c = 1.5$ then $cv = [1.5*2\ 1.5*5] = [3\ 7.5]$. The geometric interpretation of scalar multiplication is to stretch the vector v by a factor of c (figure 10.3). The scalar c doesn't change the direction of the vector; it just makes v longer, shorter, or, if c is negative, point in the opposite direction.

Vector multiplication has two forms: the inner product and the outer product (figure 10.4). The outer product is a rank 1 matrix. It is not used in any analyses in this book, but it's good to know of its existence—if you are trying to compute the inner product and get a matrix instead, you know there is an error somewhere.

The inner product is also called the dot product and is very important in data analysis. It is the "brick" from which many signal processing techniques and data analyses are built, including the Fourier transform, convolution, and correlation. The dot product is powerful because it is a single number that tells you something about the relationship between two vectors.

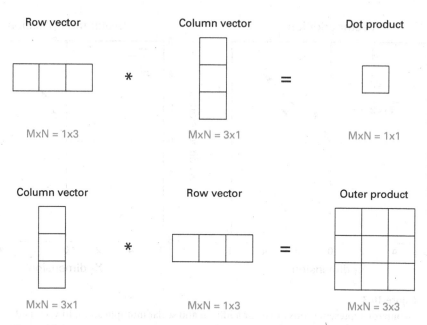

Figure 10.4
Illustration of the inner product (also called dot product) and outer product. It is easy to accidentally compute the outer product by swapping the order of the row and column vectors, so be careful with this during analyses.

The algebraic interpretation of the dot product is a single number that is obtained by multiplying corresponding elements in two vectors and then summing the element-wise multiplications. The dot product is indicated with a dot between the vectors. If v = [3 2 6] and w = [–1 3 2], then $v \cdot w$ is [3*–1 + 2*3 + 6*2] = [–3 + 6 + 12] = 15. Two important things to notice about the dot product are that it is defined only between vectors of the same length and that it produces a single number regardless of the lengths of the vectors. The formula for the algebraic interpretation of the dot product is

$$\Sigma v_i w_i$$

Before learning about the geometric interpretation of the dot product, you need to know about vector lengths. It is intuitive from the geometric interpretation of vectors that each vector has a length—it is a line, and the length of the line is simply its distance away from the origin (or wherever the vector starts). How is this length computed? If you think about a vector as being the hypotenuse of a right triangle, with the origin and a point on the x axis being the other points of the triangle, then the Pythagorean theorem tells us that the length of the hypotenuse is the square root of the summed lengths of the two sides squared. For vectors, this simplifies to the square root of the dot product of the vector with itself.

Now for the geometric interpretation of the dot product: The dot product is the multiplication of the lengths of the two vectors scaled by the cosine of the angle between them. The angle between the two vectors is critical in determining this mapping: The closer those vectors are to each other, the larger the dot product. The formula for the geometric interpretation of the dot product is:

$$|v||w|\cos\theta$$

where θ is the angle between the vectors. This formula and the previous one give identical results; they are just two different ways of looking at the same procedure. Note that when the angle between two vectors is 90° ($\pi/2$ radians) the cosine of that angle is 0, making the dot product 0 regardless of the lengths of the vectors. In this case, these two vectors are said to be *orthogonal* to each other.

In MATLAB, there are several ways to compute the dot product.

```
sum(v.*w)
dot(v,w)
v'*w
```

The last expression with the apostrophe is the matrix multiplication implementation, which will make more sense after reading the next section. Its accuracy requires you to know that both vectors are column vectors; otherwise you get the outer product rather than the inner product, or you might get a MATLAB error. In practice, it's best to avoid this construction.

10.3 Matrices

Of course you are familiar with 2D and 3D matrices already from images and data sets you may have worked with. Now that you are primed to think about vectors, you can think of a matrix as a collection of column vectors stacked next to each other (or as a collection of row vectors stacked on top of each other). This is a natural way to think about matrices in neuroscience, because data are often stored as channels by time. Matrices can have any number of dimensions, but in this chapter we will work only with 2D matrices. The sizes of matrices are given as M-by-N, where M is the number of rows and N is the number of columns (remember MR. NiCe guy!). There are many special matrices, three of which will be highlighted here (figure 10.5).

Square matrices. These matrices are shaped like a square, which means they have the same number of rows and columns, or $N = M$, or N by N (or M by M if you prefer). Nonsquare matrices are called rectangular matrices (or vectors, if N or M is one). The diagonal of a square matrix is the set of elements going from the top-left to the lower-right corners, where the row index is the same as the column index. Above and to the right of the diagonal is called the upper triangle of the matrix, and below and to the left of the diagonal is called the lower triangle.

Symmetric matrices. A symmetric matrix is a square matrix where the upper triangle is the same as the lower triangle. In other words, the matrix is a mirror image of itself across the diagonal. More formally, a symmetric matrix is one in which the matrix equals its transpose: $A = A^T$. Symmetric matrices have a lot of great properties, particularly regarding matrix factorizations such as eigendecomposition (more about this in chapter 17).

Identity matrix. This is a square matrix that contains all zeros on the off-diagonals and ones on the main diagonal. It has this name because the identity matrix times any matrix or vector is that same matrix or vector. This is the matrix equivalent of the scalar value 1 (1 times any number is that number). In text, the identity matrix is written as I or sometimes I_N, where N is the size of the matrix. The MATLAB function

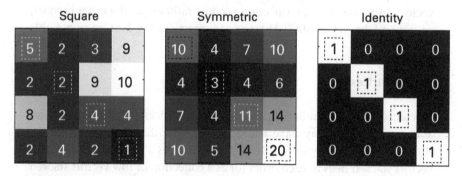

Figure 10.5
Three important types of matrices. The numbers and grayscale intensities indicate the value of each matrix element (color scaling is different for each matrix). The diagonal elements of each matrix are highlighted. Exercise 8 at the end of this chapter will ask you about this figure.

to create an identity matrix is called, cleverly enough, eye(N), where N is the number of elements in both dimensions.

Two matrices can be added together only if they are exactly the same size (both M by N). In this case, you just add the corresponding elements from the two matrices. Scalar multiplication is also straightforward: To multiply a matrix A by a scalar c (written cA), just multiply each element in A by c. Matrix-matrix multiplication is slightly trickier; we'll deal with this soon.

The transpose function is also defined for matrices. In the matrix transpose, the rows turn into columns, and vice versa. Thus, an N-by-M matrix is transposed to an M-by-N matrix.

10.4 Finding Your Way around a Matrix

When dealing with very small matrices, it's enough to point and say "that number there." But matrices can be very large, and you need to be able to access specific parts of a matrix using a more precise method.

Matrix locations are always encoded as *row, column* (MR. NiCe guy) (I know, I wrote this twice already, but it's really important and a bit of repetition never hurts). If you want to access the element in the third row and fifth column of matrix A, type A(3,5). You can use the colon operator to access multiple elements: A(2:4,3:6). And you can use square brackets to access noncontiguous elements of a matrix: A([1 3 2],[9 2]).

By convention, row 1 column 1 is at the *top left* of a matrix. Columns then go to the right and rows go down (figure 10.6). MATLAB also allows *linear indexing*, which means you can use a single number to access an element in a multidimensional matrix (also illustrated in figure 10.6). Linear indexing can be confusing and is a common source of errors even for seasoned MATLAB programmers. I recommend avoiding linear indexing whenever possible.

Here is the rule of thumb for proper indexing of matrices: If you have an N-dimensional matrix, make sure there are N − 1 commas when accessing elements of the matrix. For example, indexing a 2D matrix should be done like A(idx1,idx2) not A(idx). I have seen many students make many errors leading to incorrect matrix indexing. This is the worst kind of MATLAB error, because MATLAB might not give an error message, but the result will be incorrect. Please always remember the "N − 1 commas" rule when matrix indexing. And just in case you are sleeping while reading this, I'll mention it a few more times throughout the book.

Matrix indexing   Linear indexing

1,1	1,2	1,3	1,4
2,1	2,2	2,3	2,4
3,1	3,2	3,3	3,4

1	4	7	10
2	5	8	11
3	6	9	12

Figure 10.6
This figure illustrates two ways of identifying elements in a 3 × 4 matrix. Matrix indexing reduces the possibility of errors and should be preferred over linear indexing.

Having row 1 at the top of a matrix may seem intuitive, but images typically are drawn the other way, with row 1 at the bottom of the plot. MATLAB refers to this distinction as *xy*-format (row 1 at the bottom) versus *ij*-format (row 1 at the top). The *xy*-format is more often used for images, while the *ij*-format is more often used for matrices. This distinction was introduced in the previous chapter (`imagesc` vs. `contourf`) and will be discussed again in chapter 19 when plotting time-frequency results.

10.5 Matrix Multiplication

Matrix multiplication is not as straightforward as scalar multiplication. For one thing, not all matrices can be multiplied. Two matrices can be multiplied only when the sides "facing each other" (inner dimensions) have the same sizes. The size of the product matrix is the sizes of the sides "facing away" (outer dimensions) (figure 10.7).

Matrix multiplication is noncommutative. That is, **AB** is not **BA** (occasionally this equality holds, but these are the exceptions). You can already see this just by looking at the sizes of matrices. If **A** is N by M and **B** is M by N, then **AB** is N by N while **BA** is M by M. And if **A** is N by M and **B** is M by K, then **AB** is N by K and **BA** is not defined because the inner dimensions do not match (unless $N = K$). Two matrices are said to be *conformable* if matrix multiplication between them is valid. Now that you know the rules for matrix multiplication sizes, look back at figure 10.4 to understand why the order of multiplying vectors determines whether the result is the inner or outer product.

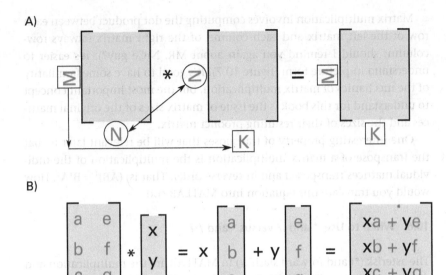

Figure 10.7
Illustration of procedures for matrix-matrix multiplication. Panel A illustrates the rules for when multiplication is valid (inner dimensions N must be equal) and the size of the resulting matrix (the outer dimensions M and K). Panel B illustrates the mechanics of matrix-vector multiplication. The resulting matrix is a weighted sum of the input columns, and the weights are defined by x and y. Panel C illustrates matrix-matrix multiplication.

Matrix multiplication involves computing the dot product between each row of the left matrix and each column of the right matrix (always row-column; should I remind you again about MR. NiCe guy?). It's easier to understand in picture form (figure 10.7). It's good to have some familiarity of the mechanics of matrix multiplication, but the most important concept to understand for this book is the issue of matrix sizes of the original matrices and the sizes of their resulting product matrix.

One interesting property of transposes that will be relevant later is that the transpose of a matrix multiplication is the multiplication of the individual matrices transposed and in reverse order. That is, $(\mathbf{AB})^\mathrm{T} = \mathbf{B}^\mathrm{T}\mathbf{A}^\mathrm{T}$. How would you translate this equation into MATLAB code?

10.6   When to Use .* and ./ versus * and / ?

The asterisk (*) and forward slash (/) in MATLAB are the multiplication and division operators. But multiplication works differently for scalars versus vectors and matrices. This can cause confusion because MATLAB needs to know whether you are trying to perform a scalar multiplication or a matrix multiplication.

To resolve this confusion, MATLAB has slightly different operators for these two procedures. The asterisk by itself means matrix multiplication. If you type A*B in MATLAB, you will get the result of matrix multiplication. For scalars, you can use * or .*. For example, if c is a scalar, v is a vector, and A is a matrix, then c*v and c*A are the same things as c.*v and c.*A.

The MATLAB operator for scalar multiplication is .* ("dot star"). For matrices, .* indicates element-wise multiplication. If A and B are both 3 × 4 matrices, then A*B is not defined and MATLAB will give an error, while A.*B produces a 3 × 4 matrix in which each element is the product of row i and column j in matrices A and B. And if v is a 1 × 5 vector, then v*v' is the dot product (a single number), v.*v is a 1 × 5 vector of each element squared, and v.*v' is undefined and will produce an error.

For many neuroscience data analyses, the scalar or element-wise multiplication (.*) is used more often than the matrix multiplication (*). This can be particularly confusing with square matrices, because if A is a 4 × 4 matrix, A*A and A.*A are both valid operations and both produce 4 × 4 matrices, although those matrices will be very different from each other. For this reason, it is good practice always to use .* and then change it to * only when you are certain that you want to perform matrix multiplication.

10.7 Linear Independence and Rank

Linear independence is a property of a set of vectors (individual vectors or as columns or rows of a matrix) and has an algebraic and a geometric interpretation. The algebraic interpretation is that if one vector can be produced from a linear combination ("linear combination" means scaling and adding) of other vectors, then that set of vectors is called *linearly dependent*. For example, if you have the vectors $\mathbf{d}$ = [4 5], $\mathbf{e}$ = [2 8], and $\mathbf{f}$ = [–11 –11], then vector $\mathbf{f}$ is linearly dependent on vectors $\mathbf{d}$ and $\mathbf{e}$, because $-3\mathbf{d} + 0.5\mathbf{e} = \mathbf{f}$. In other words, vector $\mathbf{f}$ provides no unique information: if you know $\mathbf{d}$ and $\mathbf{e}$, you can produce $\mathbf{f}$.

The geometric interpretation is that a set of vectors is linearly independent if each vector points in a different geometric dimension. For example, the vectors [1 2] and [2 4] are *dependent* because they are both on the same line; the second vector does not point off in a different geometric direction. The vectors [1 2] and [2 5], however, do point in different directions, so they are *independent*. Adding a third vector to that group would necessarily be dependent, because the first two vectors already cover the two dimensions of a 2D space. That is, any third vector in two dimensions can always be created by adding and subtracting scaled versions of the other two independent vectors.

The extreme case of linear independence is called *orthogonality*, which you learned about previously. Algebraically, two vectors are orthogonal if the dot product between them is zero. Geometrically, two vectors are orthogonal if they meet at a 90° angle.

The term *rank* refers to the number of linearly independent columns in a matrix (or independent rows in a matrix; rank is a property of a matrix regardless of whether you think about columns or rows). If all of the columns in a square matrix form a linearly independent set of vectors, that matrix is said to be "full rank." If the rank is less than the number of columns, the matrix is "reduced rank." Full-rank matrices are great to work with, particularly if they are square matrices. The MATLAB function to compute the rank of matrix A is, unsurprisingly enough, rank(A).

Let's think about matrix rank in neuroscience. Electrophysiology data are typically stored as an electrodes-by-time matrix. If each electrode measures activity from different neural populations, then the rank of matrix is the number of electrodes. It doesn't matter if neighboring electrodes record overlapping activity (e.g., due to volume conduction of the electrical potentials) as long as at least part of the activity is different (remember: linear independence does not require orthogonality). Now imagine that

the electrode wiring was damaged such that one electrode became electri-
cally bridged with another electrode. Now those two rows in the matrix
are identical, and the rank of the matrix is the number of electrodes minus
one. In practice, data matrices often become reduced-rank during processing, for example when removing sources of activity that are likely to
be noise.

Reduced-rank matrices can sometimes cause headaches for some types
of analyses. In part this is because the matrix inverse is defined only for
full-rank square matrices.

10.8   The Matrix Inverse

If you want to "neutralize" a scalar, you multiply it by its inverse to produce
the innocuous number 1. For example, 5 times 1/5 is 1 (1/5 can also be
written as 5^{-1}).

The analogous procedure in matrices is called the *matrix inverse* and is
written A^{-1}. However, the matrix inverse is a bit more complicated than
the scalar inverse, in part because the "neutral" matrix is not a matrix that
contains all ones. The neutral matrix was already introduced: It is the
identity matrix with all zeros on off-diagonals and ones on the diagonal
(see figure 10.5).

The inverse of a matrix A is the matrix A^{-1} that multiplies matrix A to
produce the identity matrix. That is, $AA^{-1} = I = A^{-1}A$. However, you cannot
simply invert each element of A to get A^{-1} (actually, this would work for
"diagonal" matrices that have non-zero elements only on the diagonal, but
the vast majority of matrices we work with in practice are not diagonal).
Computing the matrix inverse is simple for small matrices that contain
only integers, but quickly gets complicated for larger matrices. It is beyond
the scope of this book to explain how matrix inverses are computed, but
that's okay—we'll let MATLAB do the hard work for us.

The matrix inverse exists only for square matrices. Even among square
matrices, not all matrices have an inverse. In fact, the vast majority of
matrices are not invertible. The only matrices that have an inverse are those
that are square and have full rank. Noninvertible matrices are also known
as "singular." If MATLAB produces an error about "Matrix is singular or
poorly scaled," it means that MATLAB was trying to invert a matrix that
didn't want to be inverted. If the values of the matrix are very small (e.g., of
the order 10^{-13} or smaller), MATLAB may still fail to invert a theoretically
invertible matrix, due to a loss of numerical accuracy when dealing with
very small numbers. If this happens, you can try scaling up the matrix by

many orders of magnitude, although this is not guaranteed to produce a perfectly accurate solution.

In MATLAB, you can use the function inv(A) to compute the inverse of matrix **A**. The exercises will show several examples of matrices that are and are not invertible.

If the matrix is a bit "poorly behaved" but is not terribly problematic, you can substitute the pseudo-inverse (sometimes called the Moore-Penrose pseudo-inverse) in place of the full inverse (MATLAB function pinv). The pseudo-inverse is the closest approximation to a true inverse for that matrix, and equals the true inverse when the matrix is square and invertible.

10.9 Solving Ax = b

The matrix inverse is important for many reasons. One of the most important reasons is to move matrices from one side of an equation to the other. This is the case when solving systems of linear equations and, in particular, computing least-squares solutions. Consider the easy case of scalars. In the equation below, solve for x.

$ax = b$

Of course, the solution is simple: $x = b/a$. To solve this equation for x, you had to move the a from the left side to the right side, and this you did by multiplying both sides of the equation by a^{-1} (perhaps you would have explained it as dividing by a, but it's the same operation). Now we get to matrices, and the idea is the same: Multiply both sides of an equation by the inverse of the matrix.

This leads us to the least-squares equations for solving systems of linear equations. The least-squares approach is perhaps the most important algorithm in many branches of statistics and applied mathematics. You'll learn more about implementing the least-squares algorithm in MATLAB in chapter 28, but the basic idea is simple: *Solve* **Ax** = **b**. **A** is a matrix of predictors or independent variables (columns are variables, rows are trials), **x** is a vector containing the coefficients, also called regression weights, and **b** is a vector of observed data points. In other words, **Ax** = **b** is asking the question: What linear combination of independent variables can explain the observed data?

The goal of this analysis is to find the vector of unknowns, which is the vector **x**. The way to solve it is straightforward: Move **A** to the other side of the equation. Now you see where the matrix inverse is needed. We can left-multiply both sides of the equation by A^{-1}.

$Ax = b$

$(A^{-1}A)x = A^{-1}b$

$Ix = A^{-1}b$

Because the identity matrix does nothing to x, we reach our final destination of $x = A^{-1}b$. Standard matrix multiplication can be used to reduce $A^{-1}b$ to a single vector, because A is $N \times N$ and b is $N \times 1$.

10.10   Making Symmetric Squares from Rectangles

Typically, however, A is not invertible, simply because it is not square. In statistics, the matrix A is a design matrix comprising trials by independent variables. If you have three independent variables and 100 trials, then you have a rectangular matrix that cannot be inverted. How can you make this a square matrix so it can be inverted? It makes little sense to add 97 independent variables, and it makes even less sense to throw out 97 trials.

The solution is to use the transpose. It turns out that a matrix times its transpose is always a square matrix, because an N-by-M matrix times an M-by-N matrix produces an N-by-N matrix. And it gets even better than that—a matrix times its transpose is guaranteed to be a symmetric matrix. I'm not supposed to burden you with proofs, but the proof that a rectangular matrix times its transpose is square symmetric is elegant and short (recall from earlier in this chapter that $(AB)^T = B^T A^T$, and that $A^{TT} = A$).

$A^T A = (A^T A)^T = A^T A^{TT} = A^T A$

Square symmetric matrices have many great properties, for example having orthogonal eigenvectors (you'll learn about this in chapter 17). Most relevant for least squares is that square symmetric matrices are invertible as long as the columns of A are linearly independent (that is, if the rank of the matrix is equal to the number of columns).

So, computing a least-squares solution when the matrix A is rectangular involves first multiplying both sides of the equation by A^T, and then proceeding with the formulation from the previous section. The mathematics are shown below.

$Ax = b$

$A^T Ax = A^T b$

$(A^T A)^{-1}(A^T A)x = (A^T A)^{-1}(A^T)b$

$x = (A^T A)^{-1}(A^T)b$

Linear least squares is so important and is implemented so often that MATLAB has a dedicated operator for this purpose: the backslash ("\"). That final line of math above is implemented in MATLAB using the following code:

```
x = (A'*A)\A'*b;
```

If you find this really exciting, then you can look forward to working your way through chapter 28.

10.11 Full and Sparse Matrices

A *full matrix* is the normal matrix that you are used to working with. A 64-by-10,000 matrix, for example, has 640,000 elements in the matrix. That's an awful lot of elements. What if most of those elements were zeros? Image an extreme case with 639,999 zeros and a single non-zero element. It would be much more efficient simply to store the value and location of that non-zero element. That gets us from 640,000 pieces of information down to 3 (matrix coordinates, and value at that coordinate). A matrix with a lot of zeros is called a *sparse matrix* (figure 10.8). MATLAB has a special way of storing sparse matrices.

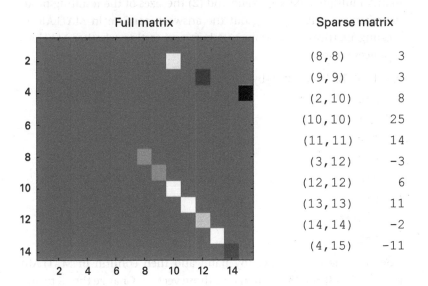

Figure 10.8
A matrix in its full (left) and sparse (right) representations.

```
A = sparse(64,10000,1);
B = full(A);
```

The function `full` will turn the sparse matrix into its full representation. Check out how much less memory the sparse matrix takes, although it contains the same information as the full matrix.

```
>> whos A B
  Name      Size            Bytes  Class     Attributes
  A         64x10000         80024  double    sparse
  B         64x10000       5120000  double
```

Otherwise, sparse matrices can be used in code in the same way as full matrices (e.g., addition, scalar multiplication, and matrix multiplication). In neuroscience, data sets that include images, local field potentials, and EEG are typically stored as full matrices, whereas action potentials are typically stored as sparse matrices. It's good to know about sparse matrix representations even if you won't use them.

10.12 Exercises

1. Given the following three matrices and their sizes, figure out (1) which matrix multiplications are valid and (2) the sizes of the resulting product matrices. After figuring out the answers, confirm in MATLAB by creating matrices of random numbers and testing whether MATLAB produces errors.

    A: 4-by-4; **B:** 4-by-7; **C:** 5-by-7

    a. A*A
    b. B*A
    c. C*A
    d. B*C'
    e. C*B*A
    f. B*B'
    g. C*B'*A
    h. C*C

2. Compute the dot product (by hand and then confirm in MATLAB) between [1 –3] and [3 1]. Then plot those vectors. Change the –3 to any other number and recompute and replot. Is it possible to get orthogonal vectors by changing the –3 without changing the other numbers?

3. Here's a trick to round a number to a specified number of decimal points:

```
aNumber = randn;
round(10*aNumber)/10
```

Work through that second line of code so you understand what it's doing. How can you modify it to round to the nearest thousand?

4. After completing the previous exercise, type `format  bank` in the Command window. Is there a difference between this formatting option and using the code in the previous exercise? Read the help file for `format` to learn about different numerical formatting options and to learn how to reset the numerical formatting to the default setting.

5. Each of the following lines of code will produce an error. Find and fix the errors without MATLAB. Then confirm your answer in MATLAB.

```
plot3([0 v3(1)],[0 v3(2)])
a=rand(2,3,4); a'
inv(rand(2,3))
```

6. A reduced-rank matrix is a matrix that has at least one column, or at least one row, that is linearly dependent on other columns or other rows. Create a 2-by-2 reduced-rank matrix by making the second column be a multiple of the first column. Then try to compute its inverse using the `inv` function. What is the result, and what is the warning message? Then compute its pseudo-inverse using the `pinv` function. For an invertible matrix A, $AA^{-1} = I = A^{-1}A$. Does this hold for the pseudo-inverse? (In linear algebra terminology, you would say that this is the closest projection to the identity matrix.) Try this exercise again using a full-rank matrix.

7. Create a rectangular matrix of 2×3 random numbers. Compute AA^T and A^TA. Before doing the computation, figure out what the sizes of AA^T and A^TA will be. Confirm by visual inspection and by plotting images of those matrices that they are symmetric. In time series analysis, if A is a time-by-channels matrix, then A^TA is called the *channel covariance matrix*.

8. One number in figure 10.5 is incorrect. Which one is it, and how can you tell it's wrong?

9. Write code that generates random 3D vectors and then computes the dot product of the vector with itself. Compute the dot product using your own custom-written code, and confirm that the function `dot`

gives the same output. Is the dot product ever negative? What if you force all elements of the vector to be negative? Why is this?

10. Below is a 3-by-3 matrix. Enter this matrix into MATLAB and compute its rank. Are you surprised by the answer? Can you change one number in that matrix to change its rank?

```
A = [ 1   3    3;
      0  -1   -2;
      7   2  -17];
```

11. Design an algorithm for obtaining linear indices from matrix indices (consult figure 10.6). Your code should work for any arbitrarily sized 2D matrix. Next, learn how to use the sub2ind function in order to sanity check your code.

12. You know it's illegal to multiply a 3-by-4 matrix by a 1-by-4 vector (the linear algebra police will come after you). But what if you want to point-wise multiply each row in the matrix by the same vector? You can write a for-loop, but this is not a good idea. Better would be to use MATLAB's repmat function, which can repeat a matrix. Use the help file for repmat to figure out how to expand a 1-by-4 vector to a 3-by-4 matrix with three identical rows (what is the rank of that matrix?). Then point-wise multiply that by the original 3-by-4 matrix.

13. An even better solution to the previous problem is to use the function bsxfun. This function is easier and faster than repmat, but the input is a bit different from what you have seen so far. Use the bsxfun help file to figure out how to do the same row-wise multiplication in the previous exercise.

14. The function diag creates a *diagonal matrix*, which is a matrix with zeros everywhere except on the diagonal. Try, for example, diag([2 1 5 6]). Create the 10-by-10 identity matrix using the functions diag and repmat (don't use the function eye). What happens when you input that I_{10} matrix into the function diag? Also try diag(rand(4)).

15. Compute the dot product between the vectors [1 3 2] and [5 2 3]. First do it by hand and then in MATLAB using matrix-product notation (x^Ty). Make sure you get the vector orientations correct in MATLAB!

11 The Fourier Transform

The Fourier transform is one of the most important signal processing tools in engineering and information technology, not to mention neuroscience data analysis. In neuroscience, the Fourier transform is used to perform frequency and time-frequency analyses and is the basis of filtering time series and many forms of image processing.

The Fourier transform implements the Fourier theorem, which states that any signal can be represented as a sum of sine waves, each having its own frequency, amplitude, and phase. In math textbooks, the Fourier transform is defined using integrals and continuous sine waves. However, in the digital world, we have only discrete time points to work with. This chapter does not discuss the continuous Fourier transform, but instead shows how the discrete Fourier transform can be implemented in MATLAB.

To understand how the Fourier transform works, you need to understand three mathematical concepts: sine waves, complex numbers, and the dot product. I know it's a bit frustrating to wait one-half of a chapter before learning about the Fourier transform, but trust me, it's worth the wait.

11.1 Sine Waves

I'm sure you already know what a sine wave is, and probably you also know that the formula to create a sine wave looks something like this:

$$a \sin(2\pi ft + \theta)$$

where a is the amplitude (height) of the sine wave, π is 3.14..., f is the frequency, t is time (if time is specified in seconds, then f can be specified in hertz), and θ is the angle offset. In MATLAB, a sine wave is generated as follows, and an example is shown in figure 11.1.

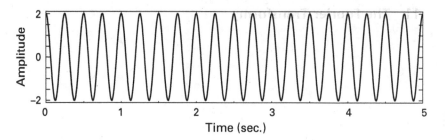

Figure 11.1

Example sine wave at 4 Hz. Count the number of cycles between any two of the seconds intervals on the *x* axis.

```
srate = 1000; % sampling rate in Hz
time = 0:1/srate:5; % units of seconds
f = 4; % units of Hz
a = 2; % arbitrary units
th = pi/2; % in radians
sinewave = a*sin(2*pi*f*time+th);
plot(time,sinewave)
```

The importance of sine waves as the fundamental basis for the Fourier transform cannot be understated. Before moving on, take some time to play around with the code above—change each of the parameters to see the effects on the sine wave. Try to develop an intuitive feel for what happens to the sine wave when you change the frequency and the amplitude.

The code above produces a *real-valued sine wave* because it contains no imaginary part. To perform the Fourier transform, we need *complex sine waves* that contain both a real part and an imaginary part. To create a complex sine wave, you need to know about complex numbers.

11.2   The Imaginary Operator and Complex Numbers

The imaginary number is the result of taking the square root of –1. "But that doesn't make sense," you are probably thinking, "no number multiplied by itself can produce a negative number; you can only take the square root of positive numbers, and preferably nice positive numbers like 4, 16, and 49." The square root of –1 does not exist, and that's why mathematicians came up with a special term for this quantity—the *imaginary operator*, which is commonly abbreviated as *i* (in electrical engineering they use *j* because *i* indicates current).

Don't lose sleep over worrying about *what it means* for a number to multiply itself and produce a negative number; arguably it is not a quantity that actually exists. Mathematicians have discovered that by using the imaginary operator, they can solve problems more easily and more efficiently than if they used only real-valued numbers. The Fourier transform is one example of this, and so you only need to worry about learning how to use complex numbers to extract frequency and time-frequency information.

A complex number is a number that contains both a real part and an imaginary part. The real part is the set of all the numbers you have been comfortable with your entire life (1, 2, −50, π, and so on). The imaginary part is any of these numbers, but multiplied by *i*.

Complex numbers look like this: 5 2*i*, −18 0.4*i*, 4 −10.2*i*. You might initially think that I wrote six numbers grouped into three sets such that the first set is the two numbers "5" and "2." But in fact I wrote only three numbers, because each complex number comprises two components. All of the real numbers that you've been using your whole life are really just a subset of complex numbers for which the imaginary part is zero. That is, the number 7 is shorthand for the complex number 7 0*i*. So when someone asks how many children you have, you can reply that it's complex, and then explain that you have non-zero real children and/or non-zero imaginary children.

Real-valued numbers can be represented on the number line. The number 7 is to the right of 5 and to the left of 7.3. Is the number 7 4*i* to the left or to the right of 7 6*i*? You guessed it—it's neither. Part of the power of complex numbers comes from having two components. They cannot be represented on a *line*; instead, they are represented on a *plane* (figure 11.2).

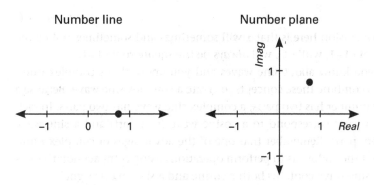

Figure 11.2
A one-dimensional (1D) number line and a two-dimensional (2D) number plane.

It is a 2D plane where the left-right axis corresponds to the real part and the up-down axis corresponds to the imaginary part. Each complex number corresponds to a coordinate in the plane.

"But Mike," you are probably thinking, "if a complex number has two parts, can we think of it as a two-element vector?" A very astute observation, reader! Indeed, it is very useful to conceptualize a complex number as a vector that starts at the origin of the complex plane and ends at the coordinate defined by the real and imaginary components of the complex number. Therefore, mathematical operations using complex numbers, such as addition and multiplication, are accomplished using the same rules that you learned about in the previous chapter (although there are additional rules such as Hermitians and complex conjugates; you'll learn some of the relevant rules later). Thinking about complex numbers as vectors is important, because we will use concepts from vector geometry to extract length and angle from complex numbers, and those quantities are the power (length) and phase (angle) in the Fourier transform.

In MATLAB, complex numbers are created in one of two ways:

```
x = 4+2i;
x = complex(4,2);
```

Warning! Although you can technically use i or j to indicate the imaginary operator in MATLAB, this is dangerous and should be avoided. Many people use i and j as counting variables in small loops. Instead, you should use 1i or 1j, which cannot be used as variables. Evaluate the following code in MATLAB:

```
i=1
x2 = 4+i
x3 = 4+1i
1i=1
```

The conclusion here is that i will sometimes and sometimes not be the square root of –1, while 1i will *always* be the square root of –1.

Now you know about sine waves and you know about complex numbers. Let's combine these concepts to create a complex sine wave. Because a complex number has two parts, a complex sine wave has two parts. In fact, these two parts correspond to a cosine wave (real part) and a sine wave (imaginary part). Remember that one of the advantages of complex numbers is that they allow us to perform operations using compact notations. A complex sine wave contains both a cosine and a sine in one signal.

Creating a complex sine wave is slightly different from a normal real-valued sine wave, because the sine wave formula is embedded inside Euler's formula. In math, the formula for a complex sine wave is $e^{i2\pi ft}$. In MATLAB, this translates to:

```
csw = exp(1i*2*pi*f*time);
```

Because a complex sine wave (csw) is a 3D function (time, real part, imaginary part), it is best represented in three dimensions, as you can see in figure 11.3.

```
plot3(time,real(csw),imag(csw))
```

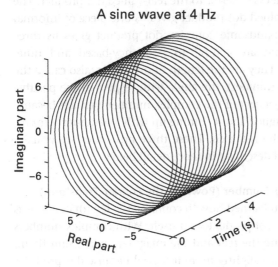

Figure 11.3
The complex sine wave in three dimensions (no special glasses required).

11.3 The Complex Dot Product

The dot product, as you know from the previous chapter, is a single number that reflects a relationship between two vectors. It is computed as the sum of point-wise multiplications between corresponding elements of the two vectors. The sign of the dot product tells you about the angle between the two vectors: a positive dot product means the angle between the vectors is acute (less than 90°), a negative dot product means the angle between the vectors is obtuse (greater than 90°), and a zero dot product means the vectors are orthogonal to each other (exactly 90° angle).

All of the dot products in the previous chapter were real-valued. Now that you know about complex numbers, it's time to upgrade your knowledge to incorporate complex dot products. A complex dot product is the dot product when one or both vectors contain complex numbers. Because complex numbers have two components, a complex dot product also has two components.

You can already see where this is going—the complex dot product is conceptualized as a coordinate point in a 2D space, where the two dimensions are the real axis and the imaginary axis. And just like with complex numbers, we can think about this coordinate as the endpoint of a vector that goes from the origin to that point.

The complex dot product is superior to the real-valued dot product. The cute but ineffectual real-valued dot product gives us one piece of information. The tall, dark, and handsome complex dot product gives us three pieces of information that are central to frequency-based and time-frequency–based analyses. They are the length of the vector (also called the *magnitude* of the complex number), the angle with respect to the positive real axis (also called the *argument* of the complex number, perhaps because ancient mathematicians argued about coming up with a better term for it), and the projection onto the real axis. These three properties are visually depicted in figure 11.4 and described in more detail below.

Magnitude of the Complex Number (Power in the Fourier Transform)
In the previous chapter you learned how to compute the length of a line as the square root of a vector dot product with itself. With complex numbers it's a similar concept: square the real and the imaginary parts, sum them, and take the square root. It's slightly more involved because dot products involving complex numbers are computed using complex conjugates. This is because $i^2 = -1$, and the length of a line needs to be positive. More on this in a few paragraphs. In the mean time, you can compute dot products by squaring the *projections* onto the axes, rather than squaring the imaginary numbers.

In MATLAB, the real and imaginary components of a complex number can be extracted using the functions `real` and `imag`, leaving us with the following implementations to compute the length of the vector defined by a complex dot product (the variable `cdp` is the complex dot product).

```
linemag = sqrt(real(cdp).^2 + imag(cdp).^2);
linemag = abs(cdp);
```

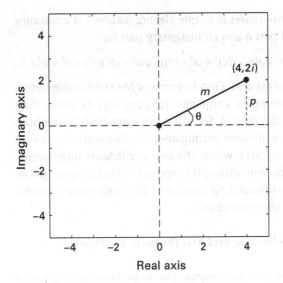

Figure 11.4
A complex number (e.g., the result of a complex dot product) provides three pieces of information that are used in frequency and time-frequency analyses: the distance from the origin (*m*), the projection onto the real axis (*p*), and the angle with respect to the positive real axis (θ).

The function abs is the same function that returns the absolute value of a real-valued number (abs(-4) is 4). It may initially seem strange that the same function gets the absolute value and the length of a line in the complex plane, but it makes sense if you think about abs more generally as computing the distance of a point away from the origin, regardless of whether that point is on a number *line* (real-valued numbers) or a number *plane* (complex-valued numbers).

In frequency-based and time-frequency–based analyses, the magnitude of this line is also called the amplitude. However, in frequency analyses, people most often plot and analyze power. Power is amplitude squared; thus, the power at each frequency is the squared magnitude of the line.

One final point about computing power. It can also be implemented as the multiplication of the Fourier coefficient with its complex conjugate. The conjugate of a complex number is a number with the same real component and an imaginary component of opposite sign. For example, the complex conjugate of the number 3 8*i* is 3 –8*i*. It is often indicated using an asterisk in the superscript (the complex conjugate of *c* is *c**). The reason why this can be used to compute power is simple algebra (below, c is a complex variable comprising a real part a and an imaginary part b):

$$cc^* = (a\ bi)(a - bi) = a^2 + abi - abi - (bi)^2 = a^2 - (bi)^2 = a^2 - (b^2)(i^2) = a^2 + b^2$$

I point this out for three reasons. First, in terms of MATLAB implementation, multiplication by complex conjugate (`c*conj(c)`) is faster than squaring the magnitude (`abs(c)^2`). Second, some data analysis methods such as spectral coherence involve conjugate-multiplication of different Fourier coefficients (`c*conj(d)`), where the two coefficients might come from different electrodes or from different frequencies. Finally, some publications use the conjugate formulation instead of the magnitude formulation, so it's good to know the equivalences.

Angle with Respect to the Positive Real Axis (Phase in the Fourier Transform)

Basic trigonometry tells us how to compute the angle between the vector and the positive real axis: The tangent of that angle equals the ratio of the lengths of the opposite to the adjacent sides, which is the ratio of the imaginary (sine) part to the real (cosine) part. Then, the angle in radians can be computed as the arctangent of that ratio.

```
lineangle = atan(imag(cdp) / real(cdp));
```

The astute reader may spot a potential problem with this implementation (if not, consult the exercises at the end of chapter 9). If the real component is zero, we are dividing by a zero. MATLAB has a function `atan2` that has an exception for this situation. In fact, MATLAB has a function to extract the angle directly from a complex number.

```
lineangle = atan2(imag(cdp),real(cdp));
lineangle = angle(cdp);
```

These phase angles have several uses in neuroscience. In time-frequency analyses, they are used to quantify the consistency of the timing of rhythmic activity over repeated trials. In synchronization analyses, they are compared across different electrodes to determine whether two brain regions are coupled (functionally interacting). In spike-field coherence, they are used to determine whether the timing of action potentials is locked to the population-level field potentials.

Projection onto the Real Axis (Band-Pass Filtered Signal in Time-Frequency Analysis)

This quantity is easy to extract using the function `real`. The projection onto the real axis is not used in the Fourier transform, but it will be used in chapters 19 and 20.

```
realpart = real(cdp);
```

11.4 Time Domain and Frequency Domain

Before learning about the mechanics of the Fourier transform, it's good to have an idea of the purpose of the Fourier transform and what the results will look like. Time series signals can be represented in the time domain or in the frequency domain. Figure 11.5 shows a signal represented in both domains. The time-domain representation of this signal is something you are used to seeing: The y axis shows the fluctuations in signal values over time, which is indicated by the x axis. The frequency-domain representation of the same signal shows the number of cycles in the signal that take

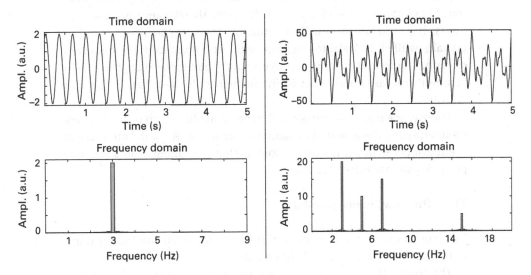

Figure 11.5

A signal viewed in the time domain (top panels) and in the frequency domain (bottom panels). When signals contain rhythmic components, the signal is often more easily interpreted in the frequency domain (bottom panels) than in the time domain (top panels).

place within a period of 1 second. If the time-domain signal has three cycles per second, the frequency-domain plot shows a bar at 3 Hz.

You should try to see the time domain and the frequency domain as being different ways of looking at the same thing. That is, imagine there is some Platonic ("Platonic" in the sense of Plato's Cave, not in the sense of "we're just friends") sine wave, and the time and frequency domains are simply different ways of looking at the same "pure" object. Jules Verne poetically explained the idea of interpreting the same object in different ways in *20,000 Leagues Under the Sea*. In response to the question "What is a pearl?" Professor Aronnax replied, "To the poet, a pearl is a tear of the sea; to the Orientals, it is a drop of dew solidified; to the ladies, it is a jewel of an oblong shape, of a brilliancy of mother-of-pearl substance, which they wear on their fingers, their necks, or their ears; for the chemist, it is a mixture of phosphate and carbonate of lime, with a little gelatine; and lastly, for naturalists, it is simply a morbid secretion of the organ that produces the mother-of-pearl among certain bivalves."

There are two major advantages of viewing and conceptualizing signals in the frequency domain. One is that many signals are easier to interpret in the frequency domain than in the time domain. This point is also illustrated in figure 11.5: Even for a relatively simple signal comprising four sine waves, it is difficult to visually disentangle the components in the time domain, while it is trivial to disentangle the components in the frequency domain. The second advantage is that many signal processing goals can be accomplished faster and better in the frequency domain than in the time domain.

There is one major advantage of the time domain, which is that it is best for interpreting the temporal dynamics of the signal. Temporal information is not lost per se in the frequency domain, but it is "hidden" in a way that precludes an easy interpretation.

11.5 The Slow Fourier Transform

Now you know all the foundational bricks from which the Fourier transform is built. It's time to put them together.

The Fourier theorem states that any signal can be represented as a sum of sine waves, each having its own frequency, amplitude, and phase. Here is how the discrete Fourier transform works: Construct a complex sine wave and compute the complex dot product between that sine wave and the signal. The resulting dot product is called a *Fourier coefficient*. Next, construct another complex sine wave at a different frequency and compute the complex dot product again. This process repeats for as many frequencies as there are time points in the data. The number of frequencies in the discrete Fourier transform is defined by the number of time points.

```
N = length(signal);
fTime = (0:N-1)/N;
for fi=1:N
    fSine = exp(-1i*2*pi*(fi-1).*fTime);
    signalX(fi) = sum(fSine.*signal);
end
signalX = signalX / N;
```

Notice that we divide by N after the loop ends. This is a scaling factor that puts the Fourier coefficients in the same scale as the original data. It is necessary because the loop involves summing over N points for each frequency. You can try running the Fourier transform code without that scaling factor—all of the relative magnitudes will be the same, but the magnitudes won't match those of the time-domain data.

How does the Fourier transform "know" which sine waves correspond to which frequencies in hertz? The vector defining the sine wave frequencies range from zero to just below one, and the data sampling rate does not appear anywhere in the formula for the Fourier transform. In fact, the Fourier transform is the same regardless of whether the data are sampled once each millisecond or once each 1,000 years. The "time" vector that defines the sine wave doesn't even need to be time; it could be distance in micrometers or inches or light-years.

Converting the frequencies of the sine waves from arbitrary units to hertz is critical for interpreting and analyzing data. Would you like to learn how this conversion is done? Keep reading ...

11.6 Frequencies from the Fourier Transform

After computing the Fourier transform, you will want to interpret the resulting Fourier coefficients in terms of frequencies in hertz. A hertz is the inverse of time (1/time), meaning that 1 cycle each second is 1 Hz, 4 cycles each second is 4 Hz, and 1 cycle every 4 seconds is 0.25 Hz.

The frequencies in the Fourier transform are in normalized units, where "1" is the data sampling rate. To convert the Fourier coefficients to frequencies in hertz, we need to think about the lowest and highest frequencies that can possibly be extracted out of a signal. Let's start with the lowest frequency. As frequencies decrease, the sine wave gets slower and slower.

When the sine wave is infinitely slow, one cycle takes an infinite length of time. At any finite time scale, this is a straight line and corresponds to a frequency of 0 Hz. In electrical engineering this is usually called DC for direct current. The DC or 0 Hz component of a signal is the part of the signal that is captured by a flat line, which is simply the mean offset (you'll see an example of this below). So the frequencies from a Fourier transform will always start at 0 Hz, regardless of the data sampling rate or the number of time points.

Now let's think about the highest-frequency sine wave that can be measured in a signal. Because a sine wave is defined by fluctuations, we need a minimum of two measurements per cycle. At two points per cycle, the highest frequency that can be measured is one half of the data sampling rate. This is called the Nyquist frequency, a term you may have heard before. Note that measuring frequencies up to the Nyquist is a *theoretical* limit; in *practice*, it is a bad idea to interpret frequencies above around 20% of the sampling rate because subsampling can still produce some strange or artifactual results, some of which are shown in figure 11.6.

Computing the Nyquist frequency in MATLAB is really trivially easy. I probably shouldn't even write it, but I will:

```
nyquist = srate/2;
```

Now we have figured out that the Fourier coefficients range from 0 Hz to the Nyquist frequency. Next we need to know how many frequency steps are in between 0 and Nyquist. This is called the frequency resolution, and it is defined by the number of time points in the data. In particular, the number of steps between 0 and Nyquist is $N/2 + 1$, where N is the number of time points in the signal, $N/2$ is the number of sine waves up to the Nyquist frequency, and +1 is for the DC component.

If you look in the Fourier transform for the code corresponding to $N/2 + 1$, your search will be unsuccessful. The Fourier transform defines frequencies all the way up to N. This may seem to violate the law that frequencies above the Nyquist are aliased into lower frequencies. Curiously enough, this is intentional in the Fourier transform. The frequencies above the Nyquist are indeed aliased, and they appear as sine waves traveling "backward" (think of the wheel hub of a moving car that appears to spin backward at high speeds). These are called the *negative frequencies*. For real-valued signals, the negative frequencies mirror the positive frequencies and the amplitude gets split between the positive and negative frequencies. For this reason, it is common practice to ignore the negative frequencies and double the amplitude of positive frequencies in between but not including the DC

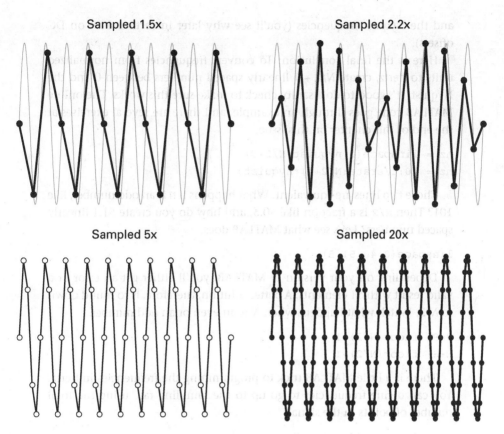

Figure 11.6
This figure illustrates some difficulties when interpreting results close to the Nyquist frequency of digitized time series. Imagine that the gray sine wave is an analog wave and that our recording equipment collects data at time points indicated by the black dots. The black lines illustrate the signal that will be observed in the data. When sampling is too sparse relative to the frequency of the sine wave, the signal is aliased into a slower frequency (top left panel). Even when the sampling is technically sufficient (more than two times the Nyquist frequency; top right panel), artifacts can appear in the data. A rule of thumb is to make sure the sampling rate of the data acquisition is more than five times the fastest frequency you hope to recover in the data.

and the Nyquist frequencies (you'll see why later in section 11.9 on DC offsets).

Here is the final conclusion: To convert frequencies from normalized units to hertz, create $N/2 + 1$ linearly spaced numbers between 0 and the Nyquist. It's good to do a sanity check to make sure this works. The online MATLAB code goes through an example, and there are several exercises at the end of this chapter on this issue.

```
hz = linspace(0,nyquist,N/2+1);
hz = (0:1/srate:N/2+1)*nyquist;
```

Those two lines are equivalent. What happens if N is an odd number, like 101? Then N/2 is a fraction like 50.5, and how do you create 51.1 linearly spaced numbers? Let's see what MATLAB does.

```
linspace(0,40,51.5)
```

Depending on your version of MATLAB, you'll either get an error or a valid result with 51 elements. A better solution, therefore, is to round down to $N/2$. Watch what happens when N is an even or an odd number.

```
N=9; floor(N/2)+1
N=8; floor(N/2)+1
```

There is a little MATLAB trick to programming the frequencies in hertz. You can define frequencies to go up to the sampling rate using the total number of points in the signal.

```
hz = linspace(0,srate,N);
```

This is dangerous code because the second half of the frequencies in this vector are inaccurate; without understanding the Fourier transform, someone might get the impression that they can reconstruct frequencies above the Nyquist. I would never admit to using such code myself.

11.7   The Fast Fourier Transform

It is important to understand how the discrete Fourier transform is implemented, because that's the best way to understand how the Fourier transform works. In practice, however, you should use the fast Fourier transform (FFT), which can save your analysis code hours or even days of computation time. If you were to decompose the Fourier transform to each individual multiplication and addition, you would find that many of the elementary steps in the Fourier transform can be reordered to be computed more efficiently; the fast Fourier transform eliminates these redundancies.

There are several algorithms for computing the fast Fourier transform, but in general they work by putting all of the sine waves into one matrix rather than using a for-loop, using matrix factorizations to break up that large matrix into many simpler matrices with many zeros (sparse matrices), and then reconstituting the Fourier spectrum.

The MATLAB function for performing the fast Fourier transform is `fft`. The FFT is not an approximation—it is a perfect implementation of the Fourier transform. That means the result of the FFT is identical to the result of the "slow" Fourier transform implemented in the previous section.

The function `fft` can take three inputs. The first input is the signal to which the FFT is applied. The second input is the N parameter of the FFT, which is discussed in section 11.10. The third input is the dimension of the data (the first input) over which to apply the FFT. The `fft` function can handle multidimensional matrices and will apply the FFT separately to each column (by default; or whichever dimension is specified by the third input) in the matrix. This is useful, for example, when computing the FFT of many trials at the same time without writing a loop over trials.

11.8 Fourier Coefficients as Complex Numbers

I'd like to spend a bit more time discussing how to work with the Fourier coefficients—and complex numbers more generally—in MATLAB. We'll start by simulating some data with known characteristics. Below are three cosine waves at 6 Hz with known amplitudes and phase offsets. The goal will be to measure these properties using the Fourier transform.

```
cos1 = 3 * cos(2*pi*6*time + 0);
cos2 = 2 * cos(2*pi*6*time + pi/6);
cos3 = 1 * cos(2*pi*6*time + pi/3);
```

Let's see how to retrieve these properties from the Fourier coefficients that are provided by the `fft` function. First we need to define the frequencies in hertz and then figure out which coefficient corresponds to 6 Hz.

```
hz = linspace(0,srate/2,floor(length(time)/2)+1);
freq = 6;
hz6 = dsearchn(hz',freq);
```

The function `dsearchn` will be explained in more detail in chapter 18, but briefly: It returns the index in vector `hz` that is closest to `freq`; in this case, it will tell us which element in the Fourier coefficient series corresponds to 6 Hz. The corresponding complex coefficients are as follows:

```
cos1: 1.5004 + 0.028257i
cos2: 0.85684 + 0.51595i
cos3: 0.24192 + 0.4374i
```

Can we make sense of these complex numbers in terms of the amplitude and phase of the original cosine waves? Not really. Perhaps things will become clearer if we think about these coefficients as vectors in the complex plane. This time the vectors are shown in a polar plot using the `polar` function (figure 11.7). Now you can see the Fourier coefficients in all their glory: The lengths of the lines reflect the amplitude of the cosine wave, and their relationship to the positive real axis reflects their phase-angle offset.

11.9   DC Offsets in the Fourier Transform

The zero-frequency component, or DC component, of a signal is simply the mean offset. We can demonstrate this by taking the same signal, adding mean offsets, and then examining the resulting power spectrum. We start with a simple "signal" (just a few numbers that I made up), with a mean

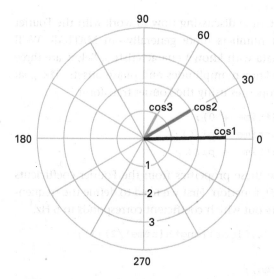

Figure 11.7
Fourier coefficients are best conceptualized as vectors in a complex or polar plane. The length of the line and the angle with respect to the positive real axis ("0" in this polar plot) are taken as amplitude and phase of the Fourier coefficient. The radial axis depicts amplitude, and the circular axis depicts angle.

value of 0.2143. Figure 11.8 shows the original signal, the signal with the mean subtracted, and the signal with a value of one added to all points (a mean offset). You can see from the time-domain plot that the temporal characteristics of the signal are unchanged; the mean offset simply shifts the signal on the y axis. Now inspect the frequency-domain plot. All of the power values are identical for all frequencies (you can zoom-in on your own MATLAB figure to verify) except for those at 0 Hz. Thus, the mean offset of a signal affects only the DC component. This makes sense, because shifting a signal on the y axis doesn't change any of its rhythmic components.

But careful examination of this plot also reveals something more insidious: The DC power value doesn't seem to match. I added a value of one to the signal, but the plot shows a power of around 2.4. I have to admit, I cheated a bit here. To recover amplitudes correctly, it is necessary to double the amplitudes of the positive frequencies, *but not the DC or Nyquist frequencies*. You can see in the online code that I doubled the amplitudes of all frequencies, so the DC value was incorrect.

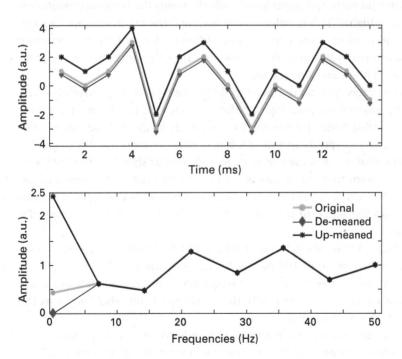

Figure 11.8
The mean offset of a time series is captured exclusively by the 0 Hz component of its Fourier transform.

Sometimes in neuroscience data analysis, this little cheat doesn't matter, because the signals are mean-subtracted before applying the FFT. When signals are mean-subtracted, the DC component is zero, and $0 \times 2 = 0$. Of course, you should strive always to have accurate code. Particularly if you plan on interpreting the DC component, you should be careful to double the amplitudes of only the frequencies above zero and below the Nyquist.

11.10 Zero-Padding the Fourier Transform

As mentioned above, the frequency resolution of a Fourier transform is defined by the number of time points in the signal. If you want a higher frequency resolution, you simply add more time points to the signal. But it is not always possible to add more time points, because you might not have any more data.

The solution is to add zeros. Zeros add no new information to the signal, but they do make the signal longer, which means the frequency resolution becomes higher. This is called "zero-padding." The extra zeros are concatenated to the end of the signal. Zero-padding on the end can get confusing because when you learn about convolution in the next chapter, you will add zeros *before and after* the signal.

Fortunately, you do not need to do the zero-padding yourself; you can specify this in the second input to the fft function. This parameter is commonly called "nfft" for the number of points in the FFT. Be careful: The parameter you specify in the fft function is the total number of points, not the number of zeros you want to pad. If your signal is 20 points long and you want to perform a 24-point FFT, the nfft parameter should be 24, not 4. If the nfft is less than the number of points in the signal, MATLAB will crop the signal to the specified value, which means removing valid signal.

Figure 11.9 shows a 14-point signal and its frequency spectrum after zero-padding by 10 or 100 points. In the time domain, the extra zeros go after the end of the signal; in the frequency domain, the effect of adding extra zeros is to sinc-interpolate the power spectrum (also known as the "zero-padding theorem").

Now that you know *how* to zero-pad, let's discuss *why* you should zero-pad. There are four reasons why you would want to zero-pad the FFT. First, zero-padding allows you to extract a specific frequency from the signal. If you want to extract exactly 10.0000 Hz but the length of your data does not

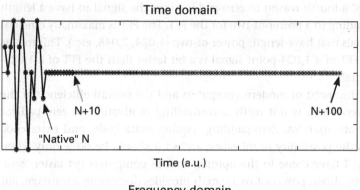

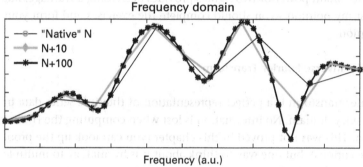

Figure 11.9
Adding zeros after a time series produces a smoother (interpolated) Fourier representation.

produce this exact frequency, you can zero-pad the right number of points to get this frequency. You'll learn more about this in chapter 18.

A second reason to zero-pad is to produce a smoother estimate of the frequency spectrum; in other words, to interpolate. You already saw this in figure 11.9, and you'll learn in the next section how this concept can be used to obtain an interpolated time series.

A third reason to zero-pad is to obtain an equal frequency resolution for two signals, which is a necessary part of convolution. You will learn about convolution in the next chapter, but briefly, one of the steps of convolution is to compute the point-wise multiplication between the Fourier spectra of two signals. Point-wise multiplication is defined only between two signals of the same length; if the signals do not have the same length in the time domain, they will not have the same length in the frequency domain. Therefore, zero-padding is necessary.

Finally, a fourth reason to zero-pad is to get the signal to have a length corresponding to a power-of-two for the FFT. The FFT is maximally efficient with signals that have length power-of-two (1,024, 2,048, etc.). This means that the FFT of a 1,024-point signal is a bit faster than the FFT of a 1,023-point signal.

With the speed of modern computers and the overall efficiency of the FFT, power-of-two is not really a compelling motivation to zero-pad for off-line data analyses. Zero-padding requires extra code, and extra code increases the possibility of mistakes, and so it should be done only when necessary. I have come to the opinion that as computers get faster, zero-padding to obtain power-of-two signals provides decreasing advantage. But that's just my opinion—you should complete the exercises and form your own opinion.

11.11   The Inverse Fourier Transform

The Fourier transform is a perfect representation of the time series data in the frequency domain. No information is lost when computing the Fourier transform. This was not proved in this chapter (you can look up the proof if you are curious), but one way to think about it is by analogy to multiple regression. In a multiple regression, the goal is to explain variance in the dependent variable (the time series) using a set of independent variables (the complex sine waves). The more independent variables you have, the more variance will be explained. If you have as many independent variables as you have data points, the model has zero degrees of freedom, and 100% of the variance is accounted for.

For those with a linear algebra background, another way to interpret this is the Fourier transform is simply a change-of-basis operation. The original signal is in R^N and the Fourier representation is in C^N. Because the Fourier transform is an invertible linear transformation, the Fourier basis is simply a way of using a different coordinate system to look at the same space, and the inverse Fourier transform can be applied to return us from the C^N Fourier basis back to the original R^N basis.

Because no information is lost in the Fourier transform, it is possible to reverse the Fourier transform and go from the frequency domain back to the time domain. This is called the inverse Fourier transform. Recall that in the (forward) Fourier transform, you start with template sine waves and use dot products to compute the Fourier coefficients. In the inverse transform, you use the Fourier coefficients to scale the template sine waves, and the sum of the scaled sine waves is the original time series.

Another difference is that the Fourier transform has a negative sign in the complex sine wave whereas the inverse does not. This negative sign cancels the imaginary part of the sine waves created in the forward Fourier transform (because $i - i = 0$), thus allowing the original real-valued signal to be reconstructed without a superfluous imaginary part.

In the exercises, you will have the opportunity to implement the "slow" discrete inverse Fourier transform. As with the Fourier transform, in real data analyses you should only use the function ifft.

You might be wondering why you should go through the trouble of taking the Fourier transform if you will apply the inverse Fourier transform to get back to the time domain. It turns out that because the FFT is so fast, many signal processing techniques can be accomplished faster and more efficiently (using less code) by doing the important computations in the frequency domain and then going back to the time domain. You will see several examples of this, including in the next chapter.

11.12 The 2D Fourier Transform

Once you understand how the 1D Fourier transform works, moving up to the 2D Fourier transform is straightforward: First, compute the Fourier transform over all the rows in the 2D matrix, then compute the Fourier transform over all the columns of that coefficients matrix. It's straightforward enough that it takes only two lines of code; nonetheless, MATLAB provides a 2D FFT function for you. Actually, the MATLAB fft2 function calls fftn, which allows for computing the FFT over any number of dimensions.

```
pic = imread('saturn.png');
picX = fft2(pic);
picX1 = fft(pic);
picX1 = fft(picX1,[],2);
```

The variables picX and picX1 are the same, although MATLAB's fftn function is a bit faster than calling the fft twice. The resulting Fourier coefficients come in a 2D matrix. It's a bit confusing at first to interpret the results, but keep in mind that the power spectrum shows the power along each dimension separately. For this reason, it is useful to think about the 2D power spectrum not as going from left to right, but as going from the corners of the matrix to the center of the matrix. And for this reason, it is common to shift the coefficients around so the four corners (corresponding to low spatial frequencies) meet at the middle, and the original center (corresponding to high spatial frequencies) is shifted to the corners. This shifting is so common that MATLAB has a built-in function to accomplish this, called `fftshift`.

You must be curious what the power spectrum of the image of Saturn is, right?! It's shown in figure 11.10. The power spectrum has a beautiful appearance, reminiscent of artists' renditions of light swirling around a black hole. Unfortunately, however, there is no cosmic connection here.

11.13 Exercises

1. Generate 10 seconds of data at 1 kHz, comprising 4 sine waves with different frequencies between 1 and 30 Hz, and different amplitudes. Add (1) a little bit of noise and (2) a lot of noise to make two time series. Compute the power spectrum of the simulated time series (use the function `fft`) and plot the results separately for a little noise and a lot of noise. Show frequencies 0 to 35 Hz. How well are the frequencies reconstructed, and does this depend on how much noise there is? Is it easier to distinguish the signal from the noise in the time domain or frequency domain?

2. The following line of code is not incorrect but is bad programming. What's wrong with this line, and how would you fix it (assume variable

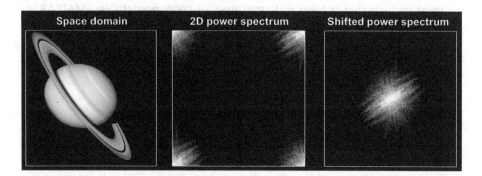

Figure 11.10
A 2D image has a 2D Fourier representation. The original power spectrum is shown in the center plot, with low spatial frequencies at the corners and high spatial frequencies in the center. It is common practice with Fourier representations of images to shift the spectrum so that the low spatial frequencies are at the center while the high spatial frequencies are at the corners (right plot).

data is a time series vector and variable n is the number of points in that vector)?

```
power = abs(fft(data)/n).^2;
```

3. Let's say you want to create the complex number 8 4i. I wrote earlier that you should use 1i instead of i in MATLAB. Is the following code correct? If not, how do you fix it?

```
z = 8+41i;
```

4. The following lines of code contain zero, one, or a few errors (occasionally, there are errors that produce no MATLAB errors). Find and fix the errors, first without MATLAB, and then confirm your answers in MATLAB.

```
x = complex(4,2*1i)
sw = sin(2*10.*(0:.001:1));
linemag = sqrt(real(cdp).^2 + real(cdp.^2))
hz = linspace(0,nyquist/2,floor(N/2)+1);
x=randn(100,1); xX=fft(x,10);%zero-padding by 10
pic=imread('2Dpicture.png'); picX=fft(pic);
```

5. Use the code that produces the right side of figure 11.5 to write a function that reproduces this image using sine waves defined by frequencies and amplitudes that are specified as inputs into the function. Make sure the code allows specifying any arbitrary number of sine waves to sum.

6. Use the function pair tic/toc to compare the computation times of the "slow" discrete-time Fourier transform and the function fft. Create signals from random numbers, and have the signals increase in length from 10 to 10,000 data points in 20 linearly spaced steps. Make a plot of the clock times (y axis) as a function of signal lengths (x axis). This exercise may take a while to run, so be patient.

7. Using the FFT (please don't use the slow Fourier transform anymore!), test the computation time for computing power of a 10,000-point random signal via abs(fc).^2 versus fc*conj(fc) (where fc is the vector of Fourier coefficients). Because this is a scientific experiment, it's best to run the procedure many times on many different vectors, and then average the results together. Show the resulting clock times in a bar plot with error bars reflecting standard deviations.

8. Create a 3-second sine wave at 6 Hz. Compute its Fourier transform and plot the resulting power spectrum. Next, recompute the Fourier

transform after zero-padding with three additional seconds (thus, the total signal will be 6 seconds). Plot the power spectrum after dividing the Fourier coefficients by the number of time points in the original sine wave, then plot the spectrum after dividing the coefficients by the number of time points including the padded zeros. Which normalization returns the accurate amplitude? When you use 6.5 Hz instead of 6 Hz, some results are the same and some results are different. What changed, and why did you get these results?

9. Here is a neat function: $\ln(x + yi)$. Translate this into MATLAB code and evaluate it for x and y from -10 to $+10$ in steps of 0.1. The result is a 2D matrix of complex numbers. Use the `surf` function to explore the magnitude, real part, and angle of this function. Use some plotting options you learned about in chapter 9 to make the images look better than with the `surf` defaults. Now try making a surface of the sum of the magnitude, real part, and imaginary part; does this image look familiar? (For bonus points, create and use a circular color scale.)

10. One easy mistake to make with the 2D FFT is to use the function `fft` instead of `fft2`. It is easy to get confused because when you input a 2D matrix into the `fft` function, it will return a matrix output, but that output will contain the 1D FFT on the columns of the input. The solution to this is—you guessed it—sanity check your code. This is easily done by plotting an image of the inverse FFT of the Fourier spectrum. Make two images of Saturn using `imagesc(ifft2())`, one after correctly applying the 2D FFT, and one after incorrectly applying the 1D FFT. Now try these combinations using `ifft` instead of `ifft2`. I guess even sanity checking is never foolproof.

11. In physics or engineering textbooks, you might see frequency-domain plots with DC in the middle and the negative frequencies to the left. (This comes from the integral expression of the Fourier transform, which has integration bounds from minus infinity to plus infinity; hence, zero is in the middle.) Use the MATLAB function `fftshift` to make one of the power spectra in figure 11.8 have the DC frequency in the center. You'll also need to adjust the vector of frequencies. In practice, you will rarely (or never) see this convention in neuroscience for time series data.

12. Write code for the "slow" discrete inverse Fourier transform. Start with the code for the forward Fourier transform. Here are three hints: The complex sine wave gets a positive i, the complex sine waves are multiplied by the corresponding Fourier coefficient that was obtained from the `fft` function, and the signal is multiplied by N after summing all

modulated sine waves. Check your result against the output of the `ifft` function.

13. The apostrophe in MATLAB actually implements something called the Hermitian transpose, which means the transpose of the conjugate. Inspect the result in MATLAB of the complex number 7 9i and its transpose. Redo figure 11.5 but show the power spectrum of `coefs` and `coefs'`. In another figure, show the phase spectra from `coefs` and `coefs'`. Then repeat this exercise using the function `transpose` instead of using the apostrophe. The lesson here is that the transpose should be applied with caution to a complex matrix. If you need Fourier coefficients to be in a different orientation, it's safer to transpose the time series first. Or transpose the complex coefficients and then take their conjugate.

14. Using paper and pen, draw a complex axis. Identify and mark the following coordinates: [2 1i], [−2 3i], [0 i], [−2 0], [$e^{i\pi/3}$]. Second, draw lines from the origin to each coordinate. Finally, reproduce this plot in MATLAB.

15. If you perform a Fourier transform of a signal that contains 200 time points sampled at 100 Hz, what is the highest frequency (in hertz) that you can reconstruct? What would the highest frequency be if you had 400 time points?

16. What are the sizes of matrices `data_fft1` and `data_fft2`? First come up with your answer, then confirm in MATLAB.

```
data = randn(100,1) + 100;
data_fft1 = fft(data);
data_fft2 = fft(data,200);
```

17. The following lines of MATLAB code extract information from the variable `fcoefs` (Fourier coefficients). Which of the following will extract (1) power, (2) phase, or (3) neither?

```
angle(fcoefs);
fft(fcoefs);
abs(fcoefs).^2;
angle(fcoefs).^2;
real(fcoefs);
```

18. Modify the "slow" Fourier transform code to put the division-by-N inside the loop instead of after the loop.

12 Convolution

Convolution is a procedure in which two signals are combined to produce a third signal that has shared characteristics of the two input signals. For example, one use of convolution is to apply a narrow-band filter to time series data; the third signal (the band-pass filtered signal) contains only the frequency characteristics of the original time series that are shared with the filter kernel.

In the parlance of convolution, the two input time series are called the signal and the kernel. By convention, we call the "interesting" time series (such as the EEG data, spike train, or brain image) the signal and we call the filter (such as a Gaussian or wavelet) the kernel. But this just helps to clarify the intention of convolution; the mechanics and results of convolution do not depend on these terms. For the remainder of this chapter, the letter N will indicate the length (number of points) of the signal, and the letter M will indicate the length of the kernel.

Convolution can be conceptualized and implemented in the time domain or in the frequency domain. It is important to understand both, but in practice you will most likely use only the implementation of convolution in the frequency domain. This chapter mainly focuses on 1D convolution for time series; the extension to 2D convolution is straightforward and will be introduced at the end of this chapter.

In mathematical texts, convolution between a signal S and a kernel K is written with an asterisk: S*K. This can cause some confusion in MATLAB because the asterisk is used for multiplication. When writing down your analysis methods, be clear about whether you are writing a mathematical formula (in which case * means convolution) or MATLAB code (in which case * means multiplication).

12.1 Time-Domain Convolution

You can think of convolution in the time domain as "smearing" the kernel across the signal. Figure 12.1 shows an example of a boxcar signal before, during, and after convolution with a Gaussian kernel.

The basic building block of convolution is the dot product, with which you are already familiar. To run convolution, flip the kernel backwards and position it such that the right-most point of the kernel is aligned with the left-most point of the signal, and then compute the dot product between the two. "But wait," you are probably interjecting, "why do we flip the kernel backwards?!" The answer is: That's just how it's done. If the kernel is not flipped backwards, the result is called the cross-covariance. In frequency-domain convolution, which you will learn about later in this chapter, the kernel is not flipped.

In order to make the dot product a valid computation, the signal needs to be zero-padded at the beginning by $M - 1$ points, and at the end by $M - 1$ points (minus 1 because the kernel and signal overlap by one point; figure 12.1). The extra zeros do not interfere with the result of convolution,

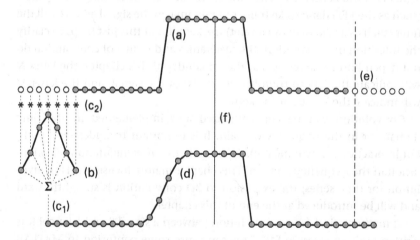

Figure 12.1
Time-domain convolution involves a signal (a) and a kernel (b). The right-most point of the kernel is aligned with the left-most point in the signal, and the dot product (point-wise multiplication and sum) is computed (c_1). The signal must be zero-padded (c_2) for this operation to be valid. The kernel is moved over by one point and the dot product is repeated. The result of the convolution when it is halfway through is shown (d; the vertical line is where convolution was paused). Convolution ends when the kernel reaches the end of the signal (e).

because zeros times the kernel is still zero. This completes the first step in time-domain convolution. The resulting dot product from this step of convolution is plotted in a location corresponding to the center of the kernel relative to where it is lined up with the signal. For this reason, it is useful to make sure the kernel has an odd number of points.

```
N = length(signal);
M = length(kernel);
halfKern = floor(M/2);
dat4conv = [ zeros(1,M-1) signal zeros(1,M-1) ];
```

The rest of convolution is simple. At each step of convolution, keep the signal fixed and slide the kernel one time step to the right. Repeat the dot product computation. Slide, dot product, repeat. Over and over again, until the left-most point of the kernel is aligned with the right-most point of the data.

```
for ti=M+1:N-M+1
    tempdata = dat4conv(ti:ti+M-1);
    conv_res(ti) = sum(tempdata.*kernel(end:-1:1));
end
```

The result of convolution starts before the signal begins and ends after the signal ends, meaning that the result of convolution is longer than the original signal. I call these the "wings" of convolution. How long is the result of convolution? The result of convolution is half the length of the kernel too long at the beginning and half the length of the kernel too long at the end. You might initially think the result is length $N + M$, but it's actually $N + M - 1$. To understand why the "minus 1" is necessary, it's easiest to see it visually explained, which I hope is clear enough in figure 12.2. $N + M - 1$ is an important quantity, and you should not forget it.

Technically, convolution is now finished. However, it is useful to cut off the wings for simplicity.

```
conv_res = conv_res(halfKern+1:end-halfKern);
```

If the convolution is not done with $N + M - 1$ points, the convolution wings wrap around such that the end of the time series sums into the beginning of the time series, and vice versa. This is called circular convolution (using $N + M - 1$ is called linear convolution). If the time series is long relative to the length of the kernel, this awkward wraparound effect has minimal impact except at the edges. You'll see an example of circular convolution in chapter 20.

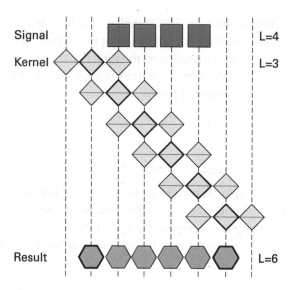

Figure 12.2

Illustration of why the result of convolution between a signal of length N and a kernel of length M is $N + M - 1$ points. In this example, the kernel is three points long, and the signal is four points long. Each kernel "row" illustrates each successive step of convolution. When the result of convolution is positioned at the center of the kernel, the result must be $N + M - 1$.

12.2 The Convolution Theorem

Time-domain convolution is okay, but (1) it's slow, and (2) it's not very intuitive *why* it is so powerful as a signal processing technique (at least, not to me). Convolution can also be implemented in the frequency domain, and this solves both problems. Before learning how to implement convolution in the frequency domain, you need to know about the convolution theorem.

The convolution theorem states that convolution in the time domain is equivalent to multiplication in the frequency domain. This means that we can implement convolution in two distinct ways and arrive at the same result (figure 12.3). The full proof of the convolution theorem is not presented here, but briefly, it involves demonstrating that the Fourier transform of a convolution between signals X and Y can be reduced to the multiplication of the Fourier transform of X by the Fourier transform of Y.

In short, the following two procedures produce identical results:

1. Time domain. Compute repeated dot products between the kernel and temporally corresponding points in the signal. At each repeat, shift the kernel δt or one time step relative to the signal.

2. Frequency domain. Compute the ... for the signal, take the FFT of the kernel, point-wise the ... (frequencies) multiply the linear coefficients together, and compute the inverse FFT.

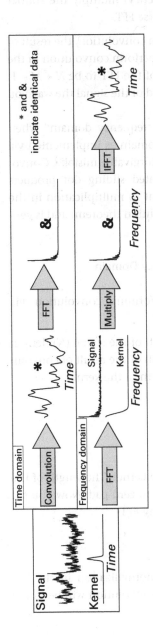

Figure 12.3
Illustration of the convolution theorem and the interchangeability of time-domain convolution and frequency-domain multiplication. The two time series with asterisks are identical, as are the two frequency spectra with ampersands. This figure is taken from figure 11.10 of Cohen (2014).

1. Time domain: Compute repeated dot products between the kernel and temporally corresponding points in the signal. At each repeat, shift the kernel over one time step relative to the signal.
2. Frequency domain: Compute the FFT of the signal, the FFT of the kernel, point-wise (i.e., frequency by frequency) multiply the Fourier coefficients together, and compute the inverse FFT.

Important! No matter which way you perform convolution, the result of convolution must have length $N + M - 1$. If you perform convolution in the frequency domain, you need to set the length of the FFT to be $N + M - 1$, but you do not need to zero-pad the start and end of the signal the way you would in the time domain.

Sometimes people say "convolution in the frequency domain" when they actually mean "convolution in the time domain as implemented via multiplication in the frequency domain." It's a forgivable mistake. Convolution in the frequency domain means repeated sliding dot products between two Fourier spectra, which is equivalent to multiplication in the time domain (this is the other half of the convolution theorem) and is generally not used in neuroscience data analysis.

12.3 Convolution Implemented in the Frequency Domain

Here is the complete five-step plan to performing convolution via frequency-domain multiplication.

1. Define the three convolution lengths: length of the signal (N), length of the kernel (M) (and now you get the length of the result of convolution for free: $N + M - 1$), and half of the length of the kernel.

```
nData = length(signal);
nKern = length(kernel);
nConv = nData+nKern-1;
nHfkn = floor(nKern/2);
```

2. Take the FFT of the signal and the FFT of the kernel. The length of the FFTs need to be $N + M - 1$, so you will need to zero-pad. Now you see one of the important reasons for zero-padding FFTs.

```
signalX = fft(signal,nConv);
kernelX = fft(kernel,nConv);
```

3. Depending on the kernel, it may require normalization in the frequency domain to ensure that the result of convolution will be in the

same units as the original signal. For wavelets (see chapter 19), this can be implemented by dividing the Fourier spectrum by its maximum.

```
kernelX = kernelX./max(kernelX);
```

4. Point-wise multiply the two Fourier spectra and take the inverse FFT. This point-wise multiplication will fail unless both FFTs are the same length, which is a good check that you did step 2 correctly.

```
convres = ifft(signalX.*kernelX);
```

5. Cut the wings off of the result of convolution: Remove one half of the length of the wavelet from the beginning and one half of the length of the wavelet from the end. The result now has length N and is directly comparable to the original signal.

```
convres = convres(nHfkn:end-nHfkn);
```

Steps 3 and 5 are not technically necessary. Step 3 is useful if you want to keep the postconvolution results in the same scale as the original data (e.g., μV for EEG data). Concerning step 5, it is difficult to imagine a situation where you would want to keep the convolution wings, and leaving them attached can only increase the possibility of confusion and mistakes.

Thinking about convolution as point-wise multiplication between the frequency spectra of the signal and that of the kernel offers an enlightening perspective on what convolution does and why it works. Let's start with a simple example of random noise (the signal) and a Gaussian (the kernel).

Figure 12.4 shows the two time series and the result of their convolution. The Gaussian acts as a low-pass filter. This happens because the shape of the Gaussian in the frequency domain is a negative exponential, and when the two spectra are point-wise multiplied, only the frequencies in the signal that match those with non-zero power in the Gaussian are preserved in the inverse FFT. More specifically, the higher frequencies are attenuated because the power spectrum of the Gaussian is nearly zero. You'll see additional examples of convolution in chapter 20.

Now you can see the importance of understanding how convolution works in the time domain and its implementation in the frequency domain. The reason why convolution is used as a filter is simple and sensible when considering how convolution is implemented in the frequency domain. And the reason why convolution produces a result of length $N + M - 1$ is sensible from the time-domain perspective.

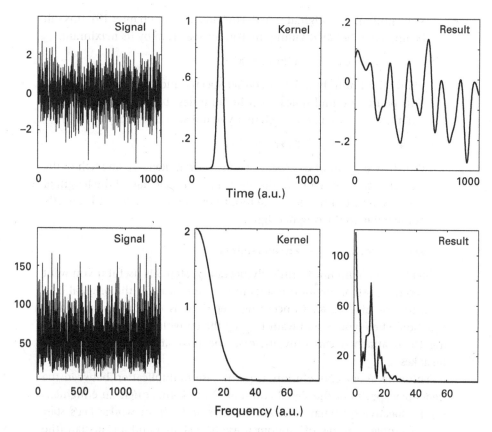

Figure 12.4

Convolution of random noise and a Gaussian. The top row shows the time domain, and the bottom row shows the frequency domain. The left-most panels show the time series signal; the middle panels show the kernel; and the right-most panels show the result of convolution. Notice that the result of convolution is a mixture of the characteristics of the signal and the kernel.

12.4 Convolution in Two Dimensions

Convolution in two dimensions is used in image processing. The principle is the same as convolution in one dimension, except that it is done over two dimensions. As with 1D convolution, 2D convolution can be done the slow way (e.g., the space domain) using two nested for-loops, or it can be done the fast and efficient way using frequency-domain multiplication (make sure to use fft2, not fft!). I'll just focus on the frequency-domain implementation here.

Let's go back to the picture of Saturn. It's very highly detailed, isn't it? How about we make it look more like a watercolor painting by smoothing it (i.e., applying a low-pass filter). Our low-pass filter will be a 2D Gaussian.

Although we have an extra dimension to deal with, the basic five-step plan of convolution is the same. There are a few minor departures from the 1D process, and selected lines of code are printed here to highlight those differences.

```
nConv = N+M-1;
picX = fft2(pic,nConv(1),nConv(2));
cr=cr(halfK(1)+1:end-halfK(1),halfK(2)+1:end-halfK(2));
```

N and M are now two-element vectors (height and width of the image) instead of one-element scalars. In the fft2 function, there are two inputs to specify the nfft parameter—one for height and one for width. And be careful that the wings must be trimmed on both dimensions. The variable cr is a 2D matrix, so how many commas should we use when indexing it?

The images in the space and the frequency domains are shown in figure 12.5. The moons disappeared in the convolution result (apologies to fans of Titan and Rhea and all the rest). Moons are small and have sharp edges, and thus they comprise spatial frequencies that are higher than the spectral characteristics of the Gaussian filter. You'll learn more about how 2D convolution is used in image processing in chapter 27.

12.5 Exercises

1. The MATLAB function for convolution is conv. Read the help file for this function, and then convolve a signal in the time domain, implement it as frequency-domain multiplication, and use the MATLAB conv function. All three results should match (you might need to skip step 3 to get the amplitudes the same).
2. Is the frequency domain really faster than the time domain for convolution? Test this yourself by convolving random signals of lengths varying from 10 to 100,000 with a Gaussian that has a length corresponding to half the length of the signal. Repeat the convolution using time-domain convolution and the frequency-domain implementation, and use tic/toc to measure computation time. Make a plot of computation times as a function of signal length.

Figure 12.5

This figure has a layout comparable to that of figure 12.4 but is 2D instead of 1D.

3. For the image of Saturn, apply a "poor-man's high-pass filter" by sub-tracting the low-pass filtered image from the original image. Does this residual image reveal the high-frequency features?

4. Can you find and fix the MATLAB error here (try to do it without looking back at the chapter)?

```
for ti=M:N-M
    tempdata = dat4conv(ti:ti+M-1);
    conv_res(ti) = sum(tempdata.*kernel(end:-1:1));
end
```

5. Do these lines of code produce the same or different results? Does it depend on whether *n* is even or odd? First think of your answer, then test it in MATLAB (e.g., using *n* = 1:10).

```
floor(n/2)
ceil(n/2)-1
round((n-1)./2)
```

6. Your friend Fred is trying to normalize kernels in the frequency domain. Would the following code work if gausX is the Fourier spectrum of a 1D Gaussian? What if gausX were the Fourier spectrum of a 2D Gaussian?

```
gausX = gausX ./ max(gausX);
```

7. Code is not provided for figure 12.5. That's your job!

8. In the code for figure 12.4, change the signal from normally distributed to uniformly distributed random numbers. What is the most notable difference in the result of convolution? Thinking back to what you learned in chapter 11, what could you change in how the signal is created to minimize this difference?

9. Write code for the "slow" 2D convolution to replicate the smoothed picture of Saturn using the same convolution kernel. You'll need to manually zero-pad and use two nested loops (if you are feeling adven-turous, you could use linear indexing to write it in one loop, but you know my opinion of linear indexing). If you are having trouble getting started, try drawing a picture of what each step in 2D convolution would look like.

10. There are two errors in the line of code below (variable cr is the result of 2D convolution). Find and fix them.

```
cr = cr(halfK(1)+1:end-halfK(2),halfK(1)+1:end-halfK(2));
```

13 Interpolation and Extrapolation

Sometimes you get data-greedy and want more data than you actually have. Or you want to guess what the data might look like if you could have measured from a broader range. Perhaps you want to synchronize camera-recorded data at 120 Hz with electrophysiologic data recorded at 2,048 Hz. In these cases, the solution is to interpolate or to extrapolate.

The distinction between interpolation and extrapolation is fairly simple (figure 13.1). *Inter*polation is when you measured points A and C and want to estimate what a point between them (B) might look like. *Extra*polation is when you measured points A and C and want to estimate what point E might look like. More formally, interpolation is when the estimated data point lies inside the boundaries of the actual measurements, whereas extrapolation is when the estimated data point lies outside the boundaries.

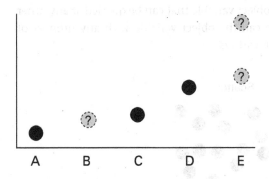

Figure 13.1
Interpolation versus extrapolation. Imagine that A, C, and D are empirically measured data points. Interpolation involves estimating the value of B. Extrapolation is the art of guessing what E (outside the measurement boundaries) might be. A linear extrapolation method might produce the lower value, while a nonlinear (e.g., spline) extrapolation method might produce the upper value.

In general, interpolated data are more trustworthy and can be estimated with fewer assumptions than extrapolated data.

13.1   The MATLAB Functions `griddedInterpolant` and `scatteredInterpolant`

Since MATLAB version 2011b, interpolation is done via the functions `griddedInterpolant` or `scatteredInterpolant`. Since version 2013, extrapolation is also done using these functions. The functions `interp1`, `interp2`, and `interpN` call `griddedInterpolant`. For users of older versions of MATLAB or Octave, `griddedInterpolant` may not be available. It is a compiled built-in function, so if you don't have these functions, simply copying the function file onto a different computer is not a viable solution. Instead, you can use the `interp*` functions, which will be explained later in this chapter.

Whether you should use `griddedInterpolant` or `scatteredInterpolant` depends on how the data were sampled. If your data were recorded on a regular lattice, that's a grid. If you recorded data from randomly scattered points, that's a scatter (figure 13.2). The inputs to these two functions are a bit different. You'll learn how to use both of these functions in this chapter.

Interpolating or extrapolating involves two steps. In the first step, you use the function `griddedInterpolant` to input the observed data and locations (e.g., time stamps or spatial locations) at which those data were measured. This produces an object variable that can be queried at any other location. The next step is to call the object variable with any number of desired new (nonmeasured) locations.

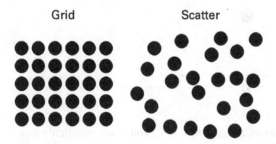

Figure 13.2

Distinction between a grid and a scatter. Imagine each dot is the physical location of an empirical measurement from which interpolation or extrapolation is estimated.

```
data = [1 4 3 6 2 19];
datatimes = 1:6;
newtimes = 1:.001:6;
F = griddedInterpolant(datatimes,data);
newdata = F(newtimes);
```

Let's piece apart this code. We start by defining six "measured" data points (variable data), sampled at regular intervals (variable datatimes). The measurements were taken on a regular lattice (i.e., the spacing between each successive measurement was the same), so we use griddedInterpolant. We want to up-sample this vector by 1,000 times, as if we were up-sampling from 1 Hz to 1 kHz. The variable newtimes specifies the time points at which we want to interpolate. The variable newtimes is regularly spaced between the first and last points of datatimes, but that was done for convenience; you could request from variable F any time point(s) you want. Note that all of the new time points are inside the boundaries of the measured time points. This means we are interpolating, not extrapolating.

A simple example like this is perfect for understanding the behavior of different algorithms for performing the interpolation. The choice of algorithm is an optional third input into the function. There are several options, including linear, three nearest-neighbor variants, and spline. Figure 13.3

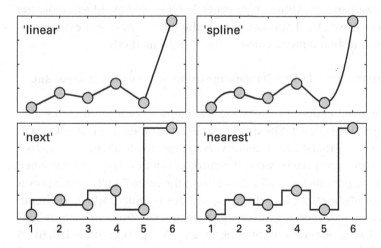

Figure 13.3
Illustration of four different algorithms for interpolation. The gray dots reflect empirical measurements, and the black lines are interpolated points.

shows results of interpolating the same data using different interpolation methods.

In practice, you will probably use linear or spline interpolation most often. The main difference between these is that the spline method is non-linear whereas the linear method is—you guessed it—linear. When interpolating over a fairly densely measured grid, the difference between linear and nonlinear interpolated results will be minor.

The more salient differences between linear and spline interpolation happen during extrapolation, particularly as the extrapolated points get further away from the measured points. Spline extrapolation can run off to extreme values and is therefore more sensitive to noise near the boundaries.

Consider figure 13.4, in which linear and spline extrapolation produced wildly different results although they were based on the same data. If you were to make important and theoretically relevant interpretations of data based on this extrapolation, you might come to completely different conclusions depending on the extrapolation method you used. (This is part of the reason why economists with different political affiliations can disagree so vehemently about how policies might affect the economy.)

Which extrapolation method is more accurate? Unfortunately, the answer is "neither and both." The accuracy of extrapolation depends on your assumptions about what the data might look like outside the boundaries of measurements. Although in general you should avoid extrapolating too far away from the boundary points, the further away you extrapolate, the more you should prefer conservative (linear) methods.

13.2 Interpolation in Two Dimensions Using `scatteredInterpolant`

There are two situations in multichannel neuroscience recordings in which 2D interpolation is used. The first is topographic plotting, in which visualization of the spatial distribution of activity is greatly facilitated by interpolating values over space between measured electrodes. The second is when an electrode provided no valid data during the recording; interpolation is used to reconstruct the activity it might have recorded, which facilitates cross-subject data pooling. Actually, topographies from human M/EEG data are 3D (because most human heads are 3D objects), but the height (z) dimension is often ignored in topographic plots for simplicity. We will ignore it here as well.

Two-dimensional interpolation requires one extra step of complexity, which is to specify a 2D grid instead of a 1D array of points. To specify a 2D

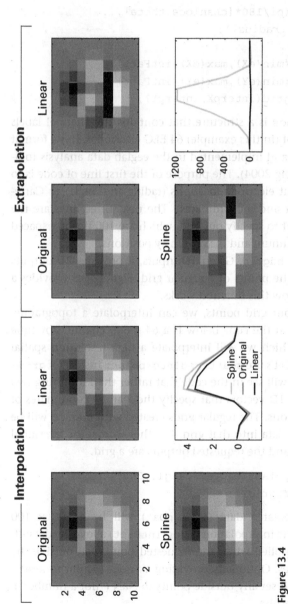

Figure 13.4

Spline and linear algorithms often produce similar results during interpolation, but can produce very different results during extrapolation. In this example, data from row 7 (interpolation) or from row 17 (extrapolation) were estimated, and the result was placed in row 7 and drawn in the line plots.

grid of points, you can use the functions meshgrid or ndgrid. The function meshgrid supports only 2D and 3D grids and swaps the columns and rows in the output (this is sometimes useful for image processing, although it can cause some confusion). The function ndgrid will be used here because it is applicable in more general situations.

```
[eX,eY] = pol2cart(pi/180*[chanlocs.theta],...
          [chanlocs.radius]);
intFact = 100;
interpX = linspace(min(eX),max(eX),intFact);
interpY = linspace(min(eY),max(eY),intFact);
[gridX,gridY] = ndgrid(interpX,interpY);
```

The variable chanlocs is a structure that contains the channel labels and locations for each of (in this example) 64 EEG electrodes. It is a format that was developed for and implemented in the eeglab data analysis toolbox (Delorme and Makeig 2004). The purpose of the first line of code is to convert the polar-format electrode locations (radius and angle) to Cartesian-format locations (x and y coordinates). The next three lines are the points at which we want to specify our grid, which are 100 linearly spaced points between the minimum and maximum xy positions.

Finally, the function ndgrid. Given 1D inputs, it provides 2D outputs that combine to form the points of a regular grid. Figure 13.5 provides a graphical overview of how this function works.

Now that we have our grid points, we can interpolate a topographic map. The variable eeg in the code below is a 64 × 640 channels-by-time matrix of data, from which we will interpolate a high-resolution spatial map at one time point. Let's use scatteredInterpolant instead of griddedInterpolant. You will see in the code that rather than inputting two 2D grids, we input two 1D vectors that specify the x and y coordinates of the measurement locations. The regular grids created with ndgrid will be used to interpolate the data into that grid. In other words, the measured locations are scattered, and the requested outputs are a grid.

```
F = scatteredInterpolant(eX',eY',eeg(:,300));
interpDat = F(gridX,gridY);
```

The resulting interpolated data (interpDat) will be a 100 × 100 matrix, as specified by the interpolation factor (intFact) above. By default, extrapolation will be applied when the requested points fall outside the measurement boundaries. Optionally providing a 'none' input to scatteredInterpolant will set any outside points to NaN ("not a number"). Figure 13.6 shows the results.

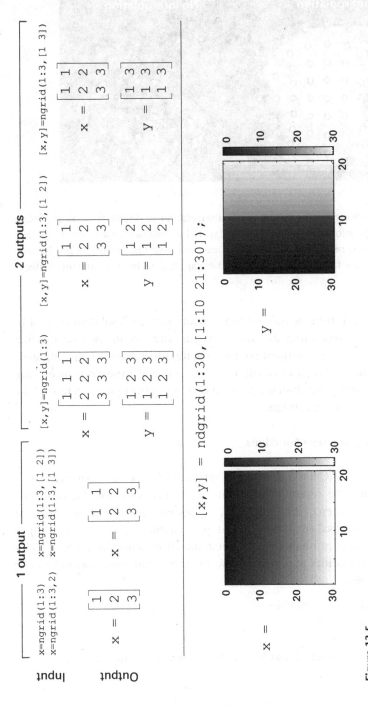

Figure 13.5
Graphical illustration of the outputs of the function ndgrid.

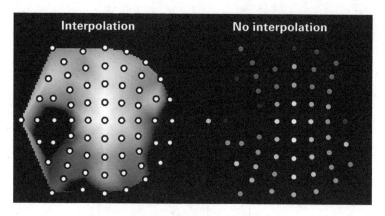

Figure 13.6
This figure illustrates the advantage of interpolation of discretely measures points (electrode positions) for interpreting the spatial distribution of activity. These data show one time point of voltage activity measured from EEG electrodes in a human. The front of the head is the right side of the map, and the left ear is the top center of the map.

At the beginning of this section, I wrote that interpolation is used in multichannel recordings for two situations. Here we focused on the first use; the second use (estimating the activity at a broken or missing electrode) is implemented in nearly the same way, with the difference being that "grid" would just be a single point corresponding to the location of the to-be-interpolated electrode.

13.3   Using `interp*` Functions

If you are using Octave or an older version of MATLAB, `griddedInterpolant` may be unavailable. In this case, you can use the functions `interp1`, `interp2`, and `interpn`. These functions work similarly to `griddedInterpolant`, but without the "middle-man" step. You input the observed measurements and their spatial or temporal locations and the specified points at which you want to estimate new data, all in the same function.

```
data = [1 4 3 6 2 19];
datatimes = 1:6;
newtimes = 1:.001:6;
newdat = interp1(datatimes,data,newtimes,'spline');
```

Again, we are performing interpolation to up-sample the data from 1 Hz to 1 kHz. The first three inputs to the function `interp1` are necessary; the fourth input optionally specifies the type of interpolation to perform, similar to the `griddedInterpolant` function.

Modifying the code so it performs extrapolation in addition to interpolation is as simple as specifying that the `newtimes` vector starts before 1 and/ or ends after 6. For example, re-run the code specifying `newtimes` from 0 to 8. Next, change the fourth input from `'spline'` to `'linear'`, and inspect the plot again. What happened? The plot shows data only from values 1 to 6, which corresponds to the boundary conditions. The `interp*` functions with linear methods do not perform extrapolation, instead setting data outside the boundaries to NaN. But fear not—you can still do linear extrapolation by adding a fifth input `'extrapolate'`.

13.4 Zero-Padding Theorem and Zero-Padding

The zero-padding theorem was mentioned in chapter 11 (Fourier transform). Briefly, adding zeros to the end of a time series produces a sinc-interpolated power spectrum, and adding zeros to the end of a Fourier spectrum produces (after taking the inverse Fourier transform) a sinc-interpolated time series.

Because frequency resolution is defined by the number of time points, the number of time points is also defined by the frequency resolution. I realize that sounds like circular logic, but it isn't, because we can manipulate the frequency resolution by zero-padding, independent of the time domain. For example, to double the sampling rate of a time-domain signal that is N points long, add N zeros to the end of its Fourier spectrum before taking the inverse Fourier transform. Fortunately, you don't need to add the zeros explicitly; you can specify the N of the inverse Fourier transform as an optional second input when calling the MATLAB function `ifft`. A simple example of applying the zero-padding theorem to perform sinc-interpolation is shown below and in figure 13.7.

```
origN=10; padN=100;
data = randn(origN,1);
datapad = ifft(fft(data)/origN, padN)*padN;
```

The key parameter for the up-sampling is `padN`. Try changing that parameter to see the effects on the resulting signal. It's a good idea to sanity check the procedure by changing `padN` to `origN`, which should have no effect.

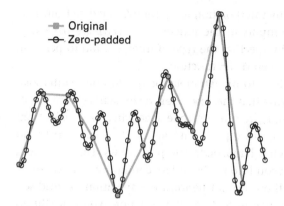

Figure 13.7
Zero-padding in the frequency domain can be used to sinc-interpolate a time series.

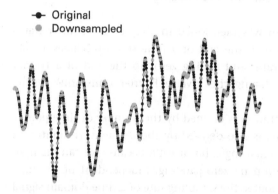

Figure 13.8
A time series in its full (black dots) and down-sampled (gray dots) forms.

13.5 Down-sampling

Down-sampling data is mostly straightforward (figure 13.8). There is one additional consideration for down-sampling, which is that higher-frequency activity can be aliased into lower frequencies. Let's take some time to unpack that statement.

Aliasing is a by-product of the fact that the highest frequency that can be measured in a Fourier transform is the Nyquist frequency (half the sampling rate). Any sinusoidal processes that have a frequency higher than the Nyquist cannot be exactly represented in the Fourier transform. But that

Box 13.1

Name: Robert Oostenveld
Position: Senior Researcher and Visiting Professor
Affiliation: Donders Institute for Brain, Cognition and Behavior, Radboud University, Nijmegen, Netherlands; NatMEG, Karolinska Institute, Stockholm, Sweden
Photo credit: Robert Oostenveld

When did you start programming in MATLAB, and how long did it take you to become a "good" programmer (whatever you think "good" means)?
I started doing simulations and data analysis in MATLAB around 1998 at the start of my PhD project. About 5 years later, I realized that my code was of sufficient quality that other researchers would benefit from me sharing it. My MATLAB code is now being used by thousands of neuroscience researchers across the world. That does not make me a good programmer per se, but it shows that my programming style is well appreciated and is efficient in contributing to science moving forward.

What do you think are the main advantages and disadvantages of MATLAB?
The advantages are that it is easy to get started with MATLAB, as the syntax is very close to the notation of linear algebra. Variables don't have to be declared, and m-files that start as scripts can easily be converted to functions. All data

in MATLAB memory can simply be saved, regardless of whether it is an array or a structure, and reused at any later point. Another advantage is that 20-year-old code still works and that transitions to newer computer hardware and operating systems have been smooth.

I consider the functional programming style of MATLAB an advantage for the often exploratory development of scientific software. The resulting collection of functions and scripts may be less well-structured compared to an object-oriented (OO) programing style, but often the design of the scientific software cannot be specified up front. MATLAB allows one to prototype the software while at the same time make scientific progress. At the end of the scientific project, there is often no need to reuse, refactor, and rewrite the prototype. However, it also one of the weaknesses that the code is often discarded at the end of the project.

Clearly the biggest disadvantage of MATLAB is that it is a commercial product. Not everyone can afford it, making the sharing and reproduction of scientific results (in the form of simulations and analyses) problematic. For those researchers who do have MATLAB, it is often problematic due to financial reasons to keep their installed version up to date. With the code that I develop and share, I consequently need to support a wide range of MATLAB versions. I cannot use the latest features of MATLAB, since these do not get back-ported to older versions.

Another disadvantage in the MATLAB language is that there is no namespace; all functions are added to the global path for the session as a whole, easily resulting in function and variable name clashes. The lack of namespaces also complicates versioning and toolbox/package management. The sharing of code is not so widespread as in Python or in Node JS. MathWorks file exchange is a valuable effort in sharing but cannot compare (yet) to "pip install somepackage" or—even better—the Node package manager (npm).

Do you think MATLAB will be the main program/language in your field in the future?

I don't think it is the "main" programming language now. It is one of the languages, besides C/C++, Java, Python, Julia, and many more, and is used in a similar way as other mathematical software (not languages per se), such as Mathematica, R, and Octave. I do expect MATLAB to become less dominant in my area of scientific computing due to the increased interest in open-source languages such as Python and Julia. Also, the development of ready-to-run open-source software applications rather than toolboxes reduces the need for the MATLAB programming environment. This is visible for example in fMRI, where FSL (C/C++ based command line tools) have taken over the dominance of SPM (MATLAB based toolbox) for standard off-the-mill functional MRI analyses.

How important are programming skills for scientists?
General programming skills reflect the skills of analytical reasoning. Therefore, they are of utmost importance. Having actual hands-on skills in programming in a specific language or environment is required to be capable of
doing new analyses. In pushing science forward, we need researchers that
don't repeat what has been done in the past, but who come up with new
research questions that also necessitate new scientific tools. Being able to
make your own software tools and to use the tools provided by others requires
researchers to understand how to use the command line, how to use scripts,
and how to program.

Any advice for people who are starting to learn MATLAB?
Write code for something that you care about. Start simple, trying to focus on
the results that the code provides. Think about reuse of the software by others,
but also by a future version of yourself. Document not only *how* to use it, but
also *why* it was implemented this way. Writing a few lines of documentation
on how a still-to-be-implemented function should behave often saves many
hours in struggling with the design of the functionality. Don't over-design
your architecture or framework if the functionality is not yet in place. Learn
from others, not only in small details regarding syntax, but also in code architecture and in the tools that they use (organization of code and data, software
version control, documentation, sharing). Get inspired by http://software-
carpentry.org, even if it does not describe the programming language that you
want to use. Discuss the code that you write with others, like you discuss your
scientific posters, presentations, and papers. Join a http://brainhack.org hack-
athon event and learn from others.

doesn't mean they won't be measured. Instead, they will be captured by
lower-frequency sine waves, as was illustrated in figure 11.6.

To prevent aliasing from occurring during down-sampling, a low-pass
filter should be applied before removing data points. The cutoff of the filter
should be the Nyquist frequency of the new down-sampled rate (not the
original Nyquist frequency). This is often called an anti-aliasing filter. The
cutoff could also be lower than the new Nyquist frequency, but it should
not be higher.

In many cases, you do not need to apply the low-pass filter yourself.
Many MATLAB functions that perform down-sampling should apply an
anti-aliasing filter prior to removing data. It's a good idea to check that a
down-sample function you want to use first applies a low-pass filter. If not,
you can apply the low-pass filter yourself before down-sampling.

If you have the MATLAB signal processing toolbox, you can use the function `resample`. If not, you can download the (free) Octave package called "signal." Many other third-party data analysis toolboxes will include functions to down-sample time series data while applying an anti-aliasing filter.

Most resampling functions do not ask for the sampling rate in hertz or any other unit. Instead, they take as inputs two integers whose ratio produces the fraction of the original sampling rate to which you want to down-sample. For example, if the original sampling rate is 1,000 Hz and you want to down-sample to 250 Hz, you would input the numbers 1 and 4.

Fortunately, you don't need to figure out these integers yourself; there is a MATLAB function called `rat` (I'm sure you immediately guessed that r.a.t. is an acronym for rational fraction approximation). Whenever possible, down-sample only to a sampling rate that is easily captured by integer fractions. That is, if the original sampling rate is 1,000 Hz, it's better to down-sample to 250 Hz than to 257.31 Hz. Down-sampling to a sampling rate that is not an easy integer fraction of the original sampling rate requires finer interpolation, which increases the risk of inaccuracies.

13.6 Exercises

1. Generate a time series of random numbers, and up-sample the time series using two methods of interpolation: the zero-padding theorem and `griddedInterpolant`. Make sure the number of time points is the same. How comparable are the results, and is it possible to find parameters to make the results more similar?

2. Simulate 3 seconds of random data (use `randn`) at 1 kHz. Down-sample the data to 200 Hz. Then up-sample the data back to 1 kHz using linear and spline interpolation. Plot all of the time courses in the time domain and their power spectra. What is the effect of down-sampling then up-sampling on the resulting time series and the power spectrum, and does linear or spline interpolation more closely match the original time series?

3. Reproduce the left panel of figure 13.6 using different values for the `intFact` parameter, ranging from 5 to 200 in 16 linearly spaced steps. Show the results in a figure with 4-by-4 subplots. Higher `intFact` values will take longer to compute but will produce smoother plots. What do you think is a good range for this parameter? Is 1,000 better than 200?

4. Rewrite the EEG interpolation code using `griddedinterpolant`. (Warning: This is a difficult problem!)

5. Do these two lines of code give identical output? Test it in MATLAB. If they differ, adjust the code so the two sets of outputs are the same.

```
[gridX,gridY] = ndgrid(interpX,interpY);
[gridX,gridY] = meshgrid(interpX,interpY);
```

6. Can you interpolate with NaNs in the data? To find out, go back to the code for figure 13.3. Replace one of the nonborder data values with NaN and re-run the code. What are the results, and do they depend on the function and/or interpolation method? Next, reproduce the figure using `interp1` instead of `griddedInterpolant`, and try again using one NaN value.

7. In the MATLAB code, down-sampling is done by an integer factor. What if you want to down-sample by a non-integer factor; for example, from 1,000 to 300 Hz? One solution is to up-sample first and then down-sample. Write code to figure out how much to up-sample in order to be able to down-sample by an integer. The code should be flexible enough to work for any two sampling rates.

8. Run interpolation using `griddedInterpolant`, `scatteredInterpolant`, and `interp1`. Before running the functions, convert the to-be-interpolated time series to various formats, such as single, int32, and so on. Which functions work with which data formats? If there are errors, what are the error messages, and do these help you debug code in the future?

9. In the code for down-sampling, I was a bit lazy and smoothed the random time series by using the MATLAB `conv` function and a kernel defined by the `gausswin` function (in the signal processing toolbox). Rewrite the code to create a Gaussian kernel and convolve it with the time series, without using the `conv` or `gausswin` functions.

14   Signal Detection Theory

Signal detection theory is a set of formal methods to describe the correspondence between pairs of binary responses (yes/no, true/false, present/absent). The "pair" can be, for example, subjective reports versus objective stimulus presentations, model predictions versus experiment conditions, or medical diagnoses versus presence of disease.

In cognitive neuroscience, signal detection theory is often applied in the context of perceptual reports in psychophysics experiments, but the concepts of signal detection theory might also be useful in other areas of neuroscience, in particular in data analysis and modeling, when trying to predict whether data were drawn from condition A versus condition B. If you are interested in a more thorough discussion of techniques in signal detection theory (though without MATLAB), see Macmillan and Creelman (2004).

14.1   The Four Categories of Correspondence

The heart of signal detection theory is that there are two perspectives of the world and that these two perspectives can match or mismatch. Imagine a human volunteer participating in a psychophysics experiment. In the experiment, a dim flash of light either appears or does not appear on the screen. This is the experiment's perspective of the world. On each trial, the human volunteer reports whether he perceived the light. This is the participant's perspective.

The participant and the experiment can agree in two different ways: They can agree that the dim light was flashed (a "hit"), and they can agree that the dim light was not flashed (a "correct rejection"). They can also disagree in two ways: The light flashed but the participant did not see it (a "miss") or the light did not flash but the participant reported seeing it (a "false alarm"). Figure 14.1 illustrates these categories.

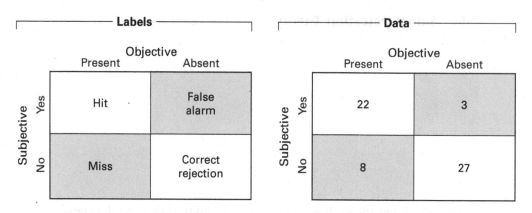

Figure 14.1

The left panel shows the four categories of responses in a binary choice situation. The right panel shows an example of a table used to compute signal detection theory variables.

The signal detection theory framework starts from these four categories and develops methods to quantify various measures of performance, such as discrimination, bias, and sensitivity.

14.2 Discrimination

In social-political discourse, most people go out of their way to avoid the act or accusation of discrimination. In signal detection theory, however, discrimination is a good thing, and you are allowed to have as much of it as possible. The signal detection theory definition of discrimination is simply your ability to distinguish between categories. Using the example situation presented earlier, discrimination is the ability to distinguish when the light was flashed versus when it was not flashed.

The standard measure of discrimination in signal detection theory is called d' (pronounced: dee-prime). The idea of d' is that accurate discrimination involves two components: saying "yes" when a stimulus was present (hit), and *not* saying "yes" when *no* stimulus was present (false alarm). To understand why false alarms must be included, consider what would happen if our light-flash participant simply responded "yes" on every single trial without even opening his eyes: His hit rate would be 100%.

D' accounts for this problem by considering that always responding "yes" also produces a false alarm rate of 100%. Subtracting false alarms from hits therefore produces a more sensitive measure of performance. A simple subtraction would be 100 – 100 (hits minus false alarms) = 0. However, a more accurate way to compute discrimination, which is the definition of d′, is the following: step 1, convert hits and false alarms from counts to proportions; step 2, convert those proportions to standard z-values; and step 3, take the difference of those z-values. The example below will be based on the counts provided in the table in figure 14.1.

```
% step 1
hitP = 22/30;
faP = 3/30;
% step 2hitZ = norminv(hitP);
faZ = norminv(faP);
% step 3
dPrime = hitZ-faZ;
```

Knowing which marginal sum to use in the denominator in step 1 can get confusing. You can say out loud (but not too loud) what the analysis should be: "How many times did the person say 'yes' given that the experiment actually flashed the light?" When you say it like this, it doesn't make sense to divide by the total number of times the person said "yes."

Let's talk about step 2 for a bit. We input the proportion to the function norminv. This function is in the statistics toolbox in MATLAB. You can also use the statistics package in Octave if you don't have access to that MATLAB toolbox. There is also a non-package-dependent solution presented in the online MATLAB code. To see what norminv is doing, consider the output of norminv as the input ranges from zero to one (figure 14.2): It transforms a

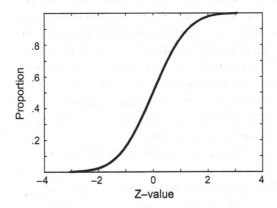

Figure 14.2
Using the function norminv to convert proportion (y axis) to z-value (x axis).

probability to a normal distribution with a peak at 0.5. You might recognize this as the same transformation that is used to convert standard deviation units to *p*-values when computing statistical significance.

Before moving forward, take a few minutes to play around with the code and see what happens to d' when the hit and false alarm rates change. What happens if you set the proportion of hits to 1 or the proportion of false alarms to 0? Uh-oh. Although proportions of 0 and 1 do not often occur in real data, it's good to protect against the possibility. This potential error can be avoided by adding a constant (e.g., 0.5) to the numerator, by adding 1 to both the numerator and the denominator, or by programming an if-then exception in case of zero.

14.3   Isosensitivity Curves (a.k.a. ROC Curves)

Did you take my suggestion and spend a few minutes exploring how various hit and false alarm rates affect the d'? If so, you may have noticed something interesting: the same d' value can be obtained for a range of different hit and false alarm rates. Imagine that person A and person B have the same d'. It is possible that person A doesn't like responding "yes" and so will have few hits but even fewer false alarms, while person B likes responding "yes" and so will have many false alarms but even more hits. This is a feature, not a bug—d' is designed to measure pure discriminability, uncoupled from the overall rate of responding and thus robust to response biases.

But this also means that we can define *ranges* of hits and false alarms that produce the same sensitivity. These are called receiver operating characteristic (ROC) curves. There are historical reasons for the term ROC, but I prefer the term isosensitivity curve (as suggested by Luce 1963) because you actually understand what it means. Let's make a plot of all d' scores using hit and false alarm rates from 0.01 to 0.99 (we'll exclude 0 and 1 for reasons mentioned in the previous section), and then draw isosensitivity curves on top of the line. How can we find the isosensitivity functions? There are analytic solutions that you can derive by applying some algebra to the d' formula. But this is a MATLAB programming book, not an algebra book, so let's figure out how to find these curves empirically.

The first step is to create a 2D matrix of d' values, with hit rate on the *y* axis and false alarm rate on the *x* axis. A trick to create this matrix is to have bsxfun expand a row vector and a column vector into a matrix, and then point-wise subtract those matrices. In other words, we'll have bsxfun perform steps 2 and 3 in the code given earlier, for many many values simultaneously. You were introduced to the bsxfun function in chapter 10, and you'll see bsxfun several more times in this book (it's one of my favorite MATLAB functions). In the exercises of this chapter, you'll be able to compare bsxfun to a double-loop.

```
x = .01:.01:.99;
dp = bsxfun(@minus,norminv(x)',norminv(x));
```

The matrix dp can be seen in figure 14.3 (it will look prettier on your color monitor). Now we want to find all of the d' values in the matrix that are equal to one. What happens when you run the code find(dp==1)?

What happens is nothing. You get no results. The problem is that there are no d' values of *exactly* 1.0000... There are, however, values that are very close to one, and these are the values we want. How close to exactly one are we willing to go? This is a parameter called tolerance. Let's try a tolerance value of 0.01; I encourage you to change this tolerance value to see the effect on the isosensitivity curves.

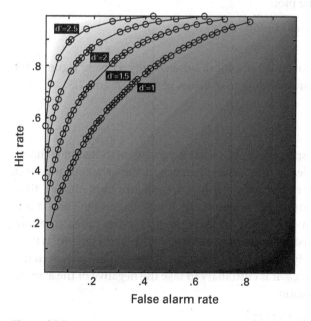

Figure 14.3
A 2D space of d' defined by myriad combinations of hits and false alarms. The lines show so-called isosensitivity curves, or regions in which the d' is the same for different hit and false alarm rates.

```
dp2plot = 1; % soft-coding
tol = .01;
idx = find(dp>dp2plot-tol & dp<dp2plot+tol);
```

The third line of code is the important line. It finds all of the indices in dp that are both above 0.99 and below 1.01; in other words, equal to 1 within the tolerance of 0.01. What are the values of idx? They are large and range up to 8,000. But how can indices be in the thousands when the matrix is size 99 by 99? You guessed it—these are linear indices. I know, I said you should avoid linear indexing when possible, but the devil always collects his due. Now that we have linear indices, let's figure out how to work with them. One way to work with linear indices is to convert them to matrix indices (see figure 10.6). MATLAB provides a function called ind2sub, which, given the size of a matrix and linear indices, returns matrix coordinates. For example, referring to figure 10.6, the result of [x,y]=ind2sub([3 4],8) is x=2, y=3.

Applying ind2sub to our dp matrix provides x and y coordinates with which to generate line plots.

```
[yi,xi] = ind2sub(size(dp),idx);
plot(x(xi),x(yi))
```

And that's how to make empirical isosensitivity curves. Figure 14.3 shows four curves corresponding to d' values of 1, 1.5, 2, and 2.5.

14.4 Response Bias

D' is insensitive to response bias, but what if you are interested in computing response bias? Clearly, there is a difference in the behavior of person A (the non-committer) and person B (the sycophant), even though they have the same d'. Fortunately, response bias is easy to compute and is just a minor change from computing d'. Response bias is also called criterion, and criterion is defined as the average of hit and false alarm rates (d' is their difference). This simply involves changing step 3 of computing d' from the difference to the average. It is customary to take the negative of the average to facilitate interpretation.

```
respBias = -(hitZ+faZ)/2;
```

The interesting feature of response bias is that it is orthogonal to d'. That is, two people can have the same response bias but different d'. This is best seen by creating a similar plot as figure 14.3 but for bias (shown in figure 14.4). Thus, to localize someone in d' space, you need to compute

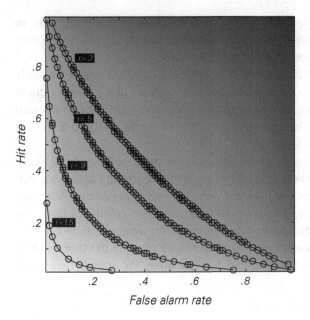

Figure 14.4
Isoresponse lines on the same signal detection "yes" space that was shown in figure 14.3.

their d' and their response bias. In a statistical sense, we might say that d' and response bias are two basis functions for the signal detection "yes" space.

14.5 Conditional Accuracy Functions

Conditional accuracy functions, also sometimes called CAFs, are a way to examine accuracy as a function of some continuous variable such as reaction time (the time it takes to indicate the response). CAFs are not traditionally included in the corpus of signal detection theory analyses, but they fit nicely within the signal detection framework. In the light-flash experiment, for example, we might ask whether subjects made more errors when they responded quickly (perhaps impulsively) compared to when they responded slowly (perhaps cautiously).

   At a coarse level, we could split the trials in the experiment into two groups according to whether they had a below-the-median reaction time or an above-the-median reaction time, and then compute accuracies for those two trial groupings. Or perhaps we could split the trials into three groups to get better precision. Ten groups? Let's first try it with seven groups just to become familiar with the MATLAB code, and then I'll provide a guideline for picking the number of discretizations.

The first thing we need to do is discretize data into N roughly equally sized groups. It's a "rough" discretization because we don't want to throw out data, and we don't want to be constrained to have perfectly equal-sized groups. For example, two groups out of $N = 11$ would be 6 and 5 (I'd call that "roughly equally sized").

To discretize a vector based on values in the data, we have two options: the easy way and the better way. The easy way is to sort the data in ascending order and then use the function `linspace` to define discretization bin edges. In this example, data from 100 trials and two columns will be discretized. You can imagine that column 1 is the response time on each trial and column 2 is accuracy (0 for error, 1 for correct); in other words, a trials-by-variable (reaction time/accuracy) matrix. We start by inventing our data.

```
ntrials = 100; nbins = 7;
d = [500+100*randn(ntrials,1) rand(ntrials,1)>.3];
d = sortrows(d,1);
```

The second line of code simulates reaction times in milliseconds with a mean of 500 and a variance of 100, and then it simulates accuracy on each trial with 1 indicating a correct answer and 0 indicating an incorrect answer (what is the average accuracy over all trials?).

The third line of code sorts the rows of the matrix. There is a meaningful relationship between the columns, so we don't want to sort just the first column; instead, we want to sort both columns according to the values in the first column. We do this with MATLAB's `sortrows` function, which sorts according to the column number specified by the second input. Now that the matrix is sorted, we can define indices for which rows should be in which bins.

```
binidx = ceil(linspace(0,nbins-1,length(d)));
discdata = zeros(2,nbins);
for i=1:nbins
    discdata(1,i) = mean(d(binidx==i,1));
    discdata(2,i) = mean(d(binidx==i,2));
end
```

Notice how we use `binidx` to create a logical vector, which we then use to compute the mean of only the columns in `d` that correspond to `binidx==i` being true. The ",1" and ",2" refer to the second dimension of d, and it is very important to get this right—typing `mean(d(binidx==i))` won't produce a MATLAB error but will give the wrong answer. To understand why, recall the distinction between matrix indexing and linear indexing. And then remember the guideline of using $N - 1$ commas when indexing an N-dimensional matrix.

So now we did something in MATLAB. How do we know it's correct? In other words, how can we sanity check this procedure? Accuracy on each trial was randomly generated, so we wouldn't expect it to show any systematic relationship with reaction times. However, the sorting was based on the reaction times, so we certainly expect reaction times to increase monotonically (and also linearly) with the discretization. A quick inspection of figure 14.5 reveals this to be the case, so I'd call that a positive sanity-check result.

Okay, that was the easy way to discretize data into N roughly equally sized groups. Now it's time to learn the better way. The better way allows us to leave the matrix in its original order—perhaps trial order is important for other analyses and you want to preserve it—while also increasing the precision of the discretization. But first, you need to learn about the function `tiedrank`.

```
>> tiedrank([pi 1 4 5 4 100000])
   ans = 2 1 3.5 5 3.5 6
```

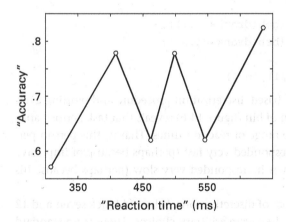

Figure 14.5
This figure shows the CAF—averaged accuracy as a function of average reaction time. The data are random, so we don't expect any relationship. But the linear increase in average reaction time provides a sanity check that the code produced a sensible result.

The function `tiedrank` returns the rank (order) of the numbers in a vector. In this example, the number 4 appears twice, and therefore its rank is repeated (we can say that the fours are tied for third place). Now let's see what to do with this function.

```
temp = tiedrank(d(:,1))/ntrials;
temp = temp*nbins;
drank = ceil(temp);
```

The first line of code computes indices of the rank-order of the reaction times, and then scales those indices to go from 0 to 1. This normalization occurred because the output of `tiedrank` is integers from 1 to `ntrials`, and then we divided by `ntrials`. Next, this 0-to-1 normalized vector is scaled up to the requested number of bins, so the values are scaled between 0 and `nbins`. Finally, we round those numbers up to the nearest integer (what would happen if I used `round` or `floor` instead of `ceil`?).

Now let's try it with real data. The online MATLAB code includes behavioral data from a published study (Cohen and van Gaal 2013) in which human volunteers had to report whether a diamond or a square was presented on a computer screen. I know it sounds like a super-easy task, but the stimuli were small, the presentation time was short, a visual mask made the stimulus difficult to see, and the human research volunteers were asked to respond as quickly as possible.

```
for i=1:12
    caf(i,1) = mean(beh(drank==i,1));
    caf(i,2) = mean(beh(drank==i,2));
end
plot(caf(:,2),caf(:,1),'o-')
```

Applying the tiedrank-based discretization procedure and plotting average accuracy as a function of bin (figure 14.6) reveals that task performance was optimal for a middle range of reaction times. That is, this person performed worse when he responded very fast (perhaps because of impulsive responding) and also when he responded very slow (perhaps because his attention lapsed).

Let's discuss the number of discretizations. Why did I pick seven and 12 bins in these examples? They were arbitrary choices. There is no standard algorithm for determining the appropriate number of bins for discretization. However, we can take a hint from histograms. One of the algorithms used to determine the appropriate number of bins when creating a histogram is called Sturges's rule (it's a bit of a misnomer because it's more of a

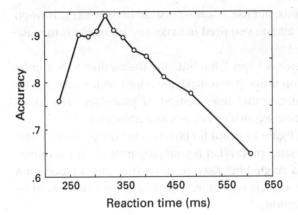

Figure 14.6
A CAF from real data. Both very fast and very slow reaction times were associated with relatively poor accuracy in this visual discrimination task. The Goldilocks zone of optimal performance was around 325 milliseconds.

suggestion than a rule). In math, Sturges's rule is $1 + \log_2 n$, where n is the number of data points. This can be translated into MATLAB as follows.

```
nbins = ceil(1+log2(n));
```

The main reason why Sturges's rule should be treated as a suggestion is that it is based entirely on the number of data points. For simple data, that's fine. But multivariate data sets often have additional considerations. For example, d' requires a sufficient number of hits and false alarms; to compute d' for each bin, you might want to use fewer bins than suggested by Sturges.

Regardless of how you select the number of bins, it is important to keep the number of bins constant across all experiment conditions, subjects, groups, and so on. This will facilitate comparison and interpretation, and it will reduce the possibility of systematic biases affecting the results.

14.6 Exercises

1. Combine figures 14.3 and 14.4 to plot d' and response bias on the same hits and false alarms space. Do the lines look like they are orthogonal? Looking at figure 14.2, it could be that the hit and false alarm rates are better conceptualized in log space rather than linear space. Adjust the plots so the axes are logarithmically scaled (hint: set). How does it look? Perhaps try having only one of the axes in logarithmic scale.

2. Repeat the 2D d′ plots, but use `imagesc` instead of `contourf`. How do the results change, and do you need to make any adjustments to make the results accurate?

3. Now reproduce exercises 1 and 2, but initialize the x-values to be z-units instead of proportion units. If you make the adjustment correctly, you should get straight diagonal lines instead of parabolas. (For bonus points, do this by deleting two `norminv`'s and adding one).

4. Create CAFs like in figure 14.6, but for bins ranging from 4 to 40. Overlay all CAFs on the same plot. What is your judgment about the importance of selecting a single bin? Keep in mind that this is based on a single subject from a single experiment; your judgment should not be taken as a generalization.

5. I used `bsxfun` in section 14.3 because I don't like loops. Reproduce the same result using two for-loops. Then use the `tic/toc` function pair to check the timing. How much extra time do you need to use for-loops instead of `bsxfun`? Now obtain the same result using `repmat` (no loops, but no `bsxfun`). How does the timing of `repmat` compare to that of for-loops and of `bsxfun`?

15 Nonparametric Statistics

The goal of most data analyses is to draw conclusions about the hypotheses or describe the patterns of results in data explorations. To reach this goal, it is necessary to determine and apply some threshold that helps you decide which results should be interpreted and which should not. There are software programs that are specifically designed for the kinds of statistics often used in psychological, clinical, and epidemiologic studies (analyses of variance, multiple regression, factor analysis, hierarchical modeling, etc.) such as SPSS, R, and SAS. MATLAB currently has no competitive toolboxes for these methods. But many statistical procedures that are often used with neuroscience data can be implemented in MATLAB. This is advantageous because it reduces the need to worry about exporting and importing data to other programs.

15.1 The Idea of Permutation-Based Statistics

Let's start more generally with the idea of statistics. Data contain signal, but they also contain noise and variability. Furthermore, it is generally not possible to measure every individual of the population of interest (people, mice, neurons, whatever), which means you need to make population inferences based on small samples. You want to know whether your findings are real effects or whether the findings could have been observed by chance even if there were no effects (no effect is also called the null hypothesis).

Generally, we assume that an effect is real if the size of the effect is big enough that it's unlikely to have occurred by chance. But how do you know what effect sizes could be obtained by chance; that is, if the null hypothesis were true? There are two ways to answer this question. Traditional statistical approaches in psychology, medicine, and related fields take the approach of making assumptions about what data would look like if the null hypothesis were true, and then evaluating the observed result relative to the assumed distributions. If the observed result is far enough away from the assumed null hypothesis distribution ("far enough" typically corresponds to being more extreme than 95% of the null hypothesis distribution), it is considered statistically significant. Statistics programs like SPSS, R, and SAS are specially designed to compute and evaluate various null hypothesis distributions.

The second way to estimate the effect sizes that could be obtained under the null hypothesis is to create an empirical null hypothesis distribution. This approach is particularly useful when the data violate assumptions used to create the null hypothesis distribution or when the distributions under the null hypothesis are unknown. These situations often arise in neuroscience data. Furthermore, because in neuroscience it is typical to collect a very large amount of data (many neurons, many electrodes, many voxels), it is somewhere between impractical and impossible to test whether all of the data conform to the assumptions used to create null hypothesis distributions.

Thus, the idea of permutation-based statistics is to avoid making assumptions that are likely to be violated. Instead of relying on what a theoretical null hypothesis distribution would look like if certain assumptions are met, the null hypothesis distribution is generated empirically, and the observed results are then evaluated relative to the empirical null hypothesis distribution (figure 15.1). This chapter focuses on such methods. If you would like to read more about the theory and justifications of permutation-based statistics in neuroscience, consider Maris and Oostenveld (2007), Cohen (2014), and Nichols and Holmes (2002).

To be clear, I am not questioning the validity or near-ubiquitous usage of the standard corpus of assumption-based parametric statistics. But permutation-based statistics are often used in neuroscience and provide a better opportunity for sharpening your MATLAB programming skills.

15.2 Creating an Empirical Null Hypothesis Test

Imagine that you recorded spiking activity from a visually responsive neuron while the research animal was shown pictures of male faces and female faces. There were 200 trials, 100 of each gender. We will simulate the firing rates as long-tailed Gaussian distributions. We'll use a "poor man's" skewed distribution by taking the log of the positive values.

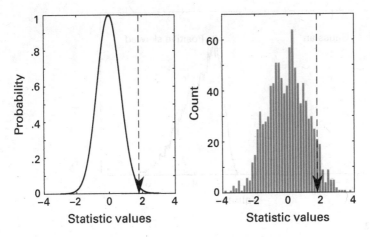

Figure 15.1
Theoretical (left) and empirical (right) null hypothesis distributions. The vertical dotted gray line is the statistic value of the observed result. If this value is far enough away (e.g., >95%) from the center of the distribution, it is considered statistically significant. In this chapter, you will learn how to create an empirical null hypothesis distribution.

```
% create Gaussian distribution
r = randn(1000,1);
% skew it
r(r>0) = log(1+r(r>0));
```

That last line of code is the crucial one, and it's also a complicated one. We want to manipulate only the positive numbers, which we access using `r>0`. If you look at the output of `size(r)` and `size(r(r>0))`, you will see that `r` is a 1,000-by-1 matrix (duh, that's what we specified in the preceding line), while `r(r>0)` is somewhere around 500-by-1. This happens because `r>0` is a logical array that is true for positive values and false for nonpositive values; `r(r>0)` accesses only the true (positive) elements. Those positive values are replaced by a modified version of themselves (figure 15.2).

Next we create distributions of firing rates for the two conditions.

```
N = 100;
% male pictures
r = randn(N,1);
r(r>0) = log(1+r(r>0));
fr_males = 26-r*10;
```

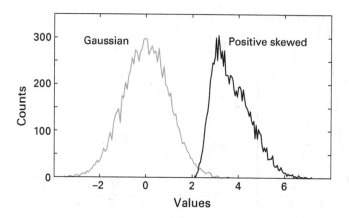

Figure 15.2

Two distributions of random numbers that will be the basis for the simulated neural
firing rates.

```
% female pictures
r = randn(N,1);
r(r>0) = log(1+r(r>0));
fr_females = 30-r*10;
```

The third line of code stretches out the numbers by a factor of 10 and
then adds an offset value of 26. The random values are turned backward
because in fact the log transform skews the data negatively, and we want a
positively skewed distribution. All of these scaling factors are arbitrary, but
they give a more plausible range of firing rates in spikes per second (let's
pretend the stimuli were on-screen for one second) (figure 15.2).

And just for fun, we'll mix these trials together as if they were randomly
presented during the experiment (randomly presented trials is a good idea
to prevent unwanted trial sequence effects).

```
allfr = cat(1,fr_males,fr_females);
% same result: allfr = [fr_males fr_females];
allfr = allfr(randperm(N*2));
```

The function `cat` concatenates the two matrices along the first dimension. Replace the "1" with a "2" in the first input and check the size of
`allfr`. The function `randperm` randomly permutes integers between 1 and
the input. Try it on a simple case by running this code several times:
`randperm(5)`.

Now the variable `allfr` has the trials from both conditions randomly
intermixed. Which trials had the male faces and which had the female

faces? Uh-oh. We didn't keep track. What a rookie mistake! Let's try it again.

```
allfr = cat(1,fr_males,fr_females);
[conds,neworder] = deal(randperm(N*2));
allfr = allfr(neworder);
conds(neworder<N+1) = 1;
conds(conds>1) = 0;
```

The function `deal` might be new to you. It's a neat function that assigns any input identically to however many output variables you want. Try a simple example:

```
[a,b,c,d,e] = deal(rand)
```

Those five variables are identically random (like teenagers trying to be nonconformists in the same way), because the output of the `rand` function was passed into the `deal` function, which dealt out the same variable to however many outputs were requested.

Next, the code assigned the first N trials in `neworder` to be one and the rest to be zero. If you don't understand why the "+1" is necessary, try to identify the numbers 1 to 3 using the following code: `1:6<3`.

Should we trust the code without sanity-checking it? Absolutely not! One easy way to check the code is by comparing mean firing rates before and after shuffling.

```
mean(fr_males)
mean(allfr(conds==1))
```

Okay, so the code seems to be working so far. Let's get back to the main task, which was to generate an effect size under the null hypothesis. The "effect size" in this case is the average firing rate difference between the two conditions. Because we simulated these data, we know that the firing rates from these two conditions reflect sampling from *two different distributions*. But that's not the null hypothesis—the null hypothesis is that there is just *one distribution* that we've accidentally labeled as two conditions (i.e., our neuron is very politically correct and thus perceived 200 faces as equals, regardless of gender). If the null hypothesis were true, there is no true difference in firing rates between male and female pictures; any non-zero difference can be attributed to an artifact of variability and sampling error.

What we want to do now is create a situation in our data that could have arisen if the null hypothesis were true. Basically, that just means randomly labeling 100 trials as "male" trials, and labeling the other 100 trials as "female" trials. We can adapt some of the code from above.

```
fakeconds = randperm(N*2);
fakeconds(fakeconds<N+1) = 1;
fakeconds(fakeconds>1) = 0;
```

The key difference from the previous code is that here we are shuffling the condition labels without changing the order of the data. So the condition labels (`fakeconds`) are misaligned with the actual data. That was the goal, of course, and by computing the difference in average firing rates with the shuffled conditions, we get a firing rate difference that could occur under the null hypothesis (remember, the null hypothesis is that both conditions were created from the same distribution).

```
[mean(allfr(conds==1)) mean(allfr(conds==0))]
[mean(allfr(fakeconds==1)) mean(allfr(fakeconds==0))]
```

The first line should give numbers around 30 and 26. The second line should give numbers that are much closer together, perhaps around 28. Taking this one small step further, we can compute the condition difference in average firing rates for the real and null hypothesis situations. That difference for the fake condition labels should be close to zero.

```
mean(allfr(conds==1)) - mean(allfr(conds==0))
mean(allfr(fakeconds==1)) - mean(allfr(fakeconds==0))
```

15.3   Creating a Null Hypothesis Distribution

The difference in firing rates in the null hypothesis simulation was not exactly zero. This is due to a combination of sampling variability and noise. Although you *expect* the difference in firing rates under the null hypothesis to be *exactly* zero, in reality it is unlikely ever to be exactly zero. This creates difficulties for statistical evaluations, because if the observed difference were 2 Hz and the simulated null hypothesis difference were 1.2 Hz, how would you know whether the observed difference was statistically significant?

The short answer is that you cannot know, at least not from one null hypothesis simulation. The solution to this problem is to examine the null hypothesis effect size not once, but many times. Perhaps hundreds or thousands of times. Each null hypothesis simulation will produce slightly different results, and all of these values together will build a distribution of null hypothesis effect sizes. This distribution is easy to create; simply put the code above into a loop.

```
nPerms = 1000;
permdiffs = zeros(nPerms,1);
for permi=1:nPerms
    fconds = randperm(N*2);
    fconds(fconds<N+1) = 1;
    fconds(fconds>1) = 0;
    permdiffs(permi) = ...
        mean(allfr(fconds==0))-mean(allfr(fconds==1));
end
```

Notice again that although the condition labels change in each iteration, the order of the data never changes. Now let's see our null hypothesis distribution and where the real condition difference value fits into this distribution.

```
hist(permdiffs,50)
hold on
obsval=mean(allfr(conds==0))-mean(allfr(conds==1));
plot([obsval obsval],get(gca,'ylim'))
```

Remember from chapter 9 that plotting a line requires specifying two points. In a vertical line, the *x*-axis value doesn't change, and we set the *y*-axis values to go from the bottom of the plot to the top of the plot.

You can see in figure 15.3 that our observed value is at the positive tail of the null hypothesis distribution. When you reproduce this plot, you might

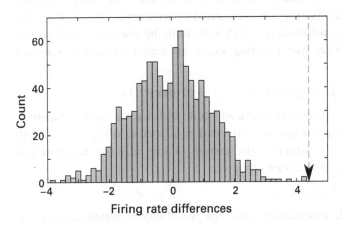

Figure 15.3
A distribution of null hypothesis statistic values from 1,000 randomization tests and the "observed" empirical value indicated by the dashed line.

get a few random condition label-swappings that produced a larger condition difference than the real condition labeling (more on these differences in section 15.7).

15.4 Evaluating Significance

The idea of statistical significance is to estimate the probability that we would have obtained a statistic (in this example, the average firing rate difference) at least as large as our observed value if the null hypothesis were true. In parametric statistics, this probability is computed based on a theoretical distribution; in permutation-based statistics, this probability is directly computed from the empirical null hypothesis distribution obtained above.

There are two options for evaluating the statistical significance of the observed effect size. One is to compute the normalized distance of the observed statistic from the distribution of the null hypothesis values. We want this to be normalized in order to compare across any measurement and any data scale. If we measured the firing rates in terms of spikes per second, spikes per hour, or spikes per decade, the numbers would be different but the relative condition differences would be the same. Needless to say, statistical significance should reflect the relative condition differences, not the scale in which the data happen to be. The normalization is done by converting to standard deviation units, otherwise known as a Z-transformation (note to readers from the United States: in Europe it's pronounced "zed transform"). It is obtained by subtracting the mean of the null hypothesis distribution and then dividing by the standard deviation of the distribution. The resulting values are called z-values or standard deviation units.

```
z = (obsval-mean(permdiffs)) / std(permdiffs);
```

For the simulation that produced figure 15.3, I got $z = 3.561$. Standard deviation units can easily be converted to a p-value using the MATLAB function normcdf, which is the inverse of the norminv function you learned about in the previous chapter.

```
p = 1-normcdf(abs(z));
```

The normcdf function evaluates the probability of obtaining a specific value in a normal (Gaussian) distribution. The value of 3.561 is at the right side of the distribution, and so the p-value would be 0.999 ... Subtracting this value from 1 gives a more statistically traditional p-value of $p < 0.001$.

Finally, taking the absolute value of the z-score ensures that the "1-" part will always work. Try this yourself by replacing z with 3 versus –3. Note also that this *p*-value is one-tailed; to obtain a two-tailed *p*-value you would need to multiply p by two.

The function `normcdf` is in the MATLAB statistics toolbox. If you do not have access to this toolbox, you can use the Octave statistics package, which has a comparable function of the same name.

The z-normalization approach is valid only if the null hypothesis distribution is roughly normally distributed, which is the case in figure 15.3. If the distribution strongly deviates from a normal distribution, you can use the second approach to obtain the statistical significance, which is simply to count the number of null hypothesis test values that were more extreme than the observed value. "More extreme" could mean larger or smaller depending on the direction of the effect. This count should be divided by the number of permutations that were tested.

```
p = sum(permdiffs>obsval)/nPerms;
```

If you have a large effect, you can get a *p*-value of exactly zero. Statistician readers may not like a *p*-value of exactly zero, but it can happen when every single permutation produced a result that was less extreme than the observed value.

15.5   Example with Real Data

Let's try this with some real data. The online code includes 200 trials of data from an EEG study in which a human volunteer made button presses in response to visual stimuli, and sometimes made errors (Cohen 2015). The first 100 trials of the data set are correct responses, and the next 100 trials are incorrect responses. Time-frequency decomposition was applied to extract the time-varying spectral characteristics (you'll learn about this in chapter 19). Our goal will be to find regions in the time-frequency plane that show statistically significant differences between correct and error responses. The variable `allpow` is a 3D time-by-frequencies-by-trials matrix. Let's start by computing the average difference in time-frequency power to see how it looks.

```
realdif = squeeze(mean(allpow(:,:,101:200),3) ...
                - mean(allpow(:,:,1:100),3));
contourf(timevec,frex,realdif,40,'linecolor','none')
```

The variable `timevec` is the vector of time points, and the variable `frex` is the vector of frequencies. The result is shown in figure 15.4 (upper left panel). There is a robust-looking condition difference around 400–700 milliseconds and around 4–9 Hz (the "theta band"). Now for permutation testing. Before reading the next paragraph, think about what exactly you want to shuffle (randomize), and how you would do it.

The answer is to randomize the trial order. If the null hypothesis were true and there were no real condition differences, we should be able to randomly assign any trial to either condition, and it would make no difference.

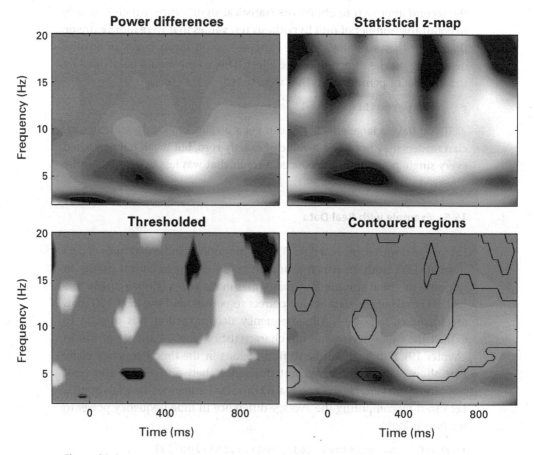

Figure 15.4
Differences in time-frequency power in scalp EEG data between making a mistake and making a correct response. Different panels show different ways of illustrating the statistical z-map.

How do we implement this in MATLAB? One possibility would be to randomize the third dimension of the variable `allpow`.

```
allpow = allpow(:,:,randperm(200));
```

But I don't like this solution, because now we've lost the original trial order. A better solution is to define a randomized order in a separate variable, like this:

```
fakeord = randperm(size(allpow,3));
fakedif = mean(allpow(:,:,fakeord(1:100)),3) - ...
          mean(allpow(:,:,fakeord(101:200)),3);
```

Now we have a shuffled condition difference without having to alter the original data. The two lines above will produce one null hypothesis time-frequency condition difference map; now we need to put those lines in a loop over permutations, and save the difference map at each iteration. For this, we populate a new 3D matrix that contains the shuffled condition difference maps for all 1,000 permutations,

```
permdif(permi,:,:) = fakedif;
```

where `permi` is the looping variable. After 1,000 iterations, we compute the mean and standard deviation maps over those permutations, and then compute the z-map. This is similar to what we did earlier in this chapter with the simulated firing rates, except here, instead of a single z-value, we have a time-by-frequency map of z-values. Each pixel contains its own statistic.

```
permmean = squeeze(mean(permdif,1));
permstd = squeeze(std(permdif,[],1));
zmap = (realdif-permmean) ./ permstd;
```

You can see the resulting z-map in figure 15.4. With a single z-value, we can determine whether it is significant by comparing it to the threshold defined by the p-value. But with a map of z-values, the picture becomes nuanced: individual pixels might be supra- or sub-threshold, and we need to visualize the entire map to interpret the results. Figure 15.4 shows this thresholded image, as well as the original condition difference map with significance regions outlined in contour lines, using the code below.

```
subplot(223)
zthresh = zmap;
zthresh(abs(zthresh)<norminv(1-pval)) = 0;
contourf(timevec,frex,zthresh,40,'linecolor','none')
```

```
subplot(224)
contourf(timevec,frex,realdif,40,'linecolor','none')
hold on
contour(timevec,frex,logical(zthresh),1,'k')
```

15.6 Extreme Value–Based Correction for Multiple Comparisons

Neuroscience data analyses often involve conducting the same statistical test on many variables. In the previous section, for example, we computed 1,225 statistical tests (25 frequencies and 49 time points). And that was only for a single electrode. In practice, you might test for condition differences over dozens of neurons or electrodes and several hundreds of time points or time-frequency points. These situations lead to statistical concerns of increased possibility of obtaining a supra-threshold result by chance (a.k.a. the multiple comparisons problem).

There are several approaches for addressing multiple comparisons concerns. Bonferroni correction may come to mind, which is a standard method for multiple comparisons in statistics. Bonferroni correction involves dividing the p-value threshold by the number of comparisons (e.g., a p-value threshold of 0.01 would be used if there were five tests to perform). However, there are situations in which Bonferroni correction is inappropriate in neuroscience. For example, Bonferroni correction assumes that the multiple tests are independent, whereas different neurons recorded from the same animal or neighboring time-frequency pixels are not independent measurements. More important, Bonferroni correction is really easy to implement, so there is little advantage to discussing it in a MATLAB book.

Instead, here we will learn about one multiple comparisons correction technique that is based on generating extreme values under the null hypothesis. The idea is that on each iteration during permutation testing, we search through the entire shuffled data set (i.e., all time points, frequencies, electrodes, image pixels, fMRI voxels, or whatever makes up the "multiple" in multiple comparisons) and find the two most extreme values—the most extreme negative value and the most extreme positive value. Those two values over all permutations create a distribution, and we then take the 2.5% value of the negative and positive tails (corresponding to a two-tailed 5% probability) as the threshold. Let's see how this is implemented in the code. The two lines below are placed inside a loop over permutations after computing the `fakedif` matrix that was created earlier.

```
exvals(permi,1) = min(fakedif(:));
exvals(permi,2) = max(fakedif(:));
```

Notice that we are not taking one value per pixel, but rather one value per map. That's how this method corrects for multiple comparisons at the map level. After going through 1,000 permutations, we have our distribution (figure 15.5). Now we need to define the threshold.

```
pval = .95;
lowerThresh = prctile(exvals(:,1),100*pval);
upperThresh = prctile(exvals(:,2),100-100*pval);
% now threshold
threshmap = realdif;
threshmap(threshmap>lowerThresh & ...
          threshmap<upperThresh) = 0;
```

In this case, the threshold is very stringent, and no results remain statistically significant. You can also see this from figure 15.5—the thresholds are beyond the distribution of data values. This can happen when the single-trial data are noisy, because noise often produces extreme values. After going through chapter 26, you will be able to implement another method

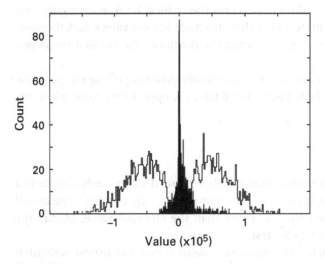

Figure 15.5
The white bars show a histogram of extreme pixel values taken from permutation testing. The black bars show the distribution of pixel values taken from the observed data with no shuffling.

for correcting for multiple comparisons based on cluster sizes (see exercise 12 in chapter 26).

15.7 Meta-permutation Tests

If you are following along in MATLAB (I'm sure you are!), you will have gotten a different result than what is reported here. That's normal—the permutation testing relies on random numbers. Run the permutation test over and over again, and even though the original data are not changing, you'll find different significance values each time you reevaluate the code. It is possible that an effect might even become nonsignificant ($p > 0.05$). This is a bit awkward. You want the significance of the test to reflect the effect size, which does not change after the data are already collected.

Unfortunately, there is no perfect solution to this. You can increase the number of permutations, but that doesn't solve the problem. This is simply a feature of permutation-based statistical testing.

There is, however, a way to ameliorate this issue, which is to perform a meta-permutation test (Cohen 2014). This involves running the entire permutation testing procedure many times (e.g., 20), obtaining many z-values, and then averaging those z-values together to obtain the final result. Averaging z-values is a second hierarchical level of averaging over data and will be more stable than the individual z-values and therefore is better than increasing the number of permutations within a single permutation test.

Running a meta-permutation test simply involves putting the permutation testing code inside a loop. You'll have the opportunity to do this in the exercises.

15.8 Exercises

1. The p-value counting method implemented in the online code is a one-tailed test, because we are counting only the number of permuted statistical values that are larger than the observed value. Modify the code to be a two-tailed test.
2. You could also use a standard parametric t-test for statistical evaluation if the data are roughly normally distributed. Perform a two-sample t-test on the simulated firing rate data from section 15.2. If you don't know what MATLAB function performs a two-sample t-test, you'll need to search for it.

3. Perform a meta-permutation test according to the description in section 15.7. Try it on the simulated firing rate data and on the EEG time-frequency data. Do the thresholds vary much over 20 meta-permutation runs?

4. Add two characters to the line of code below to plot a vertical line that goes through 50% of the size of the y axis.

```
plot([5 5],get(gca,'ylim'))
```

5. The following lines of code will produce an error. Before running the code in MATLAB, find and fix the error. Then test the original code and your revised code in MATLAB. Pay attention to the text of the error message and make sure your solution works.

```
r = randn(1000,1);
r(r>0) = log(1+r);
```

6. Although science needs humans to interpret results, computers can determine whether a particular z-value is statistically significant. Write code that reports whether a z-value is significant, given a user-specified p-value threshold (e.g., 0.05). Use the `fprintf` function to display useful information. Make sure the code works for both negative and positive z-values.

7. The online code has the following three lines. What is the purpose of the variable `colorval` and why did I scale it by 0.8?

```
contourf(timevec,frex,realdif,40,'linecolor','none')
colorval = max(abs(realdif(:)))*.8;
set(gca,'clim',[-colorval colorval])
```

8. The MATLAB function to compute a percentile score from a distribution is called `prctile`. This function is in the Statistics and Machine Learning toolbox. If you have this toolbox, use the help file for that function to figure out how to compute the 98% percentile of a distribution of 10,000 random integers uniformly distributed between 14 and 2,345 (you will also need to generate this distribution). Next, devise a solution to this problem that is not dependent on the toolbox.

9. Your friend Sue thinks the thresholding is confusing when it's done in a single line, so she splits it into two lines. But now it doesn't work. What's the problem with Sue's code?

```
threshmap(threshmap>lowerThresh) = 0;
threshmap(threshmap<upperThresh) = 0;
```

10. Compare in the online code how I computed the variables `realdif` and `fakedif`. Aside from the inclusion of the variable `fakeord`, there is one difference between the code that produces these two variables. What is it, and does it matter for the analysis? Why or why not?

11. The code for plotting the multiple-comparisons-thresholded contour map produces a warning and then an error. What are these messages, and what caused the error?

12. The following two lines differ by only one character. What is the effect of this seemingly inconsequential difference, and what does this tell you about how MATLAB concatenates inside square brackets?

```
[mean(allfr(conds==1)) -mean(allfr(conds==0))]
[mean(allfr(conds==1)) - mean(allfr(conds==0))]
```

16 Covariance and Correlation

Covariance, as you might have guessed from the name ("co" "variance") is a measure of the relationships of variances across multiple measures. Covariance is the starting point for many multichannel analyses, including principal and independent components analyses, source-space imaging of M/EEG data, and least-squares fitting.

Covariance among N channels produces an N-by-N covariance matrix. Covariance matrices preserve the original scale of the data, meaning that if you multiply the data by 1,000, the values in the covariance matrix will increase by 1,000. If the covariance matrices are normalized by the variances of the individual variables, the scaled covariances range from –1 to +1, and are then called correlation coefficients. This chapter will show you how to compute covariances and correlations and will introduce two types of correlations (Pearson and Spearman).

16.1 Simulating and Measuring Bivariate Covariance

Before learning how to compute covariance, it is useful to know how to simulate data with a known covariance structure. There are two ways to create correlated variables. One is shown below, and another is introduced in a later section and explored further in the exercises:

```
r = .6;
n = 100;
x = randn(n,2);
x(:,2) = x(:,1)*r + x(:,2)*sqrt(1-r^2);
```

where r is the desired correlation. How do you know that this procedure worked? Figure 16.1 shows that the two variables are clearly related to each other, although that doesn't prove the correlation is what r specifies. You

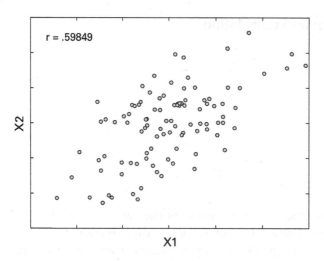

Figure 16.1
Two random variables with an enforced correlation between them.

can sanity-check it using the `corr` or `corrcoef` functions. But let's move on first; we're building up to a sanity check.

Before learning about correlations, we will learn about the "unrefined" version of a correlation: covariance. The definition of covariance is the sum of the point-wise multiplications between two vectors, divided by the number of time points minus one. Of course you recognize this as the dot product divided by $N - 1$. This is another example of how the dot product is a simple but fundamental computation for many signal processing applications. In MATLAB, this can be implemented as follows:

```
cov = x'*y/(n-1);
```

This code will produce the correct result if the variables are column vectors, because x gets transposed to a row matrix, and x'*y is the dot product x^Ty. It's a good idea to check that the output of this expression produces a single number, not a matrix. If it produces a matrix, that would be the outer product, which is not what you want here (see figure 10.4).

Did you notice a problem with that line of code? There is no error per se, but I did something I shouldn't have done. I used cov as a variable, but cov is also the name of a function (which, as it turns out, computes the covariance of the input variables). Better to use a variable name like cov1. If you ever suspect that a variable name you want to use might be a function name, type `which cov`.

Let's check to make sure that our manual covariance matches MATLAB's function to compute covariance.

```
cov(x,y)
```

If you've been following along in MATLAB, then you received an error message when typing the code above. What was the error, and why is MATLAB upset? Type `clear cov` and then re-run the preceding line. You should get identical results between the manual covariance computation and the output of the `cov` function.

Does the covariance match the value of `r` that you specified for the correlation? There will be some difference between the measured covariance and the specified `r`, because `r` is the true correlation in a theoretical distribution with an infinite number of data points, and you are sampling only a subset from this distribution. But the covariance should be reasonably close. Does this mean that covariance is the same thing as correlation?

In this specific example, the answer is yes. When all of the variables are normalized to have a mean of zero and unit variance (which is the case with outputs of the function `randn`), then covariance is indeed the same thing as correlation. Let's see what happens to the covariance when the data do not have unit variance. Try recomputing the covariance, but first multiply x and y by 100 (don't recompute the variables, just scale the existing values by 100). The relationship between the variables hasn't changed, but the covariance has changed (by how much?). The lesson here is that a covariance value on its own can be difficult to interpret, because it depends both on the relationship between the variables and on the scale of the data. Test this in the code by scaling the data to other numbers and confirming that the covariance value follows the scaling factor.

Multiplying x and y by 100 increased the variance, but the mean values were still zero (or very close to it). Now try *adding* 100 to x and y, such that their means are 100 instead of zero. Then recompute the covariance, manually using the dot-product-and-divide code above, and again using the MATLAB `cov` function. Are the two results still the same? The answer is no. The reason why is explained in the next paragraph, but try to figure it out first, perhaps by careful thinking or perhaps by looking into the cov.m file to see what it does.

The reason why the two procedures provided different values is that a covariance is valid only when it is computed between zero-mean variables. You must always subtract the mean of each variable before computing the covariance. The MATLAB `cov` function does this on line 154 (in my R2015a version; the precise line number may differ by version). This is particularly important for matrix decomposition techniques like principal components analysis, because otherwise the first component will be driven by the mean, not by the covariances. You'll see this in the next chapter and again in chapter 33.

So the best practice for computing covariance is to subtract the mean of each variable first. If the data already have zero mean, then re-subtracting zero won't do any damage. In other words, get in the habit of thinking about mean-subtracting whenever you think about computing covariance. In the code below, new variables are created to avoid overwriting x and y. This is useful in case the mean offsets are relevant for subsequent non-covariance-related analyses. The resulting cov1 should now match the output of MATLAB's cov function. And just for fun, let's use the bsxfun function here. You can see by comparing the code for mean-subtracting xx and yy that bsxfun is not necessary, but I wanted to use this opportunity to remind you about its power and utility.

```
xx = bsxfun(@minus,x,mean(x));
yy = y-mean(y);
cov1 = xx'*yy/(n-1);
```

16.2 Multivariate Covariance

Bivariate covariances are nice and whatever, but the beauty of covariances comes from multivariate data sets. As proof of this, run imagesc(cov1) and think about whether any self-respecting museum would hang this picture up on their walls. In this section, we will compute covariance *matrices* of multivariate data sets that will produce pictures you might see in a museum (well, maybe, if they were repeated, colored differently, and signed by Andy Warhol).

Let's start with a real data set. We'll compute the covariances among all pairs of channels in EEG data. There are 64 channels, so we'll get a 64-by-64 matrix (4,096 total covariances, although they are not all unique, as you'll see later). You might think that to compute an all-to-all covariance matrix, you need to have two loops over channels and then apply the dot product computation to each pair of electrodes. But this definitely violates the rule of avoiding loops whenever possible. Instead, we can use matrix multiplication. Recall from chapter 10 that matrix multiplication can be interpreted as a series of dot products between each row in the left matrix and each column in the right matrix. This is exactly what we need to compute a covariance matrix.

```
data = squeeze(EEG.data(:,:,1));
cov2 = data*data'/(EEG.pnts-1);
```

Why did I transpose the second matrix here but the first vector (x) in the code earlier? No deep mathematical reason, just that the matrix data happened to be organized as channels by time. I could have transposed data in the preceding line and then used data'*data. This is an important detail to check. Remember the rule about matrix multiplications: inner sizes must match, and the outer sizes define the size of the product matrix. The variable data is channels-by-time, and its transpose is time-by-channels. We want to end up with a channels-by-channels matrix; hence, the right matrix must be transposed.

It is also valid to turn the matrices around and end up with a time-by-time matrix. This would reflect the spatial covariance at each pair of time points, rather than the temporal covariance at each electrode pair. Such a covariance matrix is also interpretable, but we are focusing here on the channel-by-channel covariance matrix. The important lesson here is to sanity check this covariance matrix by making sure the dimensions correspond to channels.

Before having a look at this covariance matrix and thinking about how to interpret it, there is one important piece missing from the code above. Any guesses?

These data include mean offsets. So the covariance matrix also includes a mean offset. We can use bsxfun to remove the mean before computing the covariance. Because these are 2D matrices, we need to think carefully about how the mean is removed. We are computing the covariance *over time*, so we want to remove the mean *over time*. In the organization of these data, time is the second dimension.

```
data = bsxfun(@minus,data,mean(data,2));
cov2 = data*data'/(EEG.pnts-1);
```

Now we have a valid and interpretable covariance matrix. Let's see how it looks: imagesc(cov2) (figure 16.2). It's a lot nicer-looking than the bivariate covariance, but still not quite museum-ready. Maybe you can spend a few minutes applying your knowledge of color scales and the contourf and set functions to help with the final touches.

The first thing you might notice about this covariance matrix is that the upper right part looks similar to the lower left part. In fact, they are more than just similar; they are identical. All covariance matrices are symmetric (the proof of this was given in chapter 10.10, the section about matrices multiplying their transposes). For example, the element in the 4th row and

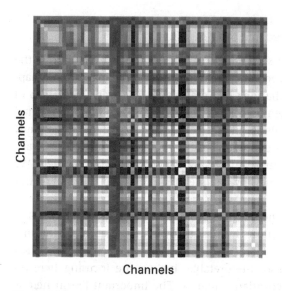

Channels

Figure 16.2
Channel-by-channel covariance matrix of scalp EEG data recorded from a human brain.

10th column is the same as the element in the 10th row and 4th column. You can verify this yourself in MATLAB by comparing `cov2(4,10)` with `cov2(10,4)`.

The second thing to notice is that the diagonal (from the top left to the bottom right) is different from the off-diagonal elements. The diagonal is the covariance of each channel with itself; in other words, the variance. If the data were normalized to produce a correlation coefficient matrix (we'll get to this topic soon), then the diagonal would be all ones.

The diagonal elements in a matrix can be extracted using the MATLAB function `diag` (see also chapter 10, exercise 14). If you type `plot(diag(cov2))`, the result will be a line, and each element in this array corresponds to `cov2(1,1)`, `cov2(2,2)`, and so on up to `cov2(end,end)`. The variances on the diagonal can be informative about the data. Interpolating these values over the topography reveals that some electrodes have higher variability than other electrodes (figure 16.3). In general, this is due to some combination of noisy electrodes plus some electrodes measuring brain activity that is more task-relevant and thus more dynamic. However, this is just a single trial of data, so we might expect the noise to be greater than the signal.

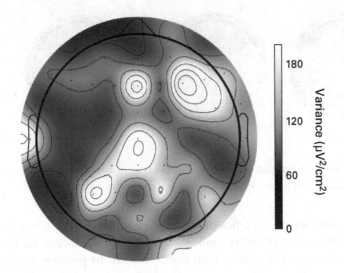

Variance (μV²/cm²)

Figure 16.3
A topographic plot of variances over space, estimated from one trial of EEG data.

Interpolation is done in figure 16.3 using a function called `topoplotIndie`. This function is the same as the eeglab function `topoplot` except that all dependencies are bundled so you can use this function without needing to download the eeglab toolbox (Delorme and Makeig 2004), which is available for free download from the Web. The small indie version is provided here because you shouldn't need to install a third-party toolbox just to reproduce this figure.

Enough of the variances; let's inspect some covariances. Each row (or column) of the covariance matrix reflects the covariance from each electrode to all other electrodes. Therefore, one way to visualize the covariances is to select one "seed" electrode and examine the covariances between that and all other electrodes. Note that covariance is a symmetric measure—a covariance value indicates nothing about directionality, it simply describes the relationship between pairs of electrodes. Figure 16.4 shows a few topographic maps of seeded covariances. You can try using different seed sites and see how the covariance changes.

16.3 From Covariance to Correlation

A correlation coefficient is similar to a covariance, except that the value is normalized between –1 and +1. A coefficient of –1 means that whenever

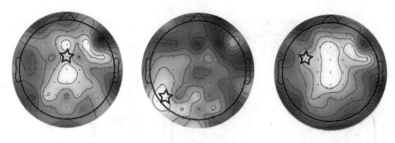

Figure 16.4

Seeded covariance topographic maps illustrate covariance between selected electrodes (see stars) and all other electrodes. Data are shown from a single trial.

x goes up, y goes down. A coefficient of zero means that x and y are completely unrelated to each other, and a coefficient of +1 means that x and y go up and down together, with no deviations. In practice, if you ever see a correlation coefficient of exactly or very close to –1 or +1 in empirically measured biological data, then you should check for mistakes. Two different components of a complex biological system like the brain will rarely show such a strong correlation, although of course there are exceptions.

There are two ways to obtain a correlation coefficient from a covariance. One is to normalize all of the variables to their standard deviations prior to computing the covariance. This involves subtracting the mean and dividing by the standard deviation per channel. In fact, you've already done half the work for this normalization by subtracting the mean. The code below does the rest.

```
data = bsxfun(@minus,data,mean(data,2));
data = bsxfun(@rdivide,data,std(data,[],2));
cor2 = data*data'/(EEG.pnts-1);
```

Note the difference in input order between the `mean` and `std` functions. The `mean` function takes the second input as the dimension over which the mean should be computed, while the `std` function takes the third input as the dimension (the second input is a normalization option, and this should be kept at empty to use the default). Now the covariance is a correlation.

The second way to obtain a correlation coefficient from a covariance is to compute the unscaled covariance as we did in the previous section, and then divide that covariance by each channel's variance. Actually, this isn't really a *different* way; either you normalize the data before computing the covariance matrix or you normalize the covariance matrix. The effect is the same. We'll start with the two variables created at the beginning of this chapter. These variables are already zero-mean with unit variance, meaning correlation and covariance are the same thing. Let's make our task more difficult by first stretching out the variables.

```
x(:,1) = x(:,1)*100 + 2;
x(:,2) = x(:,2)*20 + 10;
```

Next we compute the covariance between these variables, and then compute the variances of each variable individually.

```
cx  = x(:,1)'*x(:,2) / (n-1);
vx1 = x(:,1)'*x(:,1) / (n-1);
vx2 = x(:,2)'*x(:,2) / (n-1);
```

Finally, we divide the covariance by the square root of the product of the variances—square root because we actually want to scale by the standard deviation, not by the variance. (In case your high-school math slipped for a moment, recall that the square root of a times b is the same thing as the square root of a times the square root of b.) Let's test our normalized covariance against the output of MATLAB's corr function (you can use corrcoef if you don't have the stats toolbox). The results should be identical.

```
cor1 = cx / sqrt(vx1*vx2);
cor2 = corr(x);
```

Of course, the variables cor1 and cor2 should be identical. But, uh-oh, they're not. Our sanity checking revealed a mistake somewhere. Check my code carefully and see if you can find one small but important missing step.

Now let's figure out how to apply this covariance-to-correlation conversion to a larger covariance matrix. We need to create a channel-by-channel matrix in which each element is the product of the standard deviations of each pair of channels. Needless to say, we're not going to do this in a double for-loop when we can implement it as an outer product. To get a better idea of what this rank-1 matrix looks like, try making an image and plotting it.

```
stdMat = sqrt(diag(cor2)*diag(cor2)');
imagesc(stdMat)
plot(stdMat)
```

From the image you might not see that this is a rank-1 matrix. Plotting all rows of the matrix, however, reveals that all rows are a scalar multiple of

Box 16.1
Name: Hualou Liang
Position: Professor
Affiliation: Drexel University, Philadelphia, Pennsylvania

When did you start programming in MATLAB, and how long did it take you to become a "good" programmer (whatever you think "good" means)?

I was introduced to MATLAB way back in 1996 when I just started my postdoc career. My experience in programming prior to MATLAB was heavily Fortran, largely due to my PhD training in high-energy physics where Fortran has established itself as the lingua franca. I must admit I was a tad dubious about MATLAB when I gave it a try. I was completely blown away with the demonstrations. I do not remember how long it took to become a "good" programmer. My sense is that it's rather easy and quick to get started, though it may take years to master MATLAB.

What do you think are the main advantages and disadvantages of MATLAB?

Pros: Intuitive and easy to learn, fast prototyping, large collections of demos and toolboxes, and many more.
Cons: Not free. Lot of built-in functions, but many people do not know about them. Slow for large-scale problems.

Do you think MATLAB will be the main program/language in your field in the future?

It seems certain that MATLAB will remain the main language and will become even more powerful when used in combination with C and R, and even other languages such as Python.

How important are programming skills for scientists?

Programming skills are becoming more important, turning into the core competency for the new generation. The more you know about programming skills, the more you will be efficient, creative, and competitive.

Any advice for people who are starting to learn MATLAB?

Learning by doing and by reading the codes from experts.

the same data (the standard deviations). The final step is to scale the covariance by the standard deviation matrix.

```
cov2 = cov2 ./ stdMat;
```

The result is shown in figure 16.5. On the diagonal, you end up with the variance divided by the variance, which is 1. A sensible result, of course, because every variable is perfectly correlated with itself.

16.4 Pearson and Spearman Correlations

There are a few different types of correlations. The method described so far in this chapter produces a *Pearson correlation*. Pearson correlations are appropriate when you expect a linear relationship between normally distributed variables. Not all relationships are linear, however, and these situations can produce misleading Pearson correlations. This limitation is often illustrated using *Anscobe's quartet* (figure 16.6).

A second type of correlation is called a *Spearman correlation*, sometimes also called a *rank correlation*. The Spearman correlation tests for a monotonic relationship, regardless of whether it is linear or nonlinear. This is useful for data that have a non-Gaussian distribution, such as power spectra, neuron firing rates, and image pixel values. Neuroscience data are often

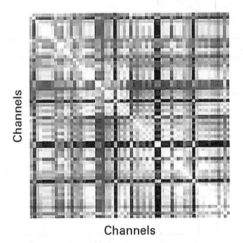

Figure 16.5
A channel-by-channel correlation matrix that is based on the covariance matrix. When generating this image on your computer, compare the color scale for this figure with that of figure 16.2.

non-normally distributed, and unless you have a specific reason to expect a linear relationship, the Spearman correlation is often preferable. If the data are roughly normally distributed and there are no outliers, Pearson and Spearman will generally produce very similar results (e.g., top left panel of figure 16.6).

The Spearman correlation works similarly as a Pearson correlation, except that the data are converted to rank order. Rank order means that, for example, the numbers [0 0.1 0.11 10,000] become [1 2 3 4]. Now you see why the Spearman tests for a monotonic and not necessarily linear relationship.

To compute a Spearman correlation, the first step is to rank-transform the data. You've already been introduced to the tiedrank function in

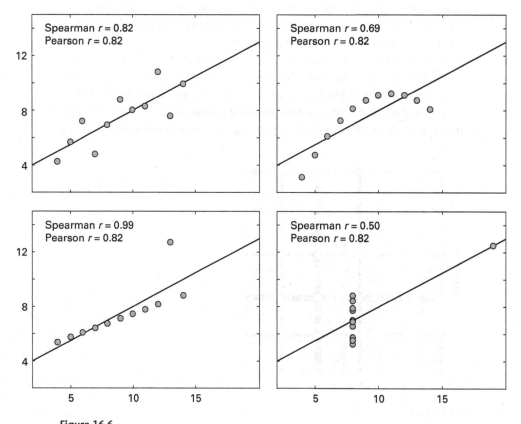

Figure 16.6
Anscobe's quartet illustrates that different data distributions can have identical Pearson correlation coefficients. In some cases, a Spearman (rank-order) correlation is preferable.

chapter 14, and now you can learn a bit more about it here. This function accepts matrix inputs and will rank-transform along the second dimension. Consider the following code. The goal is to rank the rows, but inspect the output of `tiedrank`.

```
d = [1 2 10 100; 2 10 100 1];
tiedrank(d)
```

Rather than ranking row-wise, `tiedrank` ranked column-wise. The solution is to transpose the input, and you might want to transpose the output as well to restore the original matrix size.

```
tiedrank(d')'
```

With that in mind, we can convert our bivariate data to rank order and apply the Spearman formula, which involves the number 6.

```
xr = tiedrank(x')';
c1 = 1-6*sum((xr(1,:)-xr(2,:)).^2)/(n*(n^2-1));
```

Confirm that in this case, the Spearman and Pearson correlation coefficients are similar to each other. Why is this? (Hint: What is the distribution from which these data were drawn?)

Spearman and Pearson are not the only types of correlation coefficients. There are perhaps a half-dozen different correlation methods that are designed for specific types of data. However, Spearman and Pearson are appropriate for the vast majority of correlation analyses in neuroscience.

Are you wondering why you should use these multiline implementations rather than simply using the functions `cov`, `corr`, and `corrcoef`? I have three answers to that: (1) this chapter would have been much shorter; (2) you would have learned a lot less; and (3) the mechanics of correlation might have remained a mystery. Correlation and covariance form the bases for many simple analyses like correlation, and also for many advanced analyses, including principal components analysis, source-space imaging of MEG and EEG, source separation, optimization, and clustering. It is important to understand the basic building blocks in order to apply more sophisticated analyses in appropriate ways. That said, in practice, there is nothing wrong with using the functions `corr`, `corrcoef`, or `cov`.

16.5   Statistical Significance of Correlation Coefficients

The statistical significance of a correlation coefficient can be obtained through parametric or permutation-based methods. The parametric method is computed by MATLAB in the `corrcoef` and `corr` function, and the *p*-value is given in an optional second output. The *p*-value given by these functions is based on assumptions about the distributions of correlation coefficients expected under the null hypothesis for variables with normally distributed values.

The second approach for obtaining statistical significance is to use permutation-based statistics, as discussed in the previous chapter. This involves shuffling the mapping of the two variables with respect to each other and creating an empirical distribution of correlation coefficients expected under the null hypothesis of random associations between the two vectors. This method will be explored in the exercises.

16.6   Geometric Interpretation of Correlation

There is one final point I'd like to discuss. Recall from chapter 10 that many expressions in linear algebra have an algebraic interpretation and a geometric interpretation. So far you've learned about the algebraic interpretation of covariance and correlation. There is also a geometric interpretation of the same mathematical procedure.

The geometric interpretation of a dot product between two vectors in *N*-dimensional space is the magnitude of those vectors scaled by the cosine of the angle between them. If you imagine that the two to-be-correlated data series each reflect a coordinate in an *N*-dimensional space and then imagine those two coordinates being end points of vectors from the origin, then the product of their magnitudes scaled by the angle between them is the covariance. Furthermore, if the two vectors are normalized to have length one (the linear algebra term would be unit vectors along the same direction as the original vectors), then the dot product is simply the cosine of the angle between the vectors. Now recall some basic trigonometry: When two vectors point exactly opposite each other, their angle is 180° (π radians) and the cosine is –1; when two vectors meet at a right angle, their angle is 90° ($\pi/2$ radians) and the cosine is 0; and when two vectors point in the same direction, their angle is 0° and the cosine is +1. (Don't believe me—try it in MATLAB using the `cos` function!)

Thus, correlation is the same thing as the cosine of an angle between two unit-vectors in a high-dimensional space. This interpretation is sometimes called *cosine similarity*, but don't let the fancy term confuse you.

16.7 Exercises

1. In this exercise, you will explore the effects of correlation strength and
    sample size on simulated correlated data. Write a triple-loop to create
    two signals of different lengths and correlation strengths and 100 itera-
    tions (trials) of each length-strength pair. For each simulated correla-
    tion, compute the squared distance between the true correlation that
    you specified and the estimated correlation coefficient. Plot the trial-
    averaged results in two 2D images (one image for average correlation
    coefficients, and one image for the average distance to the true correla-
    tion. How do you interpret the results—is it better to have a strong
    effect or more data?

2. So far you've learned only how to create a bivariate data set with known
    correlation. What about a multivariate data set with a known covari-
    ance structure? This can be implemented by constructing the desired
    covariance matrix and scaling random numbers by the Cholesky fac-
    torization of that covariance matrix (the Cholesky factorization is a
    way to represent a symmetric matrix using two triangular matrices).
    You can also create multivariate correlated data using eigendecomposi-
    tion, which you'll learn about in the next chapter. Below is code to
    create a 1,000-by-3 matrix of correlated random numbers. Perform a
    sanity check to test whether the data show the requested covariance
    matrix.

```
v = [1 .5 0; .5 1 -.3; 0 -.3 1];
d = randn(1000,3)*chol(v);
```

3. The biggest potential mistake when using the function `tiedrank` is to
    rank along the incorrect dimension. This was easy to check visually in
    a small matrix. How could you sanity check the output of `tiedrank`
    for a large matrix (e.g., channels by time)?

4. Use the MATLAB function `corrcoef` to sanity check the correlation
    for the channel pair [13, 46] in figure 16.5.

5. What is the difference in the code below if x and y are column vectors
    versus row vectors? What if x were a row vector and y were a column
    vector?

```
cov1 = x'*y/(n-1);
```

6. Generate random variables x and y and evaluate `corr(x,y)` with x
    and y being column vectors or row vectors. What happens in either
    case? How could you adjust the code to make sure the output always

produces a single covariance value rather than a rank 1 matrix (the outer product)? What happens when you repeat this exercise using the functions `corrcoef` and `cov`?

7. Use permutation testing to obtain a p-value for the statistical significance of the correlation produced in the first section of this chapter. What do you shuffle on each iteration? Check your significance against the p-value returned from the second output of `corrcoef` or `corr`.

8. Assume the variable `cov2` is a 64-by-64 covariance matrix. If `cov2(end)` is the same thing as `cov2(end,end)`, and `cov2(1)` is the same thing as `cov2(1,1)`, is `cov2(4)` the same thing as `cov2(4,4)`? Why or why not?

9. Compute the channels-to-channels covariance matrix that was used in figure 16.4 in two different ways. First, average all trials together before computing the covariance matrix. Second, compute the covariance matrix for each trial separately and then average together 99 covariance matrices (there are 99 trials). Plot the two results next to each other, using the same color scaling. Are there striking differences, and how would you explain them?

10. Write a function that takes 3D raw EEG data (channels by time by trials) and a number of trials as input and that outputs and plots the covariance matrix for a random selection of the specified number of trials. If there is only one trial, the input would be a 2D channels-by-time matrix; make sure your function can deal with this situation.

11. MATLAB code to produce figure 16.6 is not provided (don't worry, the data values are provided). Re-create this figure. (Hint: try the function `lsline`.)

12. Compute the covariance matrix from section 16.2 as double-loop. Make sure the result is the same as the matrix multiplication implementation.

17 Principal Components Analysis

Principal components analysis (PCA) is a matrix decomposition technique that is applied to multivariate data sets. It can be used for reducing the dimensionality of the data, for denoising the data, as a spatial filter, or as a precursor for certain analyses.

17.1 Eigendecomposition

PCA is based on a matrix factorization technique called eigendecomposition (also sometimes called eigenvalue decomposition or eigenvector decomposition). Here is the idea of an eigendecomposition: When you multiply a vector by a matrix, the matrix typically changes the direction of the vector. That is, matrix A transforms vector x into vector b (thus producing the familiar $Ax = b$), and vector b points in some direction that is different from that of vector x. However, some vectors are special and point in the same direction going into and coming out of multiplication with matrix A. Matrix A may cause vector x to get longer or shorter or it may cause vector x to face the other direction, but vector x will stay on the same line (figure 17.1). This leads to the fundamental eigenvalue equation: $Ax = \lambda x$ (the λ is the scaling factor; it's just a single number). This equation says that multiplying vector x by matrix A is the same thing as scaling vector x by the number λ. The vectors x are called eigenvectors. They are unique to matrix A in that the eigenvectors of matrix A are different from the eigenvectors of matrix B (except in a few exceptional cases such as $B = A$). The eigendecomposition of a matrix finds those special eigenvectors and their associated eigenvalues. Eigendecomposition can be applied only to a square matrix, and if the matrix has full rank, there are as many unique non-zero eigenvalues as there are columns in the matrix.

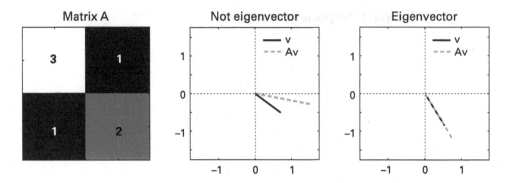

Figure 17.1
Illustration of eigenvectors in two dimensions. The left panel shows the visual representation of a 2-by-2 matrix. The other two panels show a vector **v** (black line) and vector **Av** (i.e., the result of multiplying vector **v** by matrix A). When vector **v** is not an eigenvector, **Av** points in a different direction; when vector **v** is an eigenvector, **Av** is simply a scaled version of **v**.

Interesting things happen when the matrix **A** is symmetric (i.e., when $\mathbf{A} = \mathbf{A}^T$). First, the eigenvectors are all orthogonal to each other (more precisely: the eigenvectors can be constructed in a way that they are all pairwise orthogonal), which is to say that the dot product among all pairs of eigenvectors is zero. Second, each eigenvector reveals the direction of highest variance that is orthogonal to the previous directions. That is, the first vector points in the direction of maximum variance; the second vector points in the direction of maximum variance that is orthogonal to the first direction; the third vector points toward the maximum variance that is orthogonal to the first and second directions; and so on. Finally, the eigenvalues are guaranteed to be real-valued if A is real-valued, although this property is not terribly interesting for most neuroscience applications.

PCA is one of several *blind source separation* (sometimes called BSS) techniques. These techniques are called "blind" because they are not restricted by *a priori* knowledge of patterns in the data. Other BSS techniques include independent components analysis, generalized eigendecomposition, and many other variants. BSS is a major topic in computer science and data mining. PCA is a fundamental technique that is the starting point or inspiration for many other BSS techniques. If you are interested in BSS and related multivariate spatial filtering techniques, eigendecomposition is a good place to start.

17.2 Simple Example with 2D Random Data

We're going to generate a plot that you may have seen before when reading about PCA (figure 17.2). This figure illustrates the concept that an eigende-composition of a covariance matrix (which, you will remember from the previous chapter and from chapter 10, is always symmetric) returns vectors that produce new axes such that each eigenvector (a.k.a. principal component) captures as much variance as possible along one dimension while being orthogonal to previous eigenvectors. We can also call these "orthogonal basis vectors of the eigenspace of the data" (tip for a drinking game: trying repeating that phrase five times as fast as possible).

Now that you understand the concept of PCA, let's see how it is implemented in MATLAB. We begin, as we usually do, by simulating some 2D random data. We want the distribution to be "squashed" (the technical term would be *anisotropic*) for reasons that will soon become clear.

```
x = [ randn(1000,1)  .4*randn(1000,1) ];
```

These data are not rotated, so a PCA will tell us that the standard Carte-sian coordinates are the best basis vectors ([0 1] and [1 0]). That's pretty boring. To see the PCA in action, we want to swirl these data around and thus force them to be correlated. In the previous chapter, you learned one method of creating correlated variables; now you will learn an alternative method, which is to rotate the data distribution around the origin.

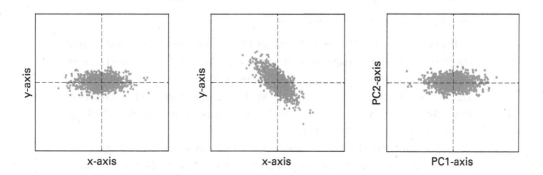

Figure 17.2
Illustration of PCA on actual simulated data. Correlated random data were created (left and middle panels). The middle panel shows the two eigenvectors of the covari-ance matrix of the data (a.k.a. principal components). The right panel shows the same data plotted in PCA space instead of the original *xy* space.

To rotate the data, we can multiply the data matrix by a rotation matrix. A 2D rotation matrix involves multiplying the data by cosines and sines of the angle (θ) to rotate. In MATLAB this is implemented as follows:

```
th = pi/4;
R = [ cos(th) -sin(th); sin(th) cos(th) ];
y = x*R;
```

Thinking back to the discussion in section 16.6 that correlation is the same as cosine similarity, it shouldn't be surprising that rotating the angle of an anisotropic distribution produces correlated variables.

Now that we have our data, the next step in performing PCA is to compute the covariance matrix (let's call it covmat). Because the data here were generated using the function randn, the mean should be zero or very close to it. Does that mean we can forget about mean-normalizing the data in this example? Nope! You should always mean-normalize the data before computing covariance. It's a good habit and promotes healthy data-analysis hygiene. It's the MATLAB equivalent of brushing your teeth before asking that special someone out to dinner.

Now we're ready for the eigendecomposition. The MATLAB eig function takes an N-by-N matrix as input and returns two N-by-N matrices that contain the eigenvectors and the eigenvalues. The eigenvalues are scalars and appear in the diagonal elements; all other elements are zeros. The input matrix can be any N-by-N matrix, but in this chapter (and most often in neuroscience analyses), the input will be a symmetric covariance matrix.

```
[vecs,vals] = eig(covmat);
```

If you get confused about the order of the outputs, the easiest sanity check is to create an image of both of them (figure 17.3). The eigenvectors matrix is all non-zero and looks like channel 900 on the hotel television. The eigenvalue matrix has all zeros on the off-diagonals and eigenvalues (typically sorted by magnitude) on the diagonals. Some people call these variables V (vecs) and D (vals).

The reason why the eigenvalues come in matrix form instead of vector form is that one commonly used matrix factorization (called *diagonalization*) requires this format. In case you are curious, that factorization is $\mathbf{A} = \mathbf{S}\Lambda\mathbf{S}^{-1}$, which means the square matrix $\mathbf{A}$ can be represented using eigenvectors ($\mathbf{S}$) and eigenvalues (Λ). This decomposition works when the eigenvectors are all independent, and is useful, among other reasons, for decoupling linear system components or for computing matrix powers. You can check yourself that covmat is the same thing as vecs*vals*inv(vecs).

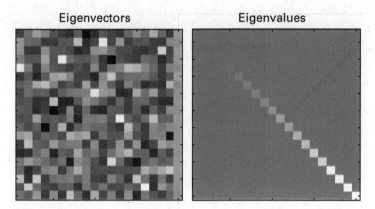

Figure 17.3
It's easy to confuse the order of the outputs of the `eig` function. An easy sanity check is to visually inspect the two outputs.

Anyway, if you want to extract the diagonal elements of a matrix, use the function `diag`. The same function can be used to create a diagonal matrix from a vector.

```
valsVec = diag(vals);
valsMat = diag(valsVec);
```

The eigenvectors associated with the largest eigenvalues are the most important. In fact, you can compute the amount of variance explained by each eigenvector by scaling the eigenvalues (you will see this in the exercises). Although there is no inherent ordering of eigenvalues and eigenvectors, the algorithms that MATLAB uses to compute eigenvalues typically results in eigenvalues (and their associated vectors) appearing in ascending order. Some people find it more intuitive to have the eigenvectors sorted in descending order and therefore will re-sort the matrices. It doesn't matter whether you have them ascending or descending, but it is important to know which order you are working with. And it is even more important to sort both matrices or neither of the matrices; don't open yourself up to confusion by sorting only one of them.

Another thing to know about MATLAB's `eig` function is that it normalizes all eigenvectors to have unit length (when the input matrix is symmetric, which is always the case for covariance matrices). We will now plot the eigenvectors on top of the scattered data, and scale the vectors to have lengths defined by the eigenvalues (figure 17.4).

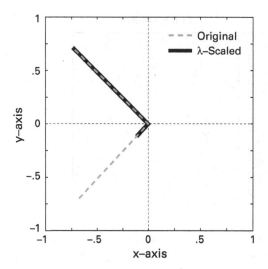

Figure 17.4

Original and scaled eigenvectors. The eigenvalues tell you how important each direction is; the vectors just tell you which way to point. The two λ's here are 0.144 and 1.01.

```
plot(y(:,1),y(:,2),'o')
plot(vals(1,1)*[0 vecs(1,1)],vals(1,1)*[0 vecs(2,1)])
plot(vals(2,2)*[0 vecs(1,2)],vals(2,2)*[0 vecs(2,2)])
```

Those are long lines of code; let's work through them. Plotting a line requires specifying two x points and two y points, as you learned in chapter 9. We want the lines to start at the origin, hence the zeros. The two pairs of square brackets define the x and y coordinates. And then the vector is scaled by the eigenvalues. Try removing the eigenvalue scaling to see that the vectors are given unit length. Displaying the scaled eigenvectors helps you appreciate the concept of finding directions of maximal variance.

Before moving on, take a few minutes to play around with the MATLAB code for this section. Try changing the amount of anisotropy (the amplitude coefficients for the two vectors) and the angle of rotation. Inspect the eigenvalues and eigenvectors to see how those are affected. Try to gain an intuitive feel for what PCA reveals about a 2D data distribution. A PCA on a 1,000-dimensional data set is exactly the same concept, just really difficult to visualize.

17.3 PCA and Coordinate Transformation

One application of PCA is to change the coordinate space from the original axes to another set of axes that better characterize the data. This was briefly introduced in figure 17.1. Our data were initially in Cartesian space, and the PCA gave us two vectors that better described the patterns of variance in the data. We can now use those eigenvectors (principal components) as new-and-improved axes with which to describe the data. This means that the data go from xy space to principal components space, and this in turn means that each data point is defined by having "scores" on each component rather than having an x and a y value. For a standard basis space (the typical Cartesian plane), the "score" (i.e., the distance along each basis vector) is simply the x and y coordinates. But PCA rotates the axes, so we want to compute the distance from each eigenvector. Scores are computed by multiplying the data by the eigenvectors, and there are therefore as many scores as there are vectors.

```
pc1 = y*vecs(:,1);
pc2 = y*vecs(:,2);
```

Remember that the eigenvalues and vectors are typically in ascending order, meaning that pc2 contains most of the variance (how could you sanity-check to make sure the second component is the biggest?). We can plot the data using their principal component scores (figure 17.2, rightmost plot). You might immediately notice that the "new" plot looks awfully similar to the original data, even though the PCA was computed on the rotated data instead of on the original data. Indeed, that's how the data began their existence before they were rotated. But if you look closely, you'll see that the data are flipped relative to the original data. The component axes are designed to capture the most variance, and the components sometimes point in different directions from the original data.

There are several advantages of such a coordinate transformation. One is to identify patterns of variance in multivariate data. Another is to perform dimensionality reduction by ignoring dimensions in the data that have little meaningful variance. In this example, if we assume that PC1 (the smaller component) is just noise, we can ignore it and perform subsequent analyses only on PC2 (the larger component). This would bring the dimensionality of the data from two to one. This dimensionality reduction can be used to group variables together (analogous to how factor analysis is used in questionnaire studies) or it can be used for cleaning the data (ignoring dimensions that contain only noise).

17.4 Eigenfaces

Now that you have an intuition of what PCA does and how to program it in MATLAB, it's time for an example. An early idea for how to get computers to recognize faces and emotional expressions was to learn a set of "eigenfaces" and categorize each face according to its score on different eigenface dimensions. In this section, we will compute eigenfaces and show that it takes only a small number of eigenfaces to recognize an individual.

The faces are stored as vectors in binary files (the full set includes more than 5,000 of them, but I took 500 for this exercise). You can find many different eigenface data sets freely available online; I downloaded a sample from http://courses.media.mit.edu/2004fall/mas622j/04.projects/ faces/. Let's start by loading in one file.

```
dat = fread(fopen('3500'));
```

In chapter 8, we had the `fopen` function on its own line with the option `'r'` (read) and a pointer (`fid`) to that file. Here, we access each file only once, and we read the entire file in one shot, so the additional complexity seems unnecessary.

The variable `dat` is a 1-by-16,384 vector. That doesn't seem quite right. We are expecting this file to contain an image, and therefore we should expect the data to be 2D or maybe 3D. This is our first time dealing with this type of file, so we should make sure that we successfully imported *something* (instead of, e.g., all zeros). Typing `plot(dat)` reveals that there are valid numbers in this variable.

Perhaps you've already guessed that this is a 2D image that has been linearized. The image is 128×128, and a quick reshape will reveal the content of the image: `imagesc(reshape(dat,128,128))`. It's easier to see when transposed and shown in greyscale (figure 17.5).

Maybe you are tempted to write a loop that will load in all pictures and store them in a 500-by-128-by-128 matrix. But before you do, let's think ahead about what we plan to do with the data. We want to compute a PCA, which means we need a covariance matrix. For images, each pixel is like a channel that measures "activity" at that location. The covariance matrix needs to be channel by channel, meaning we would want to linearize the matrix (PCA doesn't know or care about spatial locations). That is, the PCA requires a 500-by-128*128 = 500-by-16,384 matrix. Except for viewing the content of the images, it's best to leave them linearized.

Now it is time to load in all of the images in the folder. The files have no extension, so we can use `dir('*')` to list all files. Do we need to worry

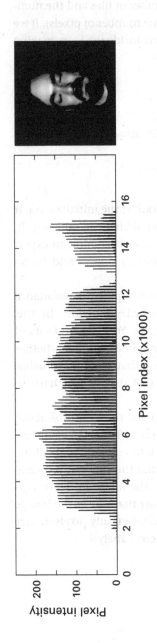

Figure 17.5
A linearized image is difficult for most people to recognize (left plot) but looks a bit better when reshaped and shown as an image (right plot).

about the file order? (No, but it's always good to ask this question when loading in many files.) Because we are using a nonselective filter, the first two entries in the output of the `dir` function will be "." and ".." (check chapter 8 if you need to refresh your memory about this situation). In order to initialize the matrix, we need to know the number of files and the number of pixels (we assume that each file has the same number of pixels). If we don't know this *a priori*, we can initialize the matrix inside the loop, making sure it runs only once.

```
for fi=1:length(filz)
    if fi==1
        allF = zeros(length(filz), ...
            length(fopen(fread(filz(fi).name))));
    end
    allF(fi,:) = fopen(fread(filz(fi).name));
end
```

The variable `allF` is for all faces. Take a close look at the initialization. It runs only in the first iteration of the loop (what would happen otherwise?), and it has five embedded functions. But you are now becoming an expert at interpreting multiple embedded functions, so this one should be no problem to figure out.

Now that we have the data, the next step is to compute the covariance matrix (mean-center first!) and then apply the `eig` function. In this example, we have 16,384 variables instead of two. You can understand why we began this chapter with 2D data. I sometimes wonder whether there is intelligent life somewhere in the universe that can conceptualize a 16,384-dimensional geometric space. The rest of us will just have to pretend.

Before computing the pixel-by-pixel covariance matrix, let's think about matrix size. The full covariance matrix will contain 268,435,456 elements. That's a really big matrix. It will take a long time to compute and will be slow to work with. And if your lab computer is from the 1980s, it probably cannot even create a matrix that big. Looking at the image in figure 17.6 reveals that the *information* we care about is smaller than the entire image. We can define a subregion that we care about and select only pixels in that subregion. We might also call this a region-of-interest analysis.

```
mask = false(128);
mask(40:100,20:120) = true;
allfaces = allfaces(:,mask);
```

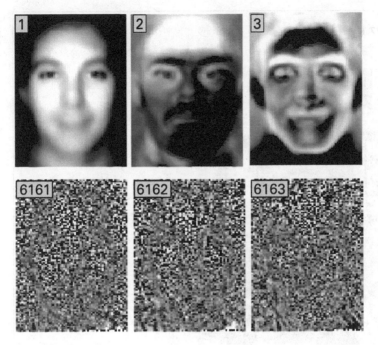

Figure 17.6
A selection of eigenfaces. The number in the boxes is the order of the eigenface. Smaller numbers capture more variance in the data.

The mask is initialized to be a matrix of Boolean falses, and then a box including pixels from row 40 to 100, and columns 20 to 120, is set to trues. This means the matrix that stores all the data is reduced. Notice that `mask` is a 2D matrix but is automatically converted to linear indices. It's a dangerous move, but I sanity-checked the result by plotting the image before and after applying the mask. Feels good to live life dangerously, doesn't it?

The eigenvectors returned by `eig` are the eigenfaces. They can be viewed like pictures, by reshaping them to two dimensions. Remember that they are not reshaped to 128-by-128, but to 61-by-101 because of the masking (the new covariance matrix will be 6,161-by-6,161). Figure 17.6 shows some eigenfaces with the largest and smallest eigenvalues. I think component no. 3 is the one most likely to enter the nightmares of small children.

One of the goals of eigenfaces is dimensionality reduction. Imagine that a computer could identify an individual's face not by analyzing all 6,161 dimensions (and this is a huge underestimate; most digital photographs have much higher resolution), but by projecting them onto a subspace of,

Box 17.1

Name: Pascal Wallisch
Position: Clinical Assistant Professor
Affiliation: Department of Psychology, New York University, New York, New York
Photo credit: Alison Wynn

When did you start programming in MATLAB, and how long did it take you to become a "good" programmer (whatever you think "good" means)?
I started programming in MATLAB in 2002. Literally the first question my PhD adviser asked me was "How is your MATLAB," and I didn't know—despite having a relatively strong background in programming and quantitative methods—what that was. I would say it took about 5 years to become good and about 10 years to become really comfortable, but I'm still learning.

What do you think are the main advantages and disadvantages of MATLAB?
MATLAB has several key advantages. First, its help function is actually helpful. Second, it is extremely beginner-friendly, as it basically takes care of all "plumbing" (e.g., memory handling) behind the scenes. Third, most of the things that are important to scientists, such as linear algebra and data visualization, are very fast end relatively effortless. Fourth, there is a tremendous user base, so if you encounter a problem, it is likely that someone else has

already encountered—and solved—it. On a related note, it is a great advantage to be able to read and understand code from other people in your lab or even other labs. Having a common lingua franca helps with that.

Most of the key disadvantages are the flip sides of these advantages. For instance, a lot of what makes MATLAB so user-friendly for beginners (e.g., no strong typing) makes it relatively slow at runtime. Another disadvantage is that MATLAB grew out of wrapper around LINPACK—a linear algebra library in Fortran—and it works great for that. But a lot of the things people now use MATLAB for feel rather tacked on; for example, GUIs (which invoke Java frames). Most important, MATLAB is still effectively stuck in pre-Internet times. If I want to deploy a program online and run it on a server, I can do that in Python, but not MATLAB. As more and more data collection is done online and even the analysis moves to the cloud, this matters.

Do you think MATLAB will be the main program/language in your field in the future?
It could go either way. We are at a turning point right now. More and more people are making the switch to Python because they feel too constrained by MATLAB—and probably because they want to join on the bandwagon. If this movement reaches critical mass, some of the key advantages of MATLAB (e.g., the large user base) go away. Then again, there are now new fourth generation languages like Julia, which are—mostly due to strong typing—extremely fast and natively support the parallel processing of big data, so the Python movement might be quenched. Moreover, MATLAB has been quite adaptive; for example, it is moving more and more toward object-oriented programming. The latest release (2016a) saw the incorporation of some Python features with the "live editor." I think if MATLAB manages to create strong online data collection and analysis capabilities, it will be around for a long time.

How important are programming skills for scientists?
Absolutely critical. As time goes on, all fields are becoming ever more quantitative, and the quantity of the data itself is also ever increasing. These are all good things, but if one cannot program, one is sidelined by these exciting developments. Importantly, if one has strong programming skills, one can *always* get an academic job (e.g., postdoc). From my students, I know that scientific programming skills (MATLAB in particular at this point) are literally the most important selling point that gets them into programs and allows them to launch an academic career.

Any advice for people who are starting to learn MATLAB?
Make sure to get a good book that focuses on the kinds of applications you have in mind. If you can, take a class from a master teacher. If possible, do all of this together with a peer. Peer-coding is becoming more and more important.

say, 50 dimensions. Let's try that—let's project a face onto the first 50 eigen-faces and see whether a 50-dimensional representation of the face is close to the full 6,161-dimensional representation.

```
nPCs = 50;
scores = allF(1,:)*vecs(:,1:nPCs);
imagesc(reshape(scores*vecs(:,1:nPCs)',facedims)')
```

The code above computes the scores from the first 50 eigenfaces (com-pare this code to computing the PC scores in the 2D case earlier in this chapter). This is done only for the first face; in the online code, the PC scores are computed for all faces, and you can select one face for plotting.

Notice that extracting the *first* 50 components (i.e., the first 50 columns of the matrix vecs) is valid only if the eigenvectors matrix has been re-sorted to go from largest to smallest eigenvalue. Otherwise, these first 50 components are actually the ones that account for the smallest amount of variance. See the online code for an illustration of this.

It's not perfect, of course, but it's "not bad." That is a subjective rating and we're not doing anything quantitative here. You can change the nPCs variable to see how many components give what level of detail. I think you could recognize some individuals with around 10 components. After around 300 components, the image is nearly perfect and there is almost no difference between using 300 versus 6,000 components. In real-world appli-cations, you can imagine that reducing a 10 million–dimensional image to a 1,000-dimensional subspace while retaining accuracy is pretty good.

17.5 Independent Components Analysis

Independent components analysis (ICA) has several conceptual similarities to PCA—it is a BSS technique that involves finding vectors that point in the direction of maximal variance, and it is often used as a data-reduction or data-cleaning technique. There are several algorithms for ICA, a few of which are commonly used in neuroscience (Makeig et al. 2004; Onton et al. 2006), including FastICA and Jade. The main distinction of ICA compared to PCA is that ICA is an iterative procedure that constrains the component vectors to be *independent* but not orthogonal. In this sense, ICA is used to unmix data, rather than to decorrelate data.

A striking example of this PCA-ICA distinction can be seen in figure 17.7. Here I generated 2D data that have a clear structure (data were gener-ated using the squash-and-rotate procedure described in section 17.2), but a structure that is clearly not orthogonal. It is obvious that there are two

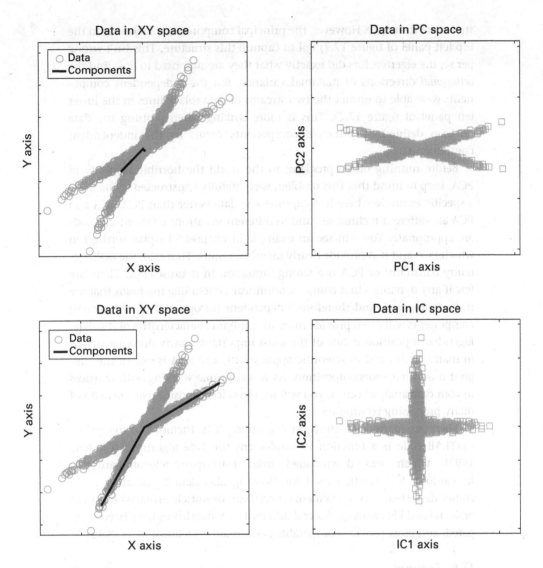

Figure 17.7

Comparison of PCA and ICA. Two data streams were generated; the goal was to separate these streams using PCA or ICA. The left-side plots show the data (gray circles) and the components (black lines). The right-side plots show the data after being transformed into PC or IC axes.

streams of data here. However, the principal components (black lines in the top left panel of figure 17.7) fail to capture this structure. This isn't wrong per se; the eigenvectors did exactly what they are supposed to do—find the *orthogonal* directions of maximal variance. But the independent components were able to unmix the two streams of data (black lines in the lower left panel of figure 17.7). This is more striking when plotting the data on axes defined by principal components versus by the independent components.

Before running off to proclaim to the world the horrible stupidity of PCA, keep in mind that this problem was carefully constructed to illustrate a specific example where ICA captures the data better than PCA. ICA and PCA are different techniques, and in different situations different methods are appropriate. You will see an example in chapter 24 (spike sorting) in which PCA and ICA provide nearly identical results. That said, the orthogonality constraint of PCA is a strong limitation in neuroscience. There are few if any dynamics in a complex nonlinear system like the brain that are truly orthogonal, and therefore independent (as opposed to orthogonal) components will often provide more meaningful characteristics of the data. Eigendecomposition is one of the most important matrix decompositions in mathematics and in scientific applications, and PCA is one of the simplest uses of eigendecomposition. As you continue working with matrices in your data analysis career, you will see eigendecomposition at the heart of many processing techniques.

There are many algorithms for computing ICA. Included in the online MATLAB code is a function to implement the Jade algorithm (Cardoso 1999), which was downloaded from http://perso.telecom-paristech. fr/~cardoso. ICA is often used for cleaning EEG data by isolating noise components (e.g., eye movements, heartbeat, or muscle artifacts) from the brain-related EEG activity. Several different ICA algorithms have been compared, and most provide comparable performance (Delorme et al. 2012).

17.6 Exercises

1. Why does the following line of code not produce a 2-by-1,000 matrix? Add one character to make it produce a 2-by-1,000 matrix, and then add one more character (without changing the previous character) to produce a 1,000-by-2 matrix.

```
x = [ 1*randn(1,1000) .5*randn(1,1000) ];
```

2. Relative eigenvalue magnitudes encode the proportion of variance explained. Divide all eigenvalues by their sum and multiply by 100 to transform the eigenvalues to percent variance explained. How much variance is explained by the first 50 eigenfaces? Write code to compute how many eigenfaces you would need to explain 73% (or any other percent) of the variance.

3. Repeat PCA on 2D random correlated data, but add a mean offset to one dimension (and don't remove it when computing the covariance matrix!). What happens to the eigenvalues and eigenvectors? Try it again with mean offsets to both dimensions. Now do you see why the data must always be mean-subtracted?

4. Your mother-in-law wants to compute a PCA on her multichannel data. She's taking things slow, starting just with the covariance matrix. Her code below produces no MATLAB errors, but it contains four problems (variable y is a time-by-channels matrix of data). What are they, what are the effects, and how do you fix them (remember to be nice—she's your mother-in-law!)?

```
cov = (y*y');
```

5. Build on figure 17.2 by repeating with 3D data. You'll need to use plot3 instead of plot, and you might need to do a bit of research to figure out how to construct a 3D rotation matrix. When plotting, use the view function to make sure the data and eigenvectors are easily visible.

6. I claimed that MATLAB returns eigenvectors with unit length. Test this in MATLAB by writing a script that will generate random square symmetric matrices of any size, and compute the lengths of all eigenvectors. If you get complex eigenvectors, you'll need to compute their magnitude. Try this again with square nonsymmetric matrices.

7. Below are two pairs of lines of code that differ slightly (assume evalsX and evecsX are 2-by-2 matrices used in the 2D PCA example at the beginning of this chapter). For each pair, decide whether the two lines produce different results, and explain why or why not. After you come up with answers, test yourself in MATLAB.

```
a1 = evalsX(1,1)*[0 evecsX(1,1)];
a2 = [0 evalsX(1)*evecsX(1,1)];
b1 = evalsX(2,2)*[0 evecsX(2,2)];
b2 = [0 evalsX(2)*evecsX(2,2)];
```

8. The MATLAB function to perform a PCA is called `princomp` (in the future it will be called `pca`; both in the statistics toolbox). For a few examples from this chapter, compare the results obtained here with the outputs of the `princomp` function. Are there any differences? If so, read the help file for this function to figure out what else `princomp` is doing that might cause the differences.

9. We can define the reconstruction error of each eigenface as the sum of squared errors of the difference between the PCA-reconstructed face versus the original face. Compute this for varying numbers of components (1 to 200) used to reconstruct each of 500 faces. Plot the reconstruction error as a function of the number of components, averaged over all 500 faces. Show the variance over faces using a patch.

10. Let's say your friend Mike really likes to have eigenvectors in ascending order. But he's not very good at MATLAB and needs your help. Write code to re-sort both the eigenvectors and the eigenvalues. What is a sanity check to make sure that the sorting is correct?

11. A PCA of a covariance matrix can be used to create multivariate random correlated data. Write code to implement this based on the following description. Then test your code by computing the correlation matrix. Create an N-by-N covariance matrix (N is the number of channels) that contains only positive values. Then take its eigendecomposition, then multiply the eigenvector matrix by the square root of the eigenvalue matrix (vectors on the left, values on the right). Finally, right-multiply the transpose of that matrix by an K-by-N matrix of randomly generated numbers (K time points).

12. How did I know the picture was 128-by-128 pixels? I have to tell the truth: There were instructions when I downloaded the data. But it would also be possible to figure it out without knowing *a priori*. You know that the images must be reshaped to integer values, and you know that pictures are generally square or close to square (assume here the images are 2D not 3D). Write some code that will report all possible sizes of reshaping a vector to two dimensions.

13. Do you agree with my choice of mask? Plot a few images before and after applying the mask. You can change the mask if you want to include more or fewer pixels.

14. Redo figure 17.2 using ICA instead of PCA. Make a figure that shows both methods, similar to figure 17.7. You can see that PCA rotated the axes to show the directions of maximum variance, whereas ICA unmixed the data into uncoupled sources. Neither approach is

"better"; they are different methods, and in some cases they can be used to obtain comparable outcomes.

15. One of the many functions of PCA is to *variance-normalize* a matrix, which is also called *sphering a matrix*. This means reconstructing the data to have equal variance in all principal directions. The formula is $\hat{y} = y\mathbf{S}\Lambda^{-1/2}$, where y is the original data, $\mathbf{S}$ is the matrix of eigenvectors, Λ is the matrix of eigenvalues, and $\Lambda^{-1/2}$ indicates the square root of the matrix inverse of Λ. Implement this formula for the data used in figures 17.2 and 17.7. Then plot the new data ($\hat{y}$) on top of the original data (y). What is the effect of sphering the data? What are the eigenvalues and eigenvectors of the sphered data?

III Analyses of Time Series

18 Frequency Analyses

Frequency-domain analyses are used to investigate processes that can be localized in the frequency domain, such as neural oscillations. The advantages of frequency-domain analyses are that they are computationally fast (thanks to a class of algorithms known as the fast Fourier transform) and are ubiquitous in many branches of science, engineering, and communications technologies.

The disadvantages of frequency-domain analyses are that the results are easily interpretable only for stationary signals (this will be demonstrated later in this chapter), and that the temporal dynamics of the signal are "hidden" in the result of the Fourier transform. "Hidden" is written with apology quotes because the Fourier transform is a perfect reconstruction of the signal; no information is lost. However, the temporal dynamics are encoded in the phases over different frequencies, and they are not easy to interpret visually or statistically. Time-varying changes in the spectral characteristics of the signal are the primary motivation for conducting time-frequency–based analyses, which are described in the next chapter.

18.1 Blitz Review of the Fourier Transform

Chapter 11 introduced the Fourier transform in detail, and you should go through that chapter before this one. Here is a quick review to get you back in the mood for thinking about the frequency domain.

The idea of the Fourier transform is to compute the dot product between the time domain data and a series of complex sine waves (sine waves that have a real part corresponding to a cosine and an imaginary part corresponding to a sine). There are as many sine waves as there are data points, and the first half-plus-one sine waves are labeled in units of hertz in linear steps from 0 (also called DC) to one half of the sampling rate (the Nyquist frequency). The resulting complex dot products per frequency are called

Fourier coefficients, and each one produces a vector in a 2D complex space from the origin to the location specified by the real and imaginary parts of the complex dot product. The squared length of that line is the power at each frequency, and the angle relative to the positive real axis is the phase at each frequency.

18.2 Frequency Resolution

The frequency resolution is the distance between successive frequencies in hertz. It is determined by the number of time points because, as you recall from chapter 11, the number of time points defines the number of frequencies between 0 Hz and the Nyquist. It can therefore be computed from the vector of frequencies. If srate is the sampling rate in hertz and n is the number of data points in the signal, the following code will compute the frequency resolution.

```
hz = linspace(0,srate/2,floor(n/2)+1);
freqres = mean(diff(hz));
freqres = hz(2)-hz(1);
```

Notice that the two computations of freqres produce identical results: The first takes the average of the derivative of the entire vector of frequencies while the second takes only the difference of the first two points. Because frequencies are linearly spaced, both implementations are correct.

Frequency resolution can also be computed without first computing the vector of frequencies, by considering that the number of frequencies is the ratio of the sampling rate to the number of data points.

```
freqres = srate/n;
```

Conceptualizing frequency resolution as a ratio of the sampling rate to the number of data points allows us to do some simple algebra and solve that equation for n. For example, if you want to extract a 35.25 Hz signal from your data, you would need 0.25 Hz frequency resolution (or 0.125 Hz, etc.). If you define the frequency resolution *a priori*, you can calculate the specific number of data points to use when computing the fast Fourier transform (zero-padding when necessary).

```
srate = 1000;
freqres = .25; % in Hz
nFFT = ceil(srate/freqres);
```

In this case, a frequency resolution of 0.25 Hz requires 4,000 data points in the fast Fourier transform (FFT). FFT lengths must be integers, which is why the code above rounds up. It's a good idea to check that your desired frequency will indeed be produced by the new nFFT.

18.3 Edge Artifacts and Data Tapering

The Fourier transform is a perfect representation of any time-domain signal, regardless of whether that signal comprises or contains sinusoidal features. This is mostly a good thing, but it can cause difficulties when interpreting the results of a Fourier transform with data that look nothing like a periodic signal. An extreme example of this is an edge.

```
ts = zeros(100,1);
ts(48:52) = 10;
plot(abs(fft(ts)/100))
```

This result, shown in figure 18.1, is completely accurate and valid. But you can imagine that having sharp edges in your time series data produces frequency-domain results that are difficult to interpret. These are called *edge artifacts*.

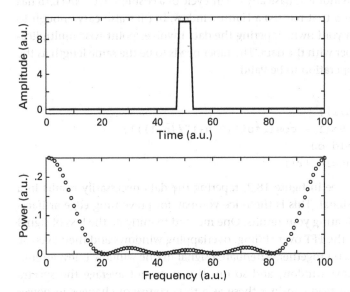

Figure 18.1
Sharp edges in the time domain (top panel) have extended representations in the frequency domain (bottom panel). If these edges are present in the data (e.g., from brief artifacts during recording), these features can impede interpretation of the results.

Edges that produce artifacts can appear in one of two places in the data. First, edges can result from an artifact during data acquisition such as amplifier saturation or interference from a mechanical device. These artifacts should be identified and removed during data cleaning. Second, edges naturally appear at the boundaries of epoched data. These edge artifacts are unavoidable and must be addressed.

The standard way to attenuate sharp edges near the boundaries of epochs is through attenuation. This is done by applying a window, also known as a taper (not *tapir*, that's something else), to the data prior to computing the Fourier transform. The purpose of the taper is to dampen the signals at the beginning and end of the data epoch. Note that tapering fixes only edges from cutting data, not edges that appear in the middle of the data like in figure 18.1.

There are several tapers that are used, including Blackman, Gaussian, Hann, and Hamming. I prefer Hann tapers because they touch zero at both sides and are relatively wide. However, the differences among different taper functions are minor. It is extraordinarily unlikely that you will get a finding with one taper that you cannot reproduce when using another.

The Hann window is basically a half-cycle of a cosine wave. MATLAB has a function hann that returns a Hann window, but it's also easy enough to compute it on your own. Tapering the data involves point-wise multiplication of the taper with the data. The taper needs to be the same length as the data for this operation to be valid.

```
n = 100;
r = randn(n,1);
hannwin = .5*(1 - cos(2*pi*(0:n-1)/(n-1)));
plot(r), hold on
plot(r.*hannwin, 'r')
```

As you can see in figure 18.2, tapering the data necessarily results in a loss of some signal. This is the price you pay for preventing edge artifacts from contaminating your results. One method to mitigate the loss of signal is to compute the FFT over sliding, overlapping windows and then average the power spectra together. Attenuated signal in one window is less attenuated in the next window, and so on. If you do not average the spectra together but instead consider these as a time course of changes in power spectra, you are computing the short-time Fourier transform, which is one method for time-frequency analysis discussed in the next chapter.

Tapering is most often applied when the time series is relatively short; that is, hundreds of milliseconds or a few seconds. If the time series is very

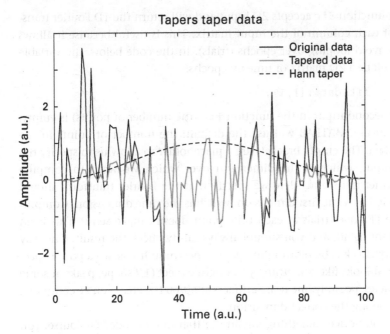

Figure 18.2
In order to attenuate edge artifacts, the data (black line) are tapered (solid gray line) before computing the FFT. The dotted black line shows the window used to taper the data. You will see an example of the effects of tapering on the power spectrum later in this chapter.

long, tapering will produce significant attenuation of real signal, while the edges might have relatively little contribution to the resulting power spectrum.

18.4 Many FFTs for Many Trials

If you have a task-related experimental design, then your data will be cut into discrete epochs (one epoch would typically correspond to one trial in the experiment). Frequency-domain analyses then involve computing the Fourier transform on each trial separately. Then, the power is extracted for each trial, and finally the average is computed over all trials. The other way around—averaging in the time domain and then taking the FFT of the trial average—is often not a useful approach, because much of the oscillatory activity in task-related data is non-phase-locked to the trial onset and therefore will be lost in the time-domain averaging process.

The function `fft` accepts 2D input and will return the 1D Fourier transform of each column of the input matrix. This is useful because it allows you to avoid looping over epochs (trials). In the code below, the variable `data` will be a 2D matrix of time by epochs.

```
dataX = fft(data,[],1);
```

The second input in the function `fft` (the number of points) is empty, which means MATLAB will use the default: the number of points in the input data. The empty brackets are a placeholder to allow us to specify the third input, which is the dimension over which we want to compute the Fourier transform. Here we specify that the Fourier transform should be applied over the first dimension, which is the time dimension. Computing the FFT over trials at each time point doesn't make sense. This is an important detail, and you should always sanity check the result. One way to sanity check is by plotting the power spectrum: If you see a power spectrum that looks like something you might expect ($1/f$ shape, peaks at some frequencies, possibly line noise artifacts at 50/60 Hz, etc.), you've computed the FFT along the correct dimension.

Did you notice something missing in that line of code? Of course, you noticed that there was no tapering. This means that the power spectrum might be contaminated by edge artifacts.

```
N = length(data);
hannwin = .5 - cos(2*pi*linspace(0,1,N))/2;
dataTaper = bsxfun(@times,data,hannwin);
dataX = fft(dataTaper,[],1);
```

Here I used the `bsxfun` function to multiply the matrix `data` by the vector `hannwin`, and `bsxfun` figured out how to expand `hannwin` appropriately over trials. Also notice that I used slightly different code to create the Hann taper compared to the code that created the Hann taper earlier in the chapter.

Perhaps you don't have trials because your data are from resting state or sleep recordings. It might still be advantageous to cut the data into many epochs, compute the Fourier transform on each epoch, and then average the power spectrum together across all epochs, rather than computing one Fourier transform of the entire time series.

There are two main advantages of epoching continuous data. First, it increases the signal-to-noise characteristics by averaging together more data. Second, it allows you to examine changes in spectral power or other features over time. One disadvantage of epoching continuous data is that the frequency resolution will be reduced by having fewer time points, but in practice the frequency resolution after epoching remains sufficient relative to the bandwidths of neural oscillations that are typical observed in EEG/LFP data. Typical epoch lengths are 1–2 seconds.

What's a good way in MATLAB to epoch continuous data? I'm glad you asked. The code below will create a continuous time series and then epoch that time series into however many epochs can be created, given the specified length of the epochs.

```
srate = 512;
n = 21*srate; % 21 seconds
data = randn(1,n);
epochLms = 2345; % epoch length in ms
epochLidx = round(epochLms / (1000/srate));
nE = floor(n/epochLidx); % N epochs
epochs = reshape(data(1:nE*epochLidx),nE,epochLidx);
```

There are several things I'd like to discuss about the code above. First, an epoch length of 2,345 milliseconds (variable epochLms) is an odd choice, and I wouldn't recommend using it in practice; it's specified like this here to show that this code will work with any arbitrary length.

The next line of code converts the requested epoch length to a number of data points, given the sampling rate. In the easy case of a sampling rate of 1,000 Hz, for example, 2,345 milliseconds would correspond to 2,345 data points.

The exact epoch duration you specify might not be possible given the sampling rate, which is why the formula is encased in the round function. If you type epochLidx*1000/srate you will see that in fact these epochs are not exactly 2,345 milliseconds long, but instead are 2,345.703125 milliseconds long. It's pretty close, though.

If you compare numel(epochs) to n, you'll notice that they differ. That's because the length of the signal does not evenly fit into an integer number of epochs. Therefore, some points were dropped at the end of the signal in order to define equal-length epochs. Although this means sacrificing some data (1,144 time points, in this case), it would be a small headache otherwise to have $N - 1$ epochs of the same length (and thus the same frequency resolution) and one epoch with a different number of data points (and thus a different frequency resolution). If you have limited data and do not want to sacrifice part of your signal, you could re-epoch the data from the end so that 1,144 points are cut off from the beginning. Results could then be averaged across the two sets of epochs.

From here you can proceed to use the `fft` function on the `epochs` matrix, as illustrated earlier for `data`. Don't forget to taper the data.

Another option for epoching continuous data, which works well for multichannel data, is to use the `reshape` function. Shall we have a primer on how the `reshape` function works? I agree; it would be a good idea. Run each line below individually and inspect the results.

```
data = 1:12;
reshape(data,1,[])
reshape(data,[],1)
reshape(data,2,[])
reshape(data,3,4)
reshape(data,4,3)
reshape(data,4,4)
```

The first `reshape` function did nothing. We requested to convert the 1-by-12 vector into a 1-by-*whatever* vector (when you use the empty bracket in the reshape function, MATLAB interprets that to mean "whatever is left over"). The variable was already a row vector, so that line simply reshaped it to itself. The line thereafter reshapes to a column vector (*whatever-by*-1). You can also reshape to a 2-by-*whatever* matrix, or a 3-by-*whatever* matrix, and so on.

Reshaping to either 3-by-4 or 4-by-3 results in valid reshapes (note that 3 times 4 is 12, which is the number of elements in the original vector), but the different reshapes make the matrix look different. The last reshape command produced an error, because you tried to reshape a 12-element vector into a 16-element matrix. That's the MATLAB equivalent of trying to push a square peg through a round hole.

Now that you understand the `reshape` function, let's get back to epoching multichannel data. You need to be careful to use the `reshape` function correctly, because you will soon see that it's easy to make horrible mistakes when reshaping.

```
nchans = 10;
data = randn(nchans,n);
epochs = reshape(data(:,1:nE*epochLidx), ...
        [nchans nE epochLidx]);
```

The data were epoched incorrectly. How do I know that? Because I sanity-checked the result.

```
plot(data(1,1:100),'r'), hold on
plot(squeeze(epochs(1,1,1:100)))
```

The first line above plots the first 100 time points from channel 1 in the continuous data, and the second line plots the first 100 time points from channel 1 and epoch 1. Those lines should perfectly overlap, but they do not (figure 18.3, top panel). That's because the epochs were reshaped over channels, not over time. So those first 100 "time points" are actually the first 10 channels looping over the first 10 epochs. Any analysis on the matrix epochs would produce junk results. But it was a legal MATLAB operation, and without our sanity check, we might never have known. The following code will epoch the data correctly.

```
epochs1 = reshape(data(:,1:nE*epochLidx), ...
          [nchans epochLidx nE]);
plot(squeeze(epochs1(1,1:100,1)),'k')
```

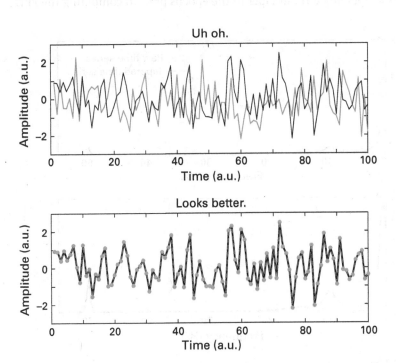

Figure 18.3

Using the reshape function to convert continuous (channels by time) to epoched (channels by time by epochs) data must be done carefully. It is easy to make mistakes with disastrous consequences, so you should always sanity check the results. In the top panel, the data were reshaped so that "time" and "channels" were swapped; in other words, the second data point comes from channel 2, not from time point 2.

Now the black line perfectly overlaps the gray line, meaning we are in fact plotting the first 100 time points of the first channel (figure 18.3, bottom panel). There are rules for how the `reshape` function works, and you could have written accurate code by following those rules. But in my opinion, trying to memorize lots and lots of MATLAB rules is tedious and is more likely to cause confusion than to prevent errors. Sanity checking is a preferable strategy that doesn't require a near superhuman capability to remember a lot of function-specific rules. And sanity checking forces you to look at and think about your data, and the closer you are to your data, the better.

Figure 18.4 shows the results of computing the power spectrum of each epoch, and then averaging the results over all epochs. You can also see the effects of applying a Hann taper to the epochs prior to computing the FFT.

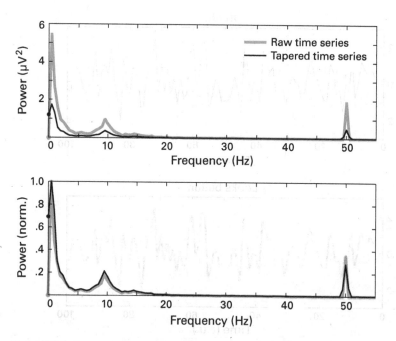

Figure 18.4
Power spectra of resting-state human EEG data showing three typical features: $1/f$ scaling, a peak at around 10 Hz (the alpha band), and line noise at 50 Hz. Data were epoched into sixty 2-second epochs, and the power spectrum was averaged across all epochs. The black line shows the power spectrum of the raw data, and the gray line shows the power spectrum of the Hann-tapered data. Because the data are attenuated from the tapering, the two results can be compared only when they are normalized (bottom plot; normalization is relative to the peak power).

Although the overall amplitude decreases, the important metric is the amplitude *relative* to other features in the frequency domain (figure 18.4, bottom plot). In this case, tapering had minimal effect (the alpha-band peak appears slightly larger), suggesting that—in these data—edges resulting from epoching did not cause significant artifacts in the frequency domain (this should not be interpreted as being generally the case).

18.5   Defining and Extracting Frequency Ranges

The main purpose of a frequency analysis is to isolate certain frequencies or ranges of frequencies. The logical step after applying the `fft` function is therefore to extract power from a frequency or range of frequencies. In most cases, it is preferable to use a range of frequencies rather than one single frequency component, in order to increase signal-to-noise ratio and to reduce the possibility of the result being driven by noise.

The question *how to define frequency ranges* can be interpreted in two ways. The simple interpretation is how to implement the definition and averaging of a range in MATLAB. The more difficult interpretation is how to determine what frequency boundaries are appropriate in empirical data sets.

We'll deal with the easy stuff first. Let's say you want to extract the power averaged between 8 Hz and 12 Hz (commonly known as the *alpha band* in humans). So far, all you have is a vector of frequency labels (variable `hz`), but you don't know *a priori* which frequencies correspond to 8 Hz and 12 Hz. It is exceedingly unlikely that this happens to correspond to the 8th and 12th entries of `hz`.

```
srate = 1000;
nFFT = 5128;
hz = linspace(0,srate/2,floor(nFFT/2)+1);
hz([8 12])
```

In this example, the indices 8 and 12 correspond to 1.3 Hz and 2.1 Hz, which is not even close to the desired alpha band. You can keep trying other indices until you happen to find an index close to 8 Hz, but this is time-consuming and not scalable to many frequencies and data sets. We need an automated way to identify the indices into the variable `hz` that correspond to desired frequencies in hertz. Trying to find the exact frequency (`find(hz==8)`) won't work (why not?).

There are three methods that will work. One is called the "min-abs" procedure (not to be confused with the exercise practice of minimizing your rectus abdominis muscle groups). It's worth taking the time to explain how this works, as an exercise in MATLAB thinking. Let's start by plotting the variable hz. You can follow the procedure visually in figure 18.5.

```
plot(hz)
```

We want to find the element in the hz vector that contains the value closest to our desired frequency of 8 Hz. The first thing we'll do is subtract the desired frequency from the vector hz.

```
desfreq = 8; % in hz
plot(hz-desfreq)
```

The plot is almost exactly the same, except that the line is shifted down on the y axis. This is useful, because we know that in this new plot, the element closest to zero contains the value closest to eight. Now our task has been reduced to finding the element in this vector (hz-desfreq) that is closest to zero. We don't care if we are a bit over or a bit under, so we can take the absolute value of this function.

```
plot(abs(hz-desfreq))
```

Now we're getting close. We want to find the minimum of this V-shaped function.

```
[minval,idx] = min(abs(hz-desfreq));
```

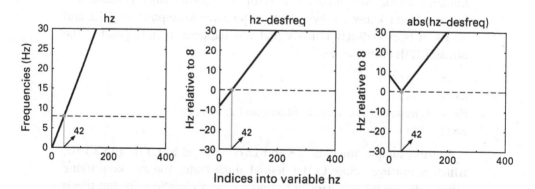

Figure 18.5
Graphical illustration of the min-abs procedure to identify the index in a vector closest to a specified point. In this example, the goal is to find the frequency in the vector of frequencies (variable hz) that is closest to 8. We start by plotting this vector (left-most panel); then subtracting the desired frequency (variable desfreq), which changes our goal to finding the index closest to zero (middle panel); then taking the absolute value of this function, which changes our goal to finding the minimum (right-most panel).

The variable `minval` is 0.0047, and the variable `idx` is 42. What do these numbers mean? Let's start with `idx`. Typing `hz(idx)` reveals that the `idx`[th] value in the vector `hz` contains the number 7.9953, which is extremely close to but not exactly 8. In fact, it is 0.0047 Hz off. When finding frequency indices, you will almost never need to know the actual minimum value, so in practice you can just request the second output.

```
[~,idx] = min(abs(hz-desfreq));
```

I like this min-abs construction because it is easy to understand how it works, and because it works well for indexing any kind of ordinal, ratio, or interval scale information (frequencies, time points, age, reaction time, etc.). It also works regardless of whether the data in the vector are linearly or nonlinearly spaced. The downside is that it works only for 1D vectors, and only for one point at a time.

The second method you can use to identify a desired frequency in the `hz` vector is the MATLAB function `dsearchn`. This function is also simple and can handle finding nearest points in *N*-dimensional space. The main annoyance is that `dsearchn` is sensitive to input formatting—it expects column-format inputs and it will produce an error if the inputs are not to its liking.

```
idx = dsearchn(hz,8);
```

The code above will produce an error message about column dimensions, and a simple transpose on the `hz` vector will give you an `idx` of 42, which is exactly the same answer as we got from the min-abs construction. That number also happens to be the answer to the ultimate question of life, the universe, and everything, as discussed elsewhere (Adams 1989).

Finally, the third method is to compute the indices directly from the frequency resolution. The MATLAB code is presented below. I think this should be the least preferred solution, because it works only for the perfectly linearly spaced frequencies of the Fourier transform. This method would not generalize to the many data analysis situations in which frequencies are logarithmically spaced, sparsely sampled, or specified based on *a priori* justifications. But it's good to have options.

```
idx = hz(1+round(desfreq/(srate/n)));
```

Let's talk about that code for a minute. The formula for the index is $1 + d/f$, where d is the desired frequency, and f is the frequency resolution. In the code above, `srate/n` is the frequency resolution, and `desfreq` is the desired frequency. That ratio is unlikely to be an integer, and indices need to be integers, so we encapsulate the code into the `round` function. You'll have to figure out on your own why the +1 is necessary, but a hint is to think about what `hz(1)` is.

That was a long but important tangent. The goal of this section is to extract specific frequency ranges from a power spectrum.

```
desfreq = [8 12]; % boundaries for averaging
idx = dsearchn(hz',desfreq');
pow = mean(abs(dataX(idx(1):idx(2))).^2);
```

That third line of code is dense. You could separate it into two lines of code, but it is good practice to be comfortable reading code with multiple embedded functions. Let's read that line of code together, starting from the most deeply embedded part. First, extract the proper range of coefficients from the Fourier series (`dataX(idx(1):idx(2))`); second, extract power from each of those coefficients (`abs().^2`); third, average the power values over all frequencies between 8 Hz and 12 Hz (`mean()`). Technically, the average is not between 8 Hz and 12 Hz; it is between 7.9953 Hz and 12.0905 Hz. But the brain is never so precise in frequency, so a little bit of simplification is fine.

It is important to extract power first and then take the average, rather than averaging the complex Fourier coefficients and then extracting power. To the extent that different Fourier coefficients have different phases (which they will), averaging the complex vectors will reduce the estimate of power.

The technical aspects of extracting a particular frequency range from the Fourier coefficients series are straightforward. The more difficult problem is knowing how to define appropriate boundaries of frequency ranges. To capture alpha-band activity, is 8–12 Hz really the best range? Or 7–12 Hz, or 8–13.3 Hz? In the literature, you will see quite some diversity in the precise boundaries of frequency ranges, not just for the alpha band, but for all frequency bands. This is because there are individual differences in peak oscillation frequencies, as well as differences between brain regions and the specific processes elicited by the experimental task if there is one (e.g., Haegens et al. 2014). How do you pick the appropriate frequency range for your data?

The best approach here is a combination of previous literature and visual inspection of your results. Look for published studies that have similar data recording and analysis pipelines, and a similar experiment setup, and draw inspiration from the frequency ranges they used. This can then be refined by inspection of the power spectral plots of your data. The key feature to look for is "bumps" or "peaks" in the power spectrum. These reflect deviations from the $1/f$ background power spectrum. The ranges you select should not be too wide (or else they will lose frequency specificity) and they should not be too narrow (or else the signal-to-noise characteristics will decrease and your results might be overly sensitive to noise).

18.6 Effects of Nonstationarities

The Fourier transform is always a perfect representation of the time-domain signal. But that doesn't mean the Fourier coefficients are always perfectly interpretable. When the time-domain signal contains nonstationarities, the Fourier coefficients become less meaningful. The more severe the nonstationarities, the weirder the frequency representation. These nonstationarities are one of the primary motivations for performing time-frequency analyses. Here we will explore how nonstationarities in amplitude and in frequency affect the frequency-domain representation of the signal. First the amplitude nonstationarity.

```
t = 0:1/.001:10;
ampl = linspace(1,10,length(t));
signal = ampl .* sin(2*pi*3*t);
```

The only difference between this sine wave and the typical sine wave that you learned about in chapter 11 is that the amplitude is a vector rather than a scalar. The frequency representation of this signal (figure 18.6) still has a clear peak at 15 Hz, but there is also non-zero power at surrounding frequencies. Those non-15-Hz frequencies were not present in the signal, but the side-lobes are necessary for the Fourier transform to capture the amplitude nonstationarities.

Now let's try frequency nonstationarities. To create frequency nonstationarities, you need to learn how to simulate sine waves with time-varying frequencies. The simplest frequency nonstationarity is called a "chirp" and involves a linear change in frequency as a function of time.

```
f = [5 15];
ff = linspace(f(1),mean(f),n);
chirp = sin(2*pi.*ff.*t);
```

Before evaluating this code, type which chirp. The result you see tells you that chirp is a MATLAB function. In fact, it's a MATLAB function that creates a chirp signal. Using this same name for the variable is bad

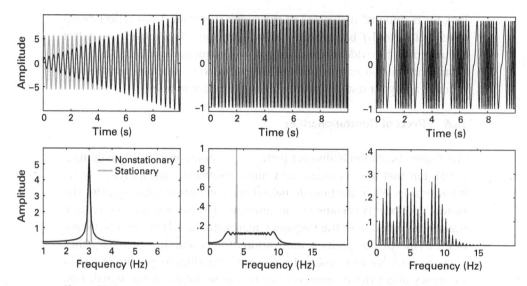

Figure 18.6

This figure illustrates the effects of temporal nonstationarities on the power spectrum.

programming. You should rename this variable (perhaps to churp) and then compute its FFT to obtain the result in figure 18.6.

It is also possible to create a sine wave with frequencies that vary arbitrarily over time. The third example in figure 18.6 was generated by defining frequencies from triangles.

18.7 Spectral Coherence

In chapter 11, you learned that power can be extracted from the Fourier coefficients by multiplying the complex coefficients by their conjugate (this is equivalent to the distance squared from the origin). To compute spectral coherence, the Fourier coefficients of one time series are multiplied by the complex conjugate of the Fourier coefficients of a second time series. (The linear algebra–inclined reader might recognize this as the Hermitian dot product.) This means the power of each of the two signals is combined, scaled by the phase angle between them. The magnitude of this multiplication is taken as the strength of connectivity. Often, the magnitude is squared, and the measure is called magnitude squared coherence.

Let's start with some simulated data to see how spectral coherence works. Two signals will be generated, each of which contains two frequency components with one frequency component being synchronous. We could generate simple sine wave functions, but you already know how to do that. Let's move one step forward and learn how to generate a signal with time-varying frequencies, similar to how FM radio uses time-varying frequency changes to encode information.

```
srate = 1000;
t = 0:1/srate:9;
n = length(t);
% create signals
f = [10 14 8];
f_ts1 = (2*pi*cumsum(5*randn(1,n)))/srate;
f_ts2 = (2*pi*cumsum(5*randn(1,n)))/srate;
f_ts3 = (2*pi*cumsum(5*randn(1,n)))/srate;
sigA = sin(2*pi.*f(1).*t + f_ts1) + randn(size(t));
sigB = sigA + sin(2*pi.*f(2).*t + f_ts2);
sigA = sigA + sin(2*pi.*f(3).*t + f_ts3);
```

Notice that signal B is defined by signal A plus unique dynamics. And then other unique dynamics are added into signal A. The effect of this code is that both signals have two frequency components, one of which overlaps (this overlap is our simulated "connectivity").

Before moving forward with coherence, let's have a look at these signals (figure 18.7). Their frequency fluctuations are too subtle to be detected by eye (you'll see more extreme frequency fluctuations later). Inspecting the power spectra reveals these nonstationarities as "carrot-shaped" frequency responses. If you have worked with neurophysiology data, you will recognize that these distributions look more physiologic than pure sine waves. You can modify the code to make the nonstationarities even larger, although we want the spectra to be fairly narrow in order to recover frequency-specific coherence.

Once we have our signals, computing spectral coherence is surprisingly simple: Compute the Fourier coefficients of both signals and point-wise multiply the coefficients of one signal by the complex conjugate of the coefficients of the other signal.

```
% spectral coherence
sigAx = fft(sigA)/n;
sigBx = fft(sigB)/n;
specX = abs(sigAx.*conj(sigBx)).^2;
specX = specX./(abs(sigAx).^2 .* abs(sigBx).^2);
plot(hz,abs(specX(1:length(hz))))
```

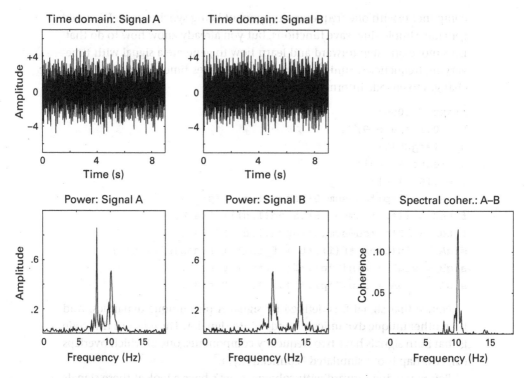

Figure 18.7

Two signals were created (A and B), each comprising two spectral components plus noise. The spectral component at 10 Hz was shared between them, which is how coherence was simulated. The lower plots show the individual power spectra and the coherence spectrum.

Notice that the coherence is normalized to the individual power of the two time series. I'm sure you can see the link between spectral coherence and correlation—both involve computing the covariance and then normalizing it by the product of the individual variances. There are two important differences between correlation and spectral coherence: (1) correlation coefficients range from –1 to +1, whereas coherence varies between zero and one—zero means no coherence and one perfect coherence; there is no such thing as "negative coherence" (although *relative* coherence between conditions or over time can be negative); (2) correlation coefficients are computed in the time domain and have no inherent spectral interpretation, whereas spectral coherence is computed in the frequency domain and can reveal frequency-specific interactions.

18.8 Steady-State Evoked Potentials

I wrote earlier that it is advisable to use fairly broad frequency ranges to boost the signal-to-noise ratio of the analyses. One noteworthy exception to this rule is the so-called steady-state response, which is a rhythmic neural response to rhythmic sensory input (typically visual, but other modalities work as well). Let me repeat that sentence with less jargon: If you look at a strobe light, a large population of neurons in your brain's visual system will fire rhythmically in sync with the flickering light. The steady-state visual evoked potential (SSVEP) is of interest to the cognitive neuroscience and vision science communities, in part because the amplitude of the SSVEP correlates with the amount of attention paid to the flickering stimulus (Norcia et al. 2015).

SSVEPs are a very narrow-band frequency feature, and therefore the analysis should isolate the stimulation frequency. Figure 18.8 shows an example of an SSVEP, in which two stimuli were simultaneously presented on the

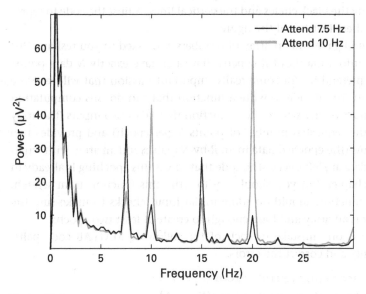

Figure 18.8
Power spectrum from an electrode over visual cortex in humans during an SSVEP experiment. The subject saw two images on the monitor, one flickering at 7.5 Hz and one flickering at 10 Hz. In different conditions, the subject was instructed to attend to the 7.5 Hz stimulus (black line) or to the 10 Hz stimulus (gray line). The difference between the power at these frequencies is taken as a measure of attention. The attention modulation can also be observed at the harmonic frequencies (15 Hz and 20 Hz).

monitor but flickered at different rates and attention was directed to one or another stimulus in different conditions. You already know how to analyze these data—compute the FFT of the data, extract power, and plot the power spectrum as a function of frequencies—so successfully completing exercise 13 should be no problem.

In SSVEP analyses, it is often useful to convert the data from the scale of the original data (microvolts or picoteslas) to signal-to-noise ratio (SNR) units. SSVEP-SNR units can be defined as the ratio of the power at the frequency peak relative to the power at surrounding frequencies. Converting to SNR units helps normalize and thus make comparable the SSVEP effect across different frequencies, individuals, and measurement devices (e.g., magnetoencephalography and EEG, which have completely different and therefore incomparable units).

18.9 Exercises

1.   The online code doesn't exactly reproduce figure 18.1. The code is missing the black circles and the vertical lines. Adjust the code to reproduce these features of the figure.

2.   Pretend I gave you an array of numbers and asked to you reshape this array into a matrix of R epochs that each have exactly N data points (also pretend it's for some really important mission that will save the galaxy). You decide to write a function that can do this computation for arrays of any size. Write a function that takes two inputs (the array and the requested number of points N per epoch) and provides two outputs (the epoched data in an R-by-N matrix and an array with whatever data are left over)—the code that solves this epoching is already in this chapter, but you should try to write this function from scratch. Your function should do some initial input checks to make sure the input is an array and long enough to create at least two epochs.

3.   There is one mistake in each of the following MATLAB code pairs. Identify and correct the errors.

```
hz = linspace(0,nyquist,floor(N/2)+1);
hz = linspace(0,nyquist,floor(N/2)-1);
freqres = hz(2)-hz(1);
freqres = hz(1)-hz(2);
hannwin = .5*(1-sin(2*pi*(1:n)/(n-1)));
hannwin = .5*(1-cos(2*pi*(1:n)/(n-1)));
[~,idx] = min(abs(timevec- -200));
[~,idx] = abs(min(timevec - 200));
```

4. If you perform a Fourier transform of a signal that contains 200 time points sampled at 100 Hz, what is the highest frequency (in hertz) that you can reconstruct? What would the highest frequency be if you had 400 time points?

5. Figure 18.1 showed the effect of an edge in the time domain on the power spectrum. Is this detrimental for the spectrum of sinusoidal components? To find out, add a pure sine wave to the variable `ts`.

6. The online MATLAB material includes a file called mouseHippocampus.mat, which contains 100 trials of LFP recordings from the hippocampus. On each trial, a visual stimulus was displayed. Show the power spectrum from this electrode for two methods of averaging over trials. First, compute the FFT and extract the power spectrum of each trial, and then average the power spectra together. Second, average the time-domain LFP traces together, and then compute the power spectrum. Show the results on the same plot.

7. How much of the data are lost when tapering? To find out, create a 100-by-120 matrix of normally distributed random numbers (120 epochs, each with 100 time points). Then, taper each epoch with a Hann window (can you do this without a loop?). Make sure you are tapering the time dimension, not the epoch dimension. At each time point over epochs, compute the sum of squared errors between the tapered signal and the original signal. The result will be a 1-by-100 vector of how much and where information is lost due to tapering. Finally, repeat this procedure using Hamming and Gaussian windows. Plot the results for different tapers overlaid on the same window.

8. The variable `data` is a 3,000-by-60 (time by trials) matrix. What is the error in the following code, and how can you fix it?

```
dataX = fft(data)/length(data);
dataPow = abs(dataX(1:length(hz))).^2;
```

9. Compute the power spectrum of the online data EEGrestingState.mat. The variable `eegdata` is a time-by-epoch matrix. Use the entire time series in one FFT, with and without applying a Hann taper. In one figure with 2 × 1 subplots, plot in the upper subplot the time-domain data before (black line) and after (red line on top) applying the Hann taper. In the lower subplot, plot the power spectrum from the entire time series without (black line) and with (red line on top) the Hann taper.

10. With the same resting-state data, cut the time series into non-overlapping epochs of two seconds. Then taper each epoch using a Hann window, take the power spectrum, and then average the power spectrum over epochs. First do this using a loop over epochs. Then get rid of the loop by inputting a matrix into the `fft` function. Check that the loop and no-loop code produce identical results.

11. Using the epoched resting-state data, compute the power spectrum using an N parameter for the FFT corresponding to the length of the entire time series. That is, zero-pad the FFT of the epochs so the frequency resolution is the same for the epochs and for the continuous time series. Then plot three power spectra on the same plot (don't taper the data for this exercise): from the continuous data, from the zero-padded epoched data, and from the non-zero-padded epoched data. You might want to scale up the power from the continuous data in order to make the power spectra more directly comparable.

12. For task-related data, it might not make sense to take the Fourier transform of the entire epoch. Using the data set sampleEEGdata.mat, compute the FFT from electrode FCz (take the power spectrum of each trial and then average the spectra over trials). First, use the entire epoch time series, and then compute again using only the data from 0 to 800 milliseconds. Plot the power spectra on top of each other. Keep in mind that the frequency resolutions will differ, and don't forget to apply a taper.

13. Reproduce figure 18.8. The data are in the ssvepdata.mat file in the online code. It might help to detrend the data before applying the FFT, and you might want to use a 4,096-point FFT to obtain fine-enough frequency resolution.

14. SSVEP data are sometimes quantified as SNR units. The frequency of the stimulus flicker is taken to be the signal (the numerator), and the surrounding frequencies are taken to be the noise (the denominator). Compute SNR at each electrode using two methods to quantify the denominator. First, take the power from frequencies 1 Hz above and 1 Hz below the peak frequency (averaged together). Second, find the minimum power values from between the peak frequency to –3 Hz, and from between the peak frequency to +3 Hz. Average those two power values to use as the denominator. Plot the results in a 1-by-3 array of topographic maps: raw power values at the peak frequency in the left-most plot, SNR using 1 Hz surrounding values in the middle plot, and SNR using the surrounding local minima in the right-most

plot. The color ranges will differ, but you can compare the spatial distributions qualitatively.

15. One way to attenuate the $1/f$ characteristic of the power spectrum is to take the FFT of the first derivative of the time series (sometimes called "pre-whitening"). The MATLAB function `diff` takes the derivative, and the result is length $N - 1$ (this makes sense: the discrete derivative of [2 1 4] is [–1 3]). In this problem, you will simulate "pink" noise with $1/f$ characteristics, add some sine waves, and then compute the power spectrum before and after pre-whitening. One method to create $1/f$ noise is to modulate the power spectrum of white noise and then take its inverse Fourier transform. The code below does most of the work for you.

```
nPnts=1000; nTrials=40;
dataX = fft(randn(nPnts,nTrials));
modfunc = linspace(-1,1,nPnts).^2;
data =
```

Make sure `data` is a `nPnts`-by-`nTrials` real-valued matrix. Next, add two sine wave components to the signal, one of 5 Hz and one of 60 Hz. Then, take the FFT of the signal on each trial and average the power spectra together. Do this before and after pre-whitening the signal. Plot the power spectra on the same plot.

How could you sanity check the result? (Hint: What should be the size of the matrix before versus after applying the `diff` function?) With this in mind, do you need to have different frequencies vectors for the raw versus pre-whitened signals?

16. Name two of the major limitations of the Fourier transform for neuro-science data analyses.

17. You perform a Fourier transform of a signal that is sampled at 100 Hz.

   a. What would be the highest frequency (in Hertz) that you can reconstruct if there are 200 time points?

   b. What would be the highest frequency if there were 400 time points?

   c. What would be the highest frequency if there were 4,000 time points?

   d. What would be the highest frequency if there were 4×10^{100} (a googol) time points?

18. What is the error in the following code? Write code to perform the following sanity check: Create a 1-second sine wave at 13 Hz using a sampling rate of 1 kHz. Then plot the power spectrum of that sine wave and identify the frequency of the peak power. If the peak is not 13 Hz, then the frequencies vector is incorrect. You should use the datacursormode tool to check the exact frequency.

```
hz = linspace(0,srate/2,floor(n/2));
```

19 Time-Frequency Analysis

The previous chapter introduced you to frequency-domain analyses of time series data. You also saw that when the time series data contained nonstationarities (meaning that the statistical characteristics of the signal such as mean, variance, frequency structure, and so forth, changed over time), the results of the Fourier transform were difficult to interpret. This is quite problematic for neuroscience, because the brain is a highly nonstationary machine. Indeed, it is no understatement to write that the vast majority of the research into the functional organization of the brain is focused on the *nonstationarities* in brain activity; that is, the changes in brain activity during sensory processing, memory, language, motor planning, and so on.

Enter time-frequency analyses. The idea of time-frequency analyses is to combine the advantages of time-domain and frequency-domain analyses while demanding only minimal sacrifices from either. Time-frequency analyses have been applied in neuroscience for decades and have been growing in popularity in recent years.

One thing to be aware of is that time-frequency–based analyses, unlike frequency-domain analyses, are *lossy* analyses. That means information is lost when going from the time domain to the time-frequency plane. This isn't a bad thing, because the goal of time-frequency analyses is to stay in the time-frequency plane. But unlike with the Fourier transform, it is not possible to reconstruct a time-domain signal from the time-frequency representation as implemented in the methods presented in this chapter (technically, it is possible, but it would entail a loss of information).

Time-frequency analysis is a big topic, much too big to cover fully in one chapter. Entire books have been dedicated to the application of time-frequency analyses in neuroscience and cognitive neuroscience (oh, just to pick one example at random: Cohen 2014). This chapter is an introduction to these methods with a stronger focus on the MATLAB programming than on the justifications of the analyses.

19.1   Complex Morlet Wavelets

A Morlet wavelet is a sine wave tapered by a Gaussian. You can probably guess that a complex Morlet wavelet is a complex sine wave tapered by a Gaussian. Let's first have a look at this wavelet before learning what to do with it (figure 19.1).

```
srate = 1000;
wavtime = -2:1/srate:2;
frex = 6.5;
s = 5/(2*pi*frex);
csine = exp(2*1i*pi*frex*wavtime);
```

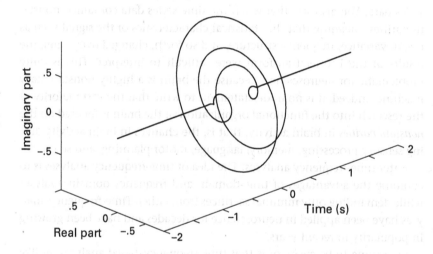

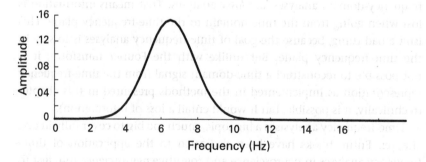

Figure 19.1
A complex Morlet wavelet is a complex sine wave tapered by a Gaussian. The top plot shows the 3D time-domain representation of a Morlet wavelet, and the bottom plot shows the power spectrum of the wavelet.

```
gaus = exp(-(wavtime.^2) / (2*s^2));
cmw = csine .* gaus;
plot3(time,real(cmw),imag(cmw))
rotate3d
```

Notice that the time vector used to create the wavelet starts and ends at the same distance away from zero. This is important because (1) the wavelet will be centered at zero, which prevents phase shifts from being introduced, and (2) the wavelet will have an odd number of points, which makes convolution convenient (this was discussed in chapter 12).

A complex Morlet wavelet involves the product of two exponentials. You might remember from your high-school algebra class that exponents of the same base can be condensed by summing those exponents. In other words, $e^x e^y = e^{x+y}$. Therefore, the implementation of the wavelet can be shortened.

```
cmw = exp(2*1i*pi*frex*wavtime-(wavtime.^2)/(2*s^2));
```

And now let's see what the wavelet looks like in the frequency domain. A neat feature of Morlet wavelets is that they have a Gaussian shape in the frequency domain.

```
cmwX = fft(cmw);
hz = linspace(0,srate/2,floor(length(wavtime)/2)+1);
plot(hz,abs(cmwX(1:length(hz))))
```

There are two parameters that define a Morlet wavelet. One is the frequency, which defines the frequency of the sine wave in the time domain and the peak frequency of the wavelet in the frequency domain. This is the variable frex in the code above.

The second parameter is the width of the Gaussian, which defines the width of the wavelet in the time domain and the width of the Gaussian in the frequency domain. This parameter determines the trade-off between temporal and frequency precisions. It is indicated by the variable s in the code above. The denominator is actually just a frequency-specific scalar; the number before it ("5" in the code above) sets the width. In publications, this number is often called the "number of cycles." The higher this number, the wider the Gaussian. You will see in exercise 3 of chapter 20 that a higher number of cycles parameter creates wider wavelets in the time domain and narrower Gaussians in the frequency domain. Higher numbers result in better frequency precision but worse temporal precision, and lower numbers result in better temporal precision but worse frequency precision. Typical values are somewhere around 4–12, depending on the goal of the analysis.

19.2 Morlet Wavelet Convolution

Assuming you haven't just started reading this book at this chapter (although it would be understandable—time-frequency analyses are great), you now know everything you need to know to perform wavelet convolution: how to create wavelets, how to perform convolution, and how to extract power and phase information from the complex dot products that result from convolution. All we need are some data. Let's start with that linear chirp from the previous chapter that produced the plateau-shaped Fourier power spectrum (see figure 18.6). We will see whether Morlet wavelet convolution is useful for characterizing nonstationary signals (spoiler: it works well providing the nonstationarities are slower than the width of the Gaussian).

The signal is the same as created in the previous chapter. We can dispense with the time-domain convolution and go straight to frequency-domain implementation, as promised by the convolution theorem.

```
nData = length(time);
nKern = length(wavtime);
nConv = nData+nKern-1;
halfwav = floor(length(wavtime)/2)+1;
as = ifft(fft(cmw,nConv) .* fft(signal,nConv));
as = as(halfwav:end-halfwav+1);
plot(time,abs(as).^2)
```

I won't reiterate the mechanics of convolution, but I would like to draw your attention to the different time vectors (and therefore different lengths) that define the signal and the wavelet. It is important, however, that they have the same sampling rate. Very important.

Figure 19.2 shows the time course of power at 6.5 Hz. Why does it peak near the center? Because that's where the chirp goes through 6.5 Hz on its way up to 10 Hz. Try re-running the code above but changing the frequency of the wavelet. You'll notice that as the wavelet frequency gets closer to 2 Hz, the power peaks earlier, and as the wavelet frequency gets closer to 10 Hz, the power peaks later. Next try changing the width of the Gaussian (the number of cycles) while leaving the frequency the same. What happens to the result, and why?

19.3 From Line to Plane

Single-wavelet analyses were hip in the 1980s, before smartphones and before Twitter (hard to believe, I know, but humans managed to squeak out

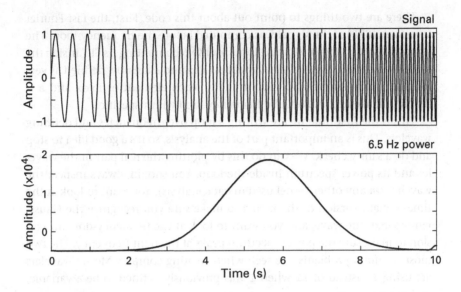

Figure 19.2

A linear chirp (top panel) and the time course of power of that signal at 6.5 Hz
(bottom panel).

a meager existence before technology saved us). To appreciate the time-
frequency dynamics of a signal, we need to see the changes over time *and*
over frequency simultaneously. This means many wavelets and a time-
frequency plane instead of a line.

There's no big mystery about how to create a time-frequency plot. You
simply repeat wavelet convolution over many frequencies and then show
the results as an image.

```
nFrex = 30;
frex = linspace(1,15,nFrex);
s = linspace(4,12,nFrex) ./ (2*pi.*frex);
sigX = fft(signal,nConv);
tf = zeros(nFrex,length(signal));
for fi=1:nFrex
    cmw = exp(2*1i*pi*frex(fi)*wavtime + ...
        -(wavtime.^2)/(2*s(fi)^2));
    cmwX = fft(cmw,nConv);
    as = ifft(sigX .* cmwX);
    tf(fi,:) = abs(as(halfwav:end-halfwav+1))*2;
end
```

There are two things to point out about this code. First, the fast Fourier transform (FFT) of the signal was computed before the frequency loop. The signal doesn't change inside the loop, and so recomputing its FFT inside the loop is redundant. Second, the variables `frex` and `s` are now defined as vectors, not single numbers. Inside the loop, each element of `frex` and `s` is called to make each wavelet be frequency- and width-specific.

This is the first time we are writing a loop over frequencies and defining wavelets. This is an important part of the analysis, so it's a good idea to stop and do a sanity check. We can do this by plotting the real part of the wavelet and its power spectrum inside the loop. You should always inspect the wavelets (or any other kernel used in data analysis). You want to look at the time-domain version of the kernel to make sure you recognize the Gaussian-tapered sine wave, and you want to look at the frequency-domain version to make sure the power spectrum peaks at the right frequency. The two most common problems I've seen when creating complex Morlet wavelets are using `i` instead of `1i` when `i` was previously defined to be a variable, and mis-specifying `s`.

```
hz = linspace(0,srate/2,floor(length(wavtime)-1)/2);
for fi=1:nFrex
    cmw = exp(2*1i*pi*frex(fi)*wavtime + ...
        -(wavtime.^2)/(2*s(fi)^2));
    cmwX = fft(cmw,nConv);
    subplot(211), plot(real(cmw))
    subplot(212), plot(hz,2*abs(cmwX));
    title([ 'Frequency = ' num2str(frex(fi)) ])
    pause
end
```

The `pause` command freezes MATLAB until you press a key on the keyboard. Run the code and keep pressing keys until the loop finishes (if you press keys and the plot doesn't update, try mouse-clicking on the figure or on the Command window; if you get bored, you can press Ctrl-c to break out of a `pause` function). If you are disappointed in what you see, then you have a good eye. There is something wrong with that code that causes a mismatch between the peak in the power spectrum and what should be the peak frequency of the wavelet.

Once you find and fix the error, you should see that these wavelets all look fine. (Hint: How do you determine the frequency resolution when creating the vector of frequencies?) Now we can move forward and look at the results.

```
contourf(time,frex,tf,40,'linecolor','none')
```

There are three striking features about this plot (figure 19.3). First of all, it looks pretty neat, like a comet careering through the sky (the color version of the figure looks better). Second, it seems to represent the chirp reasonably well, although it is smoother than the chirp itself. Here you see how wavelet convolution (and many other time-frequency methods) can smooth the results. Try changing the ranges of the number of cycles to see how this affects the smoothness of the plots. However, don't eschew the smoothing. There is noise in data, there are nonstationarities in brain signals, and there is natural variability in the precise timing and frequencies of activity over trials, conditions, and individuals. Some amount of smoothing is beneficial. I would even venture a stronger statement and say that most of the time (there are always some exceptions), having amazing precision in time and in frequency will only negatively impact your results, particularly when doing group-level analyses. By analogy, consider spatial locations: You wouldn't assume that a single neuron at some xyz coordinate has exactly the same function in every single person; instead, you assume that some patch of brain tissue has a common function, and applying a bit of spatial smoothing will help make that patch comparable across individuals.

The third striking feature of figure 19.3 is that the amplitude decreases with increasing frequency. In this case, this is not due to $1/f$ scaling, but instead is due to the wavelets not being normalized per frequency. This means that the amplitude of the original signal is incorrectly

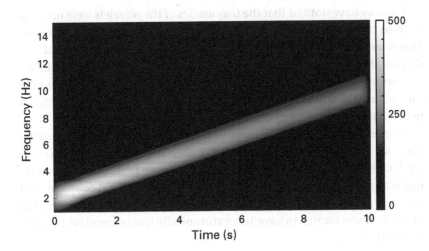

Figure 19.3
A time-frequency representation of a linear chirp, computed via Morlet wavelet convolution.

reconstructed. You can see this by adding a color bar (type `colorbar`). The color range goes from zero to somewhere around 3,000. But the amplitude of the chirp is a constant 1.0. You can also see the incorrect amplitude in figure 19.2.

The easiest way to normalize Morlet wavelets for convolution is by scaling their amplitudes to a value of one in the frequency domain. Try re-running the previous code, but this time, insert the following line after taking the Fourier transform of the wavelet.

```
cmwX = cmwX ./ max(cmwX);
```

And then re-create figure 19.3 and observe the accurate amplitude reconstruction. If you look carefully, you will observe that the reconstruction is not exactly perfect, because the frequency structure of the signal is changing faster than the width of the wavelets, which produces some inaccuracies. Still, it's pretty good. Max-value-normalizing works for wavelets that integrate to zero in the time domain. This normalization is not guaranteed to reconstruct accurate amplitudes for all kernels.

A brief tangent about amplitude versus power. Power, being amplitude squared, will exaggerate the features of the time-frequency response; that is, relatively large amplitude values will become relatively larger power values. People generally think about squared numbers as being larger than their unsquared progenitors, but remember that the square of a number between zero and one is actually smaller than the original number. Try replacing amplitude in the code above with power.

You may have noticed that the frequencies of the wavelets were not constrained the same way frequencies in the Fourier transform are constrained. This is one of the primary reasons why I wrote in the beginning of this chapter that time-frequency decomposition as presented here is a "lossy" conversion. The time-frequency plane is typically constructed using only the frequencies of interest in the data analysis, not the full possible range of frequencies.

Frequencies can be scaled linearly or logarithmically. Linearly spaced frequencies were created above, and you've seen the function `linspace` in the Fourier transform chapter as well. Now it's time to learn how to create logarithmically spaced numbers. The MATLAB function is `logspace`, and you might be tempted to use it the same way you'd use `linspace`. For example, if you wanted to have 10 logarithmically spaced numbers between 2 and 20:

```
logspace(2,20,10)
```

But this produces an unexpected result. The numbers are very large. Any idea what's going on? If you look at the first number, you might recognize that MATLAB turned your "2" into "100." 100 is 10^2. And the last number in this series is 10^{20}. The function `logspace` returns numbers as 10 to the power of the inputs. The way to use the `logspace` function as you might want to use it is to convert the inputs to base 10.

```
frex = logspace(log10(2),log10(20),10);
```

19.4 From Single Trial to Super-trial

If you have a task-related experiment, you will need to perform wavelet convolution on all trials. This means having a loop over trials, taking the FFT of the data from each trial, looping again over frequencies, and running convolution per trial and per frequency. "But Mike," you are probably thinking right now, "can't we do this without violating the avoid-loops-whenever-possible principle?" Yes we can.

This can be done by reshaping the 2D matrix of time-by-trials into a 1D matrix of time-trials. As long as the N for convolution is properly computed, the result will be comparable to the result of single-trial convolution. Let's have a look.

```
data = squeeze(EEG.data(47,:,:));
data = reshape(data,1,[]);
← convolution here →
as = reshape(as,size(data,1),size(data,2));
```

If you have multichannel data, it is also possible to concatenate data over channels to have a channel-time-trial vector. But in my experience, that ends up being such an incredibly long vector that MATLAB will run into memory problems. So you might want to keep the loop over channels.

After the analytic signal—the result of convolution between the time-trial vector and a Morlet wavelet—is reshaped back to a 2D time-by-trials matrix, the power and phase-angle time series can be extracted (figure 19.4). Trial-averaged power shows a clear stimulus-related increase in activity, as well as edge artifacts that will be discussed in the next section. Inter-trial phase clustering (ITPC) is a measure of the consistency of phase angles over trials at each time point and provides insight into the likelihood of frequency-band-specific activity taking on similar temporal configurations over trials.

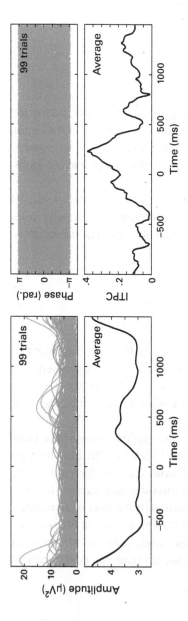

Figure 19.4

Multitrial data can be concatenated into a single vector, then convolved with a Morlet wavelet, and then reshaped back to a time-by-trials matrix (top panels; each gray line is a trial). The lower panels show trial-averaged power and ITPC (inter-trial phase clustering, which reflects the consistency of phase angles over trials at each time point).

Box 19.1

Name: Arnaud Delorme
Position: Professor
Affiliation: CerCo, CNRS, Toulouse, France
Photo credit: Arnaud Delorme

When did you start programming in MATLAB, and how long did it take you to become a "good" programmer (whatever you think "good" means)?
I learned Octave during my PhD and wrote some small programs. However, I only started to delve into MATLAB in my postdoc at the Salk Institute when I started to work on EEGLAB, which was originally only designed to streamline my own research.

What do you think are the main advantages and disadvantages of MATLAB?
I program in many other languages including Python and R. I am one of the few programmers I know who defends MATLAB.

Our users want to achieve results fast and spend the minimum time programming. Some of them have no programming experience. Python, for example, is not a language for beginners. It is too structured and is designed for professional programmers (0 for first index of matrices as in C while in MATLAB it is 1, which is more intuitive to naïve users; in Python, you have to install the numpy library to deal with matrices, which is not necessary in MATLAB; when dealing with matrices, Python also assumes object-oriented programming terminology, which can be difficult to grasp for naïve users). The indenting format of Python is also confusing for beginners (and this prevents the code from being copied and pasted to the Python command line as in MATLAB). Python is an elegant language but not one I would recommend to a novice.

R syntax is very similar to MATLAB. It is intuitive and simple. The problem with R is the quasi absence of high-performance computing solutions.

I agree that MATLAB might not be as elegant and structured as other programming languages. However, it does the job for beginners and nonprogrammers, and I think this is what counts. These users are not interested in learning to program. They want to get results with minimal programming efforts and knowledge.

Do you think MATLAB will be the main program/language in your field in the future?
In the long term, I do not think so and I do not hope so. I hope some open-source language will take precedence because the scientific community should be in charge of developing such tools.

How important are programming skills for scientists?
It depends on what level one is dwelling. The best scientists I know can program. However, they are not the best programmers either. Programming is an important skill for a scientist, but it is not the most important one.

Statistics, for example, is a skill that I think is more important than programming. Then, intuition and rigorous experimental design are also two skills that might be more important than programing. Of course, the abilities to write journal articles and to write grants are the most important skills a scientist in our society can have. This is why languages like MATLAB are important. It takes the burden off programming. It shortens the path from thinking about a way to analyze the data and implements it.

Any advice for people who are starting to learn MATLAB?
My advice would be to use your intuition and try out solutions in the MATLAB command line when you write scripts. Beginners should not start by writing a stand-alone MATLAB script or function in one shot. Beginners and even advanced users might want to copy and paste small snippets of code on the MATLAB command line to test them. When writing a MATLAB function, beginners might want to start writing a script first, then convert the script to a function once they are done. This solution makes it easier to access the variable inside the script/function while designing it.

19.5 Edge Artifacts

Where there are edges, there will be edge effects. In the previous chapter, you learned that edges can be attenuated by tapering the data. With wavelet convolution, it works a little bit differently. First of all, notice that wavelets have no sharp edges, neither in the time domain nor in the frequency domain (see figure 19.1). The edges come from the data, in particular, at the first and final data points. Whether you work with single trials or create a time-trial vector, there will be edges at the boundaries between trials.

Before talking about what to do about these edges, let's see what edge artifacts look like in the time-frequency plane. Watch what happens when the same edgy time series from the previous chapter is convolved with wavelets. Only the result is shown here (figure 19.5); one of the exercises at the end of this chapter is for you to re-create this figure. (Hint: Copy the code from earlier in this chapter and replace the data.) There are two things to notice in this figure. First of all, although this picture might make a great T-shirt, it's a horrible artifact and you definitely do not want this to contaminate your data. Depending on the size of the edge, the artifacts in the

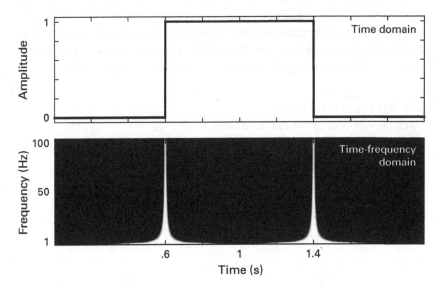

Figure 19.5
Sharp edges in a time domain have large-magnitude and multifrequency representations in the time-frequency plane. If the edges are artifacts, their time-frequency representations may overshadow the brain-related dynamics.

time-frequency plane could be orders of magnitude larger than the brain-generated dynamics in your signal.

The second thing to notice is that the artifacts do not last forever. We can use this to our advantage. Most edge artifacts subside in one or two cycles of the frequency of the wavelet (e.g., the edge artifact in a 10 Hz wavelet will last 100–200 milliseconds). One solution to edge artifacts in time-frequency analyses is to make sure they are far away from the data you want to interpret. This is accomplished by making sure the data epochs are wide enough to allow for three cycles of the lowest frequency before any time period that you care about. If the lowest frequency you want to analyze is 2 Hz, then you should make sure to have 1,500 milliseconds (500 ms per cycle times 3 cycles) before the earliest time point you want to include in your analyses (e.g., the start of the baseline period) and 1,500 milliseconds after the last time point you want to include in your analyses.

If your data are already cut into epochs that are too short, and if it is not possible to re-cut the epochs from the continuous data, you can use a procedure called reflection. Reflection is a signal processing trick to avoid edge artifacts. The idea of reflection is to make the time series three times as long by concatenating a flipped version of the time series to the beginning and to the end. It's like putting a mirror on both sides of the time series. The edge artifacts will contaminate the flipped time series, and these can then be discarded prior to analyses. For visual clarity, the exercise below will illustrate reflection using the derivative of a Gaussian (figure 19.6).

```
t = -1:1/srate:3;
x = diff(exp(-t.^2));
reflex = [ x(end:-1:1) x x(end:-1:1) ];
```

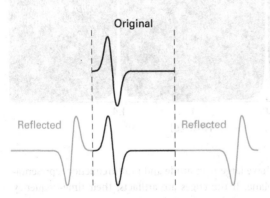

Figure 19.6
Reflection procedure that can be used to attenuate edge artifacts in short epochs.

Just make sure not to interpret any of the reflected results, lest you think that the brain works backward.

19.6 STFFT

STFFT stands for short-time fast Fourier transform (sometimes the non-furious people drop the "fast" and just go with STFT). STFFT is a different technique for extracting time-frequency information from a time series, but it produces qualitatively similar results as complex Morlet wavelet convolution. Therefore, I will review the STFFT method only briefly as an excuse to go over a bit more MATLAB.

The idea of the STFFT is to compute the FFT on a limited window of time; for example, a few hundred milliseconds. The power spectrum from that window is then stored in a separate matrix. Then the time window is shifted by some amount (e.g., 100 milliseconds), and an FFT is taken on the data from that window. This creates a time series of power spectra from temporally shifted FFTs. The code below will set up the timing parameters for the STFFT, including the window width, the number of time points around which to compute the FFT, and the frequencies vector (remember that the frequency resolution is defined by the smaller time windows, not the length of the entire time series).

```
fftWidth_ms = 1000; % FFT width in ms
fftWidth = round(fftWidth_ms/(1000/srate)/2); % in ms
Ntimesteps = 50; % number of time widths
ct = round(linspace(fftWidth+1,n-fftWidth,Ntimesteps));
% figure out Hz vector for our FFT width
hz = linspace(0,srate/2,fftWidth-1);
% Hann window for tapering
hwin = .5*(1-cos(2*pi*(1:fftWidth*2)/(fftWidth*2-1)));
```

After specifying parameters, we loop through time points (the center time points are in variable `ct`), extract a window of 1,000 milliseconds around each center time point, apply a Hann taper, and then compute the FFT. A question for you—when defining the center time points, why doesn't the code take linearly spaced points between 1 and *n*?

Now we're ready to go. Let's test the STFFT on the same chirp signal used earlier (variable `signal`). That will facilitate a comparison between Morlet wavelet convolution and STFFT. It's always a good idea to test multiple analysis methods on the same data.

```
tf = zeros(length(hz),length(ct));
for ti=1:length(ct)
    tdat = signal(ct(ti)-fftWidth:ct(ti)+fftWidth-1);
    x = fft(hwin.*tdat)/fftWidth;
    tf(:,ti) = 2*abs(x(1:length(hz)));
end
```

A few things to notice about this code. The width of the FFT provides the temporal precision, but it also constrains the frequency resolution. There is a trade-off here between having shorter windows to increase temporal precision versus having longer windows to increase frequency resolution. Common values are in the range of hundreds of milliseconds, but these windows should be sized according to your expectations about the rate of changes in the data.

The temporal resolution of the result is determined by the center time points (variable ct). This means that the time-frequency result will not have the same resolution as the original data. This usually isn't a problem because of the temporal smoothing inherent in time-frequency analyses. But this feature makes STFFT different from complex Morlet wavelet convolution, which produces an estimate of time-frequency activity at each time point.

Next is the loop over time points. This is different from wavelet convolution, in which there is a loop over frequencies. Here, the frequencies are all computed simultaneously during the FFT. At each iteration inside the loop, the data corresponding to the ti^{th} time window needs to be cut out and tapered with a Hann function (or any other tapering function), and then the FFT can be computed as you have previously done. The example above has only a single trial of data, but as you know, the fft function works on 2D data. Just make sure to compute the FFT over the correct dimension.

To plot the results, you need to compute the vector of frequencies in hertz. Make sure to compute the hz vector based on the number of time points in the FFT, not the total number of time points in the entire trial. Results are shown in figure 19.7.

Given certain parameter selection, the STFFT can produce the same results as wavelet convolution and other time-frequency analysis methods (Bruns 2004). There are, however, several important differences in implementation.

1. In wavelet convolution you get to define the frequencies, whereas the frequencies in the STFFT are defined by the number of time points in the window.

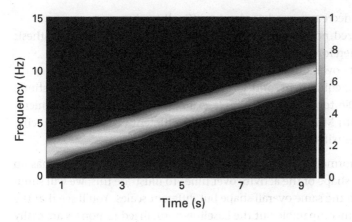

Figure 19.7
Results of the STFFT applied to the linear chirp. You can compare this result with that in figure 19.3.

2. Wavelet convolution natively provides time-frequency results with the same temporal resolution as the original data. Obtaining high temporal-resolution power estimates from the STFFT requires moving the time window forward by the smallest time step, which is time-consuming relative to wavelet convolution, though simple to implement.

3. Wavelet convolution can estimate activity at all time points, even the first and last time points. The STFFT cannot estimate activity within one half of the width of the time window at the beginning and end of the signal because each FFT has a center time point. These edges are typically contaminated by artifacts and are therefore typically uninterpretable, but having the same time points often simplifies subsequent analyses.

19.7 Baseline Normalization

So far in this chapter, we've been computing "raw" power. There are several limitations of interpreting and analyzing raw power that motivate computing and interpreting normalized power:

1. $1/f$ scaling of brain activity impedes direct comparisons of activity across frequencies.

2. Ongoing task-unrelated activity can obscure smaller (and possibly more theoretically meaningful) effects.

3. Relative decreases in power can be difficult to observe.
4. Normalized power is normally distributed under the null hypothesis and is therefore more amenable to statistical evaluation.
5. Absolute power differences across experiments, electrodes, individuals, recording equipment, and so forth, are often difficult and sometimes impossible to interpret. A simple example is comparing EEG (microvolts) with MEG (picoteslas). Relatedly, different electrode referencing montages will change the "absolute" power values.

Baseline normalization shifts and scales the data on the y axis. It has no effect on the shape of the activity over time. To illustrate this, we will simulate data with the same overall shape but different scales. You'll see that the raw data are incomparable but the baseline-normalized responses are easily visually and numerically comparable. Our two signals will be the same sinc function with different DC offsets.

```
time = -2:1/100:5; % 100 Hz sampling rate
mothersig = sin(2*pi*5* (time-3))./(time-3);
sig1 = mothersig + 2000;
sig2 = mothersig + 200;
```

Next, we apply a baseline normalization. Below I'll use decibel (dB) normalization, which is defined as $10 \log_{10}(a/b)$, where a is the activity and b is the baseline. The baseline will be defined as the average signal between −1.5 and −0.5 seconds.

```
basePow1 = mean(sig1(:,baseidx(1):baseidx(2)),2);
sig1DB = 10*log10(bsxfun(@rdivide,sig1,basePow1));
basePow2 = mean(sig2(:,baseidx(1):baseidx(2)),2);
sig2DB = 10*log10(bsxfun(@rdivide,sig2,basePow2));
plot(time,sig1DB,time,sig2DB)
```

It is clear from figure 19.8 that the signals are comparable only after baseline normalization. Zooming in to one of the signals will show that the overall shape of the function has not changed (hint to people reading this in the lab: this works better using the zoom function on your computer screen than putting this book under a microscope).

The important part of normalization is to divide the "interesting" activity by the "baseline" activity. This division provides the scale- and frequency-independent normalization; a linear subtraction of the baseline will not suffice. If you have a single time series, then decibel normalization involves dividing the entire time series by a single number. In case of multichannel or multicondition data, this can be simplified to dividing each row

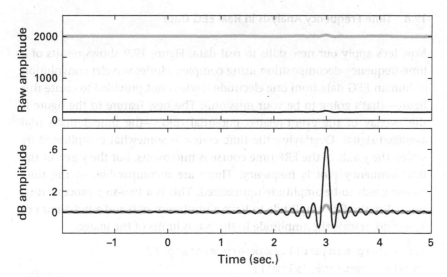

Figure 19.8
The top panel shows two time series that have large offsets. It's difficult to compare their fluctuations visually. Baseline-normalizing the time series (lower panel) puts both time series on the same scale and therefore facilitates direct comparison.

in a matrix by each element in a vector. That requires expanding the vector, which can be accomplished using the `bsxfun` function (you'll get to practice this in the exercises). Taking the estimate of the baseline activity as the average activity over a time period (e.g., –300 to –100 ms) rather than as a single time point (e.g., only –100 ms) is generally a good idea to increase the signal-to-noise characteristics of the baseline.

Decibel conversion involves taking a logarithm and therefore works only for positive values (the log of negative values is undefined, and produces `Inf`, or infinity, in MATLAB). This normalization is therefore most frequently used for power, because power values cannot be negative.

Decibel normalization is not the only valid method for baseline normalization. Percent change, for example, works the same way as decibel normalization, except the formula is different: 100*(activity – baseline)/ baseline. Notice that both decibel and percent change normalizations involve dividing by the baseline activity. Sometimes, z-normalizations are also used (subtracting the mean baseline power and dividing by the standard deviation in the baseline period). This method can be suboptimal because it is sensitive to the variance in the baseline period.

19.8   Time-Frequency Analysis in Real EEG Data

Now let's apply our new skills to real data. Figure 19.9 shows results of a
time-frequency decomposition using complex Morlet wavelet convolution
of human EEG data from one electrode (code is not provided to create this
figure—that's going to be your mission!). The new feature in the figure is
the overlay of the event-related potential (ERP)—the time-domain trial
averaged signal. Overlaying the time course is somewhat complicated by
scales: the y axis of the ERP time course is microvolts, but the y axis of the
time-frequency plot is frequency. Those are incomparable, so the time
course needs to be amplitude-normalized. This is a two-step process: first,
normalize the ERP amplitude to have a minimum of 0 and a maximum of
1; second, rescale the amplitude to the y-axis limits of the image.

```
erp = (erp-min(erp))./max(erp-min(erp));
yscale = get(gca,'ylim');
erp = erp*(yscale(2)-yscale(1))+yscale(1);
```

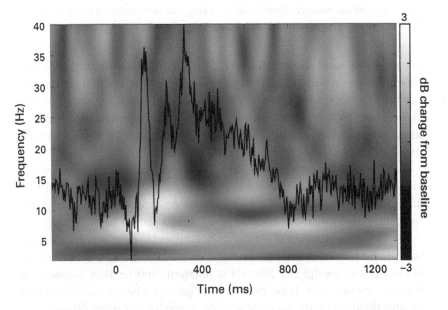

Figure 19.9
Time-frequency power map of human EEG data. Overlaid on top is the event-related
potential, the time-domain average signal.

19.9   Exercises

1. In the code for figure 19.5, uncenter the Morlet wavelet by defining wavtime from –2 to +3 seconds (also try other ranges like –2 to +2.1). How does this change the time-frequency plot, why does this happen, and what is the important lesson about constructing wavelets?

2. Write code to produce figure 19.9.

3. Adjust the code for the STFFT to estimate power at the first and last time points. (Hint: Try reflection.)

4. Adapt the code from figure 19.5 by replacing the boxcar function with a sine wave at 30 Hz. Confirm that the time-frequency result is an accurate representation. Then decibel-normalize the result. First use a baseline period of 0 to 0.2 seconds, then use a baseline of 0.2 to 0.5 seconds, and finally use a baseline of 0.8 to 1.2 seconds. Why do the results look so different for these different baseline periods?

5, Can you rescale the ERP normalization (figure 19.9) in one step instead of two? Either way, rescale the ERP to span 50% of the y axis instead of 100%.

6. In the code for figure 19.7, experiment with different values for the fftWidth parameter. What are reasonable and unreasonable ranges?

7. Reproduce figure 19.9 using decibel, percent change, and z-normalization for baseline normalizations. The numerical ranges of these different normalizations differ, so each plot will need its own color scaling. Nonetheless, they can be visually compared. Which methods are most similar to each other and why?

8. One common mistake when defining the width of the wavelet is to write 5/2*pi*f instead of 5/(2*pi*f). MATLAB interprets these two expressions, respectively, as (5*pi*f)/2 versus 5/(2*pi*f). Create a wavelet with these two pieces of code. How would you sanity check the result to make sure you've done it correctly?

9. The power spectrum of the "triangle" signal from figure 18.6 is nearly uninterpretable. Apply a time-frequency analysis of this signal. Does this better capture the dynamics in the data?

10. Perform a PCA of the human EEG data, then apply a time-frequency analysis on the first two component time series (i.e., the two components with the largest eigenvalues). Do the results depend on whether you compute the covariance matrix of the trial-averaged data versus the average of single-trial covariance matrices?

11. There is a MATLAB function called chirp, which creates a chirp. On the basis of the help information, create a chirp using the same

parameters as for the chirp we created manually in this chapter. Plot them against each other. Do they match? (The answer is no.) How much are they off, and how can you adjust your code to match the output of `chirp`?

12. Without reflection, the earliest and latest time points that can be extracted from time series data when using the STFFT is constrained by the width of the FFT window. Write code to determine the earliest and latest center time points given a specified time window width.

13. Put the two sinc signals used in figure 19.8 into a 2-by-time matrix. Then use the `bsxfun` function to apply decibel-normalization in one line of code. Check that the results are the same as when decibel-normalizing each line separately.

14. To create a Morlet wavelet, point-wise multiply a _____ by a _____. The frequency of that wavelet is determined by the frequency of the _____.

15. What are the five steps of implementing convolution in the frequency domain? Which of these are necessary and which are optional?

16. There is at least one mistake in each of the following MATLAB code paragraphs (one in part a and one in part b). Identify and correct the two mistakes.

    a.

```
fftEEG = fft(csd(7,:,10),n_conv);
fftWave = fft(wavelets(3,:),n_conv);
as = ifft(mean(fftEEG.*fftWave),n_conv);
```

    b.

```
sine_wave = exp(2*1i*pi*frequency(10).*time);
gaus_win = exp(time.^2./(2*(6/(2*pi*frex(10)))^2));
wavelet = sine_wave .* gaus_win;
```

17. In the following two pairs of code, one line is correct and one contains an error. Find and fix the errors.

```
hz = linspace(0,nyquist,floor(N/2)+1);
hz = linspace(0,nyquist,floor(N/2)-1);
n_conv = size(csd,3) + length(wavelet_time) - 1;
n_conv = length(size(csd,3))+length(wavelet_time)-1;
```

18. Take the time-frequency representation of the linear chirp using Morlet wavelet convolution, and average the power values over all time points. Then take the FFT of the entire chirp. Plot both power spectra

on top of each other (note that the frequencies vectors will be different). How similar do they look? Try changing the maximum number of cycles for the wavelets to make the two results more similar. What does this tell you about the relationship between time-frequency analyses and the FFT?

19. The online code contains the following line in the section that creates figure 19.6. What is the meaning of the first input in the plot function?

```
plot(length(g)+1:2*length(g),g,'ko')
```

20 Time Series Filtering

Time series data in neuroscience are filtered either as a pre-processing strategy (e.g., to attenuate noise in the data) or as part of data analyses (e.g., to focus analyses on frequency-specific components of the data). In the previous chapter, you learned about filtering via complex Morlet wavelet convolution; here, you will learn additional approaches for filtering time series data.

20.1 Running-Mean Filter

This filter is also variously called moving-mean, moving-average, sliding mean, and so forth. I prefer "running-mean" because it feels healthier.

A running-mean filter is a simple low-pass filter that has the effect of smoothing the time series. It involves replacing each data point by the average of the surrounding k data points. Called the *order* parameter, k is the only parameter in this filter. As the order of the filter gets larger, the smoothness increases. To illustrate this filter, we'll apply it to a sine wave with noise. And to make things more interesting, we'll use a random amplitude-modulated sine wave by interpolating a few random values.

```
srate = 1000;
time  = 0:1/srate:6;
ampl  = interp1(0:5,rand(6,1),time,'spline');
f = 8; % Hz because time is in seconds
noise = 3*randn(size(time));
signal = ampl .* sin(2*pi*f*time) + noise;
```

Before moving on to the filter, let me point out the use of .* and * in the definition of this signal. Inside the sin function, the .* was not necessary because 2*pi*f together make a scalar (they are just single numbers that multiply to make another single number). That scalar multiplies each element in the vector `time` (in other words, this is a scalar-vector multiplication). The output of the `sin` function is a 1-by-N vector, and we want to point-wise multiply that vector by `ampl`, which is also a 1-by-N vector. If you use a * instead of .*, you will get an error (to understand why, recall the rules of matrix-matrix multiplication). You could "fix" this error by transposing `ampl` or the output of the `sin` function, but this would produce the dot product or the outer product. Neither is what we want; we want point-wise multiplication. Now onto the filter.

```
k = 7;
filtsig = zeros(size(signal));
for i=1:length(time)
    filtsig(i) = mean(signal(i-k:i+k));
    %filtsig(i) = sum(signal(i-k:i+k))/(k*2+1);
end
```

Run the code. What happened? MATLAB crashed. Try to figure out why it crashed before reading the next paragraph.

MATLAB crashed because at the first iteration of the loop, MATLAB tried to access the $(i-k)^{th}$ element in the matrix, which is $1 - 7 = -2$. The solution is to start the loop at `k+1` (not at `k`, because then `i-k=0` and you would get the same error). For the same reason, the loop should end not at `length(time)`, but at `length(time)-k`. (This is another example of how filtering produces uninterpretable data at the edges.) The commented line produces the same result as the previous line and is included in case you needed a refresher on what `mean` means. The original and filtered signals are shown in figure 20.1.

In the code, k is set to 7. Does this mean the filter had an order of 7? How many points are averaged together at each time step? That's right, the filter order is actually $2k + 1$ in the code above.

Before moving forward, play around with this code a bit. Trying changing `k` and the magnitude of the noise (the hard-coded parameter "3" in the variable `noise`). The running-mean filter is very popular, in part because it is simple to implement. Do you think it works well? Does your opinion depend on `k` and on the amount of noise?

The running-mean filter can be slow for long time series, in part because it is done in the time domain. Let's think for a minute about whether it might be possible to speed this up, perhaps using the convolution theorem. Remember that the idea of time-domain convolution is to compute a time series of sliding dot products between a kernel and a signal. Now here is the million-dollar question: Can we interpret an average as a dot product, and

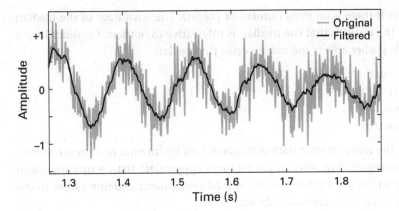

Figure 20.1
Illustration of a running-mean filter. Random noise (drawn from a normal distribution, meaning there are both positive and negative values) was added to a sine wave. The running-mean filter successfully attenuated the noise while preserving the signal.

if so, what is the kernel? (Okay, that was two questions; each can be worth $500,000).

The key realization is that all data points in an average are equally weighted, so the shape of the kernel is a boxcar with amplitude of $1/N$ (in other words: multiply each of N numbers by $1/N$ and then sum). This means we can perform the running-mean filter in the frequency domain. You'll have the chance to do this in the exercises.

Now that you appreciate that the mean over a window of time is actually just a dot product with a plateau-shaped kernel, the running-mean filter can be extended from an unweighted filter to a weighted running-mean filter. For example, the weighting can be based on distance away from the center time point. I hope you see the bigger picture here, which is that many manifestations of temporal filtering can be thought of and implemented as convolution. The differences among different types of filters lie more in the shape of the kernels than in their mechanical implementations.

20.2 Running-Median Filter

The running-median filter is very similar to the running-mean filter, except that it uses the median instead of the mean. The median of a distribution is defined as the center of that distribution (or the mean of the two center values if there is an even number of points). The advantage of the median over the mean is that the median is insensitive to outliers. Consider how a single outlier affects the mean versus the median.

```
q = randn(100,1);
q(50) = 1000;
mean(q)
median(q)
```

If the noise in your data is characterized by infrequent extreme values, a running-median filter might be more appropriate than a running-mean filter. In the example below, we will add infrequent extreme values to the same noisy sine wave used above.

```
signal = signal + 1000*isprime(1:length(time));
```

The function isprime returns a vector of trues and falses according to whether each input number is a prime number. These logical values are then treated as the numbers 0 and 1 when multiplied by 1,000. Now we apply the median filter.

```
for i=k+1:length(time)-k-1
    sortnums = sort(signal(i-k:i+k));
    filtsig(i) = sortnums(k);
end
```

Instead of explicitly using the function median, we computed the median by taking the middle value of the sorted vector. There is no major advantage of this code over the median function; I just wanted you to see that computing the median is simple. Figure 20.2 shows the unfiltered, mean-filtered, and median-filtered signals. You can see that the running-mean filter gave undesirable results.

If you have spikes in your data or other extreme values, it may not be necessary to apply the median filter to all time points as shown in the code above. Instead, you can identify only the time points that need to be filtered and take the median around those points. The exercises explore this idea further.

Is it possible to perform the running-median filter in the frequency domain? To answer this question, you need to think about whether the median is a linear function, and if so, what the kernel is. Sadly, the answer is no. The median is a nonlinear operation, while convolution is a linear operation.

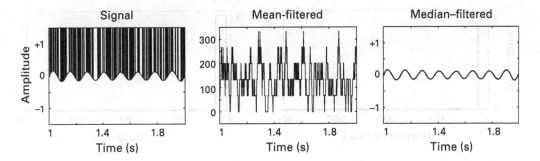

Figure 20.2

Illustration of a running-median filter. In this example, large noise spikes were added to the sine wave signal. The mean-based filter, being sensitive to outliers, does a horrible job at removing the noise. The running-median filter worked like a charm. Note the *y*-axis limits in the three plots.

20.3 Edges in the Frequency Domain

You've already learned that sharp edges in the time domain produce features in the frequency domain that can be artifacts when they overshadow more subtle effects (see figure 18.1). The reverse is also true—sharp edges in the frequency domain can cause large-amplitude features in the time domain. These features take the form of ripples. You can imagine that these features are artifacts in the context of time-frequency analyses, because oscillatory features in the data should come from oscillations in the brain, not from artifacts of filter edges. For this reason, a lot of fuss in filtering involves minimizing edge effects.

To gain an intuition of the kinds of time-domain artifacts that result from frequency-domain edges, let's create an edge in the frequency domain and then consider its time-domain representation (figure 20.3).

```
N = 400;
X = zeros(N,1);
X(round(N*.05)) = 1;
subplot(221), plot(X)
subplot(223), plot(abs(ifft(X))*N)
```

Notice how I set the frequency of the spike to be at whatever frequency is 5% of the length of the signal. What are the advantages and disadvantages of this choice?

A straight edge in the frequency domain produces an "artifact" in the time domain that is a pure sine wave. That's an interesting way to think

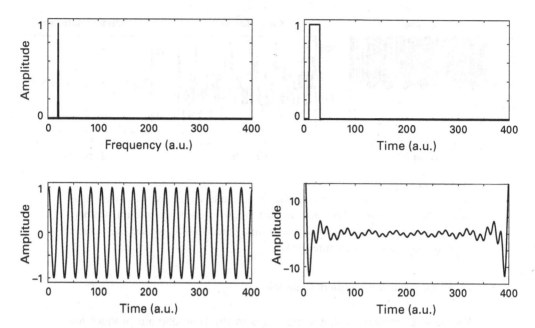

Figure 20.3
Edges in the frequency domain (top row) produce ringing in the time domain (bottom row). Although these are accurate reconstructions, if they are introduced into the data because of filtering, these ringing features could be misinterpreted as oscillations. Notice the difference in y-axis scaling in the two time-domain plots.

about sine waves—they are the artifactual manifestation of an impulse function in the frequency domain. We can also see what happens when we widen this impulse response to a boxcar shape.

```
X = zeros(N,1);
X(10:30) = 1;
```

Here I hard-coded which frequency values should be set to 1. Try evaluating the code to produce figure 20.3 and changing the parameter N. Which method is dependent and which is independent of the frequency resolution?

From these two examples, you can appreciate the trade-off that must be considered when constructing narrow-band filters: You want the filter to be narrow to maximize frequency specificity, but you want the filter to have gentle slopes rather than sharp edges, although this reduces frequency specificity. What shape might come into mind when you think about a narrow peak that has smoothly decaying sides?

20.4 Gaussian Narrow-Band Filtering

If you weren't already thinking that the answer is a Gaussian, then I'm sure you are thinking it now. Perhaps you remember from the previous chapter that Morlet wavelets have a Gaussian shape in the frequency domain. Here we will construct a similar filter, except we define the kernel in the frequency domain instead of in the time domain.

Whereas time-domain Morlet wavelets are defined by the number of wavelet cycles, frequency-domain Gaussian filters are defined by their full-width at half-maximum (FWHM). What is FWHM? It is a description of the width of a Gaussian (or Gaussian-like function). If you have a pure Gaussian, FWHM can be computed analytically, but in practice—and certainly when looking for an excuse to write a few lines of MATLAB code—it is better to measure it empirically.

FWHM is the distance between the 50% amplitude points before to after the peak (figure 20.4). If the Gaussian is normalized to have an amplitude of 1.0 (which can be obtained by dividing by the maximum), then the task is simplified by finding the point before the peak and after the peak closest to 0.5. We start by creating a Gaussian and maximum-normalizing it.

```
x = -4:.1:4;
gaus = exp(-x.^2);
gaus = gaus./max(gaus);
```

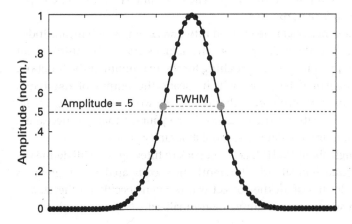

Figure 20.4
Demonstration of empirical FWHM of a Gaussian. In this case, the empirical estimate will be slightly off from the theoretical value, because there are no data points at exactly 0.5.

A few things about this code: First, before I wrote this code, I thought that gaus seemed like it could be the name of a function. Typing which gaus confirmed that it's safe to use this name for a variable. Second, what is the *s* parameter of this Gaussian? Third, truth be told, we didn't need to normalize this Gaussian (try re-running the code without that scaling line and you'll see that it's already scaled to 1), but it's good practice to remember that our empirical FWHM procedure requires a normalized Gaussian. Depending on how you define it, your Gaussian might not be amplitude-normalized.

Our next step is to find the indices before and after the peak that are closest to 0.5. You might think of using min-abs or dsearchn, but those will give only one result corresponding to whichever side of the Gaussian happens to be a tiny bit closer to 0.5 (don't trust me—try it yourself—and then look forward to learning a solution to this problem in chapter 21). A better strategy is to find the point closest to 0.5 only before the peak, and then in a separate line of code, find the point closest to 0.5 after the peak.

```
[~,pidx] = max(gaus);
prepeak = dsearchn(gaus(1:pidx)',.5);
postpeak = pidx-1+dsearchn(gaus(pidx:end)',.5);
```

The first two lines should be pretty easy to understand. We find the index of the peak of the Gaussian, and then we use dsearchn to find the point closest to 0.5 prior to that peak. The third line is a bit more complicated. Why is there a pidx-1+?

Only the post-peak data are entered into dsearchn, which means index 1 in dsearchn is really index pidx in the actual signal. Next, think about what would happen if we were searching for the maximum of the function. The maximum would be pidx, which means the output of dsearchn would be 1, which in turn means that postpeak would be pidx+1. Try re-running this code without the –1 to convince yourself that it is necessary. You can plot and use the datacursor tool as a sanity check.

Okay, enough about FWHM. Let's get back to filtering. We will define the width of the Gaussian slightly differently here compared to the previous chapter, in order to obtain the correct width when specified in hertz, and when the Gaussian is maximum-value-normalized.

```
srate=1000; N=4000;
hz = linspace(0,srate,N);
s = fwhm*(2*pi-1)/(4*pi); % normalized width
x = hz-f; % shifted frequencies
```

```
gx = exp(-.5*(x/s).^2); % gaussian
gx = gx./max(gx); % gain-normalized
```

There are two things I'd like to point out about this code. First, I specified the frequencies vector to go up to the sampling rate. As written in section 11.6 of chapter 11, this is generally something you should avoid, but it's convenient in this case because we want the Gaussian to have as many points as the data. Second, the exponential is coded slightly differently but is equivalent to what you saw previously. You can practice your math skills by converting the following two lines of code to formulas on a piece of paper, and then applying some simple algebra to prove that they are identical.

```
fx = exp(-.5*(x/s).^2);
fx = exp(-(x.^2)/(2*s^2));
```

This Gaussian will be our filter kernel, and filtering now proceeds as discussed for convolution—element-wise multiply this Gaussian by the Fourier coefficients of the data. Figure 20.5 shows the frequency-domain filter kernel and the result of applying this filter using a similar linear chirp as what we used in the previous chapter.

If this is convolution, why don't we have $N + M - 1$? Because here we are performing *circular* convolution, not *linear* convolution. The beginning and end points of the convolution get wrapped around, as if the signal were circular. You can see this in figure 20.5. This does mean that the beginning and end of the signal should not be interpreted because of the wraparound summation, but that was also the case with linear convolution. More generally, filtering methods nearly always produce uninterpretable results at the edges; you should always design your analyses with this in mind.

Applying this filter to data involves point-wise multiplying this filter by the Fourier transform of the signal and taking the inverse Fourier transform. You'll need to double the amplitudes of the result, because our filter has only non-zero values between DC and Nyquist—it's missing the corresponding shape in the negative frequencies range. Recall the discussion in chapter 11 that the amplitudes of a real-valued function get split between the positive and negative frequencies.

The Gaussian narrow-band filter has three advantages. It is easy to construct and implement in MATLAB; it produces minimal ripple artifacts in the time-domain reconstruction, because there are no edges in the frequency domain; and it is a non-causal filter, meaning it does not produce any phase distortions in the result (although it does mean that there is temporal leakage "backward" in time as well as forward in time). There are

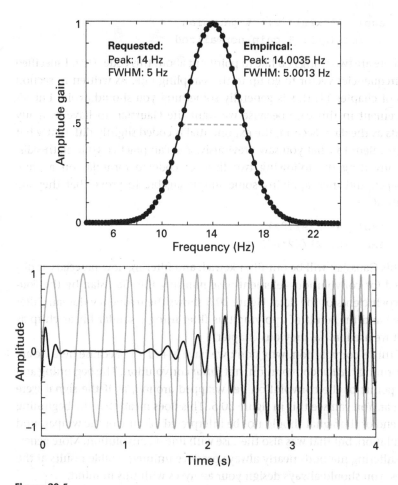

Requested:
Peak: 14 Hz
FWHM: 5 Hz

Empirical:
Peak: 14.0035 Hz
FWHM: 5.0013 Hz

Figure 20.5
The top panel shows the frequency-domain Gaussian kernel and its requested and empirically measured properties. The lower panel shows the result of a circular convolution between this Gaussian kernel and the Fourier transform of a linear chirp. The gray line shows the original chirp, and the black line shows the filtered time series.

two disadvantages. The result is real-valued, so you'll need to apply the Hilbert transform if you want to extract power and phase (more about this in a later section). Perhaps the main disadvantage is that you cannot control the shape of the filter. The Gaussian filter is—I feel silly writing this but it segues to the next section—shaped like a Gaussian. If you want one filter that spans, say, 30–60 Hz, a Gaussian filter is a suboptimal choice.

20.5   Finite Impulse Response Filter

Let's say you don't like Gaussians for some reason. Or maybe you think Gaussians are okay but you want a filter that is less restrictive than a Gaussian. In these cases, you want the freedom to define the shape of your filter. You want a finite impulse response (FIR) filter.

FIR filters can produce qualitatively the same results as wavelet convolution, Gaussian filters, and short-time Fourier transform, given certain parameter settings (Bruns 2004). Rather than implementing frequency-domain multiplication, however, FIR filters work by setting each time point to be a weighted sum of previous values (the number of previous values is called the *order* of the filter), and the weights are defined in a way that extracts certain frequencies or frequency ranges from the past values. Because each time point is a weighted sum of *previous* time points, the filter is causal, and therefore introduces phase shifts into the filtered signal. To make the filter non-causal, the filtered signal is flipped backward, filtered again, and then flipped forward again. If you want a causal filter, then the filter is applied only in one direction.

There are two steps to FIR-based filtering in MATLAB: First, specify the shape of the frequency response of the filter and the corresponding frequencies normalized to the Nyquist frequency. Second, apply the filter kernel to the data.

We start with step 1. We want our filter to have a "boxcar" shape that allows frequencies between 12 Hz and 18 Hz to pass through while attenuating frequencies below 12 Hz and above 18 Hz.

```
freqshape = [ 0 1 1 0 ];
frequencies = [ 0 12 18 srate/2 ];
```

Why are four numbers there when I wrote that we want to filter between 12 and 18 Hz? Because the shape needs to be defined across the entire frequency spectrum, going from DC to Nyquist. But this isn't a good filter shape. To see why, type `plot(frequencies,freqshape)`. This filter moves gradually from DC to 12 Hz, and then gradually from 18 Hz to

Nyquist (figure 20.6). This filter lacks frequency specificity, and the results of such a filter will be difficult to interpret. A better approach is to add two additional anchor points to make the desired frequency response look more like a boxcar. Not too boxy, though—we don't want to have sharp edges in the frequency-domain response. We will use *transition zones* of 15% of the edges. This will give us a smoother filter kernel shape.

```
tz = .15; % trans. zone, in percent
fbnd = [ 12 18 ]; % freq boundaries
freqshape = [ 0 0 1 1 0 0 ];
frex=[0 fbnd(1)*(1-tz) fbnd fbnd(2)*(1+tz) srate/2];
frex = frex./(srate/2); % norm. to Nyquist
plot(frex,freqshape)
```

Now we're getting somewhere. This frequency shape is still a bit edgy, although the edges are not so severe. We will next use this shape as an input into one of several MATLAB functions that will compute a time-domain filter kernel based on this shape. The filter kernel will be designed to maximize the gain in the frequency ranges of interest, minimize the gain in the frequency ranges not of interest (those associated with zeros), and smooth the sharp edges in between. Notice also in the code above that the vector of frequencies is now divided by the Nyquist frequency. This converts the frequencies from hertz to fraction of the Nyquist. This makes life easier for the

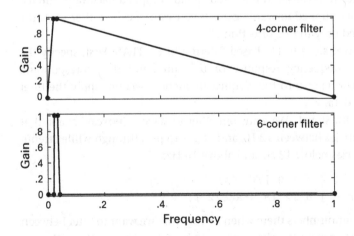

Figure 20.6
The top panel shows the frequency response of a poorly designed filter kernel. The lower panel shows a better design.

filter kernel construction algorithms, because they can optimize any filter kernel without knowing the sampling rate.

There are several MATLAB functions that will compute filter kernels, including `firls`, `firpm`, `fir1`, and so forth. Most are provided in the signal processing toolbox. If you do not have this toolbox, you can use the Octave signal package. In the interest of brevity, I'll focus only on one filter kernel method, `firls`, which stands for FIR least squares. Other filter kernel construction functions work similarly from the MATLAB implementation perspective. The function `fir1` is slightly different because you do not specify transition zones; instead, `fir1` calls `firls` with no transition zones and then smooths the resulting kernel to soften the edges.

The last piece of information we need is the order of the filter. The order determines the number of time points in the filter kernel, which means it also determines the frequency resolution of the kernel. For real-time online filtering, smaller orders are preferred because of the reduced computation time (IIR filters, for example, have small orders and are often used for real-time filtering applications). For off-line filtering, this is generally not a significant concern. The function `firpmord` estimates a filter order for the filter kernel construction function `firpm`, but these estimates are more suggestions than mathematical requirements. You might want to specify a longer or shorter order to control the filter precision, and anyway, within reasonable ranges, many different order parameter values will produce the same qualitative pattern of results (this point is discussed more at the end of this section).

We will set the order to be the number of points corresponding to three cycles of the lower bound of the filter. For a filter of 10 Hz, that corresponds to $3 \times 100 = 300$ milliseconds.

```
x = chirp(time,5,time(end),20);
ford = round((3*1000/frange(1)) / (srate/1000));
fkernel = firls(ford,frex,fshape);
```

Is three a good factor for the filter order? That's a question without a straightforward answer. Good ranges of order depend on, among other features, the frequency and the sampling rate. But you can use the code that produces figure 20.7 to gain some intuition for how this parameter controls the frequency precision of the filter. Try changing the 3 to other numbers between, say, 1 and 10 (need not be integers), and watch the effect of that parameter on the time-domain and frequency-domain versions of the figure. Also try changing the sampling rate and the frequency bounds.

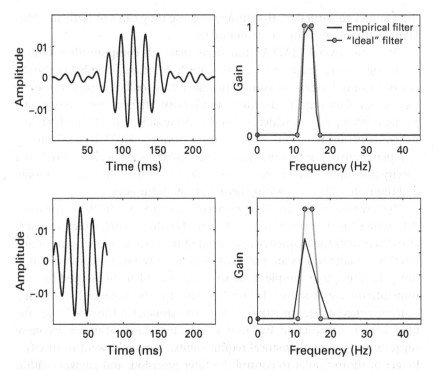

Figure 20.7
Two filter kernels were constructed that differed only in their order parameter (3 on top, 1 on bottom). The left-side plots show the time-domain filter kernels. The right-side plots show the requested (gray lines) and actual (black lines) frequency responses of the kernel.

Now that we have our filter kernel, we can use MATLAB's `filtfilt` function (also in the signal processing toolbox) to apply the filter to some data. The function is called `filtfilt` because it forward-filters the data, flips the data backward, filters again, and flips the data forward again. Forward-only filtering can be done with the function `filter` (not in the signal processing toolbox). For the sake of comparison, we'll filter the same linear chirp from earlier in this chapter (figure 20.8).

```
fdata = filtfilt(fkernel,1,x);
```

A final note about filter parameter selections. Occasionally in the psychology-EEG literature, there are strong opinions about which kernel construction functions and parameters should and should not be used. It's a strange debate because if you talk to enough people, you will hear equally strong opinions on opposite sides of the arguments ("always use causal

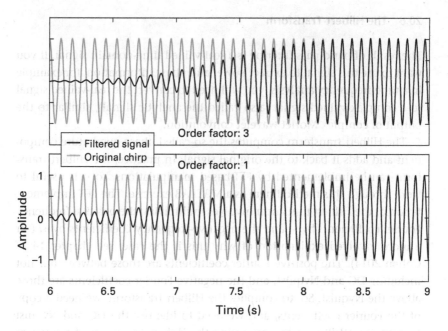

Figure 20.8
Applying a band-pass filter to the linear chirp. This is a zoomed-in view to highlight the differences. You can use the figure pan and zoom options to see that the amplitude of the filtered signal attenuates when the chirp is slower than 13 Hz and faster than 15 Hz.

filters" vs. "never use causal filters"; "filter forward and then correct using group delay" vs. "filter in reverse because group delay depends on frequency"; "always use `fir1`" vs. "use `firls` or `firpm` because you have more control" [this isn't even a sensible argument because `fir1` calls `firls`] and so on). Certainly, filter parameters are important and certainly it is possible to design bad filters that will produce bad results.

Ultimately, however, you should trust *findings*, not *parameters*. If your result disappears or qualitatively changes after minor changes to filter parameters, do not trust that result. The results you should trust, and the results you want your name associated with, are those that can be replicated regardless of who is doing the analysis, and regardless of whether you use wavelet convolution, Gaussian filter, FIR filter, an order factor of 3 versus 3.4, and so on.

If you would like to learn more about the signal processing and mathematical details of filtering as used in neuroscience, consider the book from van Drongelen (2006).

20.6   The Hilbert Transform

The result of band-pass filtering is a real-valued time-domain signal. If you want to use this signal to extract power and phase information, for example in a time-frequency analysis, you will need to convert the real-valued signal to a complex-valued signal (also called the analytic signal), similar to the output of complex Morlet wavelet convolution.

The Hilbert transform computes the so-called phase quadrature component and adds it back to the original signal. In practice, the Hilbert transform can be implemented by FFT-based manipulations, so it is useful to work through this method here as a MATLAB exercise. The FFT implementation of the Hilbert transform involves multiplying the positive-frequency Fourier coefficients by $-i$ and the negative-frequency coefficients by $+i$ (for a more involved explanation of why this is the case, see chapter 14 of Cohen 2014). The positive Fourier coefficients are those between but not including DC and Nyquist, and the negative Fourier coefficients are those above the Nyquist. So, to compute the Hilbert transform, we need a copy of the Fourier coefficients, and we need to identify the DC and Nyquist frequencies. We'll start by computing the Hilbert transform of a series of random numbers.

```
n = 20;
r = randn(n,1);
rx = fft(r);
% a copy that is multiplied by the complex operator
rxi = rx*1i;
% find indices of positive and negative frequencies
posF = 2:floor(n/2)+mod(n,2);
negF = ceil(n/2)+1+~mod(n,2):n;
% rotate Fourier coefficients
rx(posF) = rx(posF) + -1i*rxi(posF);
rx(negF) = rx(negF) + 1i*rxi(negF);
% take inverse FFT
hilbert_r = ifft(rx);
```

And voilà! We have the Hilbert transform of vector r. Identifying the DC + 1 frequency is easy—it's always the second index. Identifying the Nyquist is a bit trickier because it depends on whether the signal has an odd or even number of points. You could use an if-else construction here, but an easier method is to use mod (introduced in chapter 7; it returns the remainder of the division of the first input by the second input). The output of mod with the second input "2" tells us whether the number is odd (output: 1) or even (output: 0). To identify the negative frequencies, we just reverse this by adding the tilde.

You can compare the above results with the output of the MATLAB function `hilbert` (in the MATLAB signal processing toolbox or Octave signal package) and check that they are the same. In practice, it's easier to use the `hilbert` function than to write out the FFT manipulations above. The MATLAB `hilbert` function uses a different and slightly more efficient algorithm. If you want to challenge yourself to reproduce MATLAB's implementation, complete exercise 8 before looking at the file hilbert.m.

Once you have this analytic signal, you can extract power and phase values as you would the result of complex Morlet wavelet convolution. The results of the Hilbert transform are only interpretable for narrow-band signals, so it should be applied after band-pass filtering of the data.

20.7   Exercises

1. The following code is correct, but probably doesn't produce the desired result. What is the problem and how can you fix it? (Hint: Try plotting.)

```
x = 0:100;
gaus = exp(-(x.^2)/100);
```

2. How does a frequency-domain Gaussian compare with time-domain wavelets created in the previous chapter? Plot the real part of the inverse Fourier transform of the Gaussian. You'll need to use `fftshift`, because we haven't defined the proper phases.

3. In this exercise, you will explore the reason why the number-of-cycles parameter of the Morlet wavelet controls the trade-off between temporal and frequency precisions. Generate five wavelets at 10 Hz, changing the number of cycles from 2 to 15 in linear steps. In one figure using a 5-by-2 subplot organization, show the time-domain Gaussian that tapers the sine wave on the left plot and the frequency-domain representation of the wavelet power in the right plot. Title each subplot to indicate the number of cycles. Does this figure help you understand the time-frequency precision trade-off? (If not, make a better figure!)

4. Write a MATLAB function that applies a frequency-domain Gaussian filter. It should take five input arguments: data, sampling rate, peak frequency, FWHM, and a plotting toggle. And it should give two outputs: the filtered data and the empirical FWHM. The function should do the following steps:

a. Have a useful help information that explains what the function does and how to use it.

b. Check the inputs for accuracy and consistency, and report a useful error message if something is wrong. For example, the sampling rate input should be just a single number.

c. Compute the filter in the frequency domain based on the input parameters (number of points, peak frequency, and FWHM).

d. Apply the filter to the data.

e. Compute the empirical FWHM of the Gaussian.

f. If the plot toggle is true, produce a plot that looks like figure 20.5, although you might want to plot only a selection of the data (e.g., the first 10% of the time series).

5. I think it would be nice to have a vertical line from the peak of the Gaussian to the $y = 0$ line in figure 20.4. If you agree, then implement this in the code. If you disagree, then do it anyway, but complain about it afterward.

6. Look up the theoretical FWHM of a Gaussian. Then implement that equation in MATLAB. Empirical estimates will differ when the sampling rate is low. Compute several Gaussians with varying sampling rates and compare the theoretical and empirical FWHMs.

7. The Gumbel distribution is sometimes used for modeling skewed statistical distributions (sadly, it has nothing to do with the inspirational claymation character Gumby). Look up the formula for a Gumbel distribution and implement it in MATLAB. Write code that will produce its time-domain and frequency-domain representations. There are two parameters to the Gumbel formula; see how these affect the time- and frequency-domain responses. Do you think the Gumbel distribution would make a good narrow-band filter? Why or why not?

8. The following paragraph contains an explanation of how the Hilbert transform is implemented in MATLAB. On the basis of this explanation, write code to implement the Hilbert transform. Then, make sure it produces identical results to the code presented in this chapter. Finally, inspect the contents of the file hilbert.m to compare your solution against MATLAB's.

    Identify the positive and negative frequencies. Double the Fourier coefficients from the positive frequencies and zero-out the Fourier coefficients from the negative frequencies. Then take the inverse Fourier transform.

9. Construct a series of FIR filters (band-pass: 8–12 Hz) that vary in filter
    order from one cycle to 15 cycles of 8 Hz, keeping other parameters
    the same. Show a plot of the filter kernels in the time domain (you
    might want to add a small y-axis offset to each kernel to improve
    visibility) and their power spectra in a frequency-by-order image
    (you'll need to zero-pad to make the power spectra comparable). How
    do the frequency characteristics of the filter vary as a function of
    order? Next, create a real-valued Morlet wavelet centered at 10 Hz.
    Apply each filter to the wavelet, and measure the empirical FWHM
    of each filtered result. (Hint: Measure FWHM on the result of
    `abs(hilbert(filtsig)).`). Plot the FWHM as a function of the
    filter order. On the basis of these results, at what point would you
    draw a different conclusion about the results based on the different
    orders?

10. Here is an alternative way to compute the frequency response of a filter
    kernel. Create an FIR filter using parameters that you specify. Just keep
    the pass-band frequencies below 100 Hz. Then create a loop over fre-
    quencies, ranging from 1 Hz to 100 Hz in 1 Hz steps, in which the fol-
    lowing is done at each iteration. Create a sine wave at that frequency
    and apply the FIR to the sine wave. Then compute the FFT of the fil-
    tered result, and extract the power at the frequency of the sine wave.
    After the loop, make a plot of power as a function of the frequencies of
    the sine waves. How does this procedure differ from taking the FFT of
    the filter kernel?

11. Produce a sinc function in the time domain. Use the `fft` function and
    plot its power spectrum. Probably the plot is empty. Inspect the Fourier
    coefficients in the Command window to see what's going on. There is
    something "weird" about the time-domain function—inspect the for-
    mula for a sinc function and think about what happens at $x = 0$. Add
    some code to make sure the entire signal contains finite values (hint:
    do you know the function `isfinite`?), then inspect the frequency
    distribution.

12. Create a linear chirp and design a narrow band-pass filter. Extract time-
    varying power from the Hilbert transform of the filtered chirp. Also
    extract the power spectrum of the filter kernel. Repeat this procedure in
    a loop over filter order factors ranging from 1 to 8. Then plot all power
    time courses on top of each other and, in a separate plot, all filter
    kernel power spectra on top of each other.

13. The chirp power time series in the previous exercise will have a Gauss-
    ian-like shape. Estimate the FWHM of the power time series from each

filter order factor. Create a plot of FWHM as a function of order. What do this and the previous exercise tell you about the filter order parameter? Do you have any inspirations for ways to sanity check that your filter parameters are reasonable?

14. The running-mean filter shown in this chapter is a non-causal filter, meaning each data point becomes a weighted sum of previous and future data points. Change this to be a causal filter, meaning that each data point becomes a weighted sum only of previous (or only of future) values. Can this filter also be implemented in the frequency domain?

15. Create a low-pass FIR filter with a cutoff at 40 Hz (the lower bound is 0 Hz, meaning there is no lower bound). Guess a reasonable filter order, and inspect the time- and frequency-domain representations of the filter kernel. Then try other filter orders. On the basis of these plots, what seems to be a reasonable filter order, and what does this tell you about the amount of data you need to apply a low-pass filter? Does your answer depend on the sampling rate of the data?

16. There are many filter kernel construction algorithms, several of which come in the MATLAB signal processing toolbox. Modify the code that produces figure 20.8 to achieve similar results using `fir1` and `firpm` instead of `firls`. The results won't be exactly identical, but it should be possible to get qualitatively similar results.

21 Fluctuation Analysis

21.1 Root Mean Square to Measure Fluctuations

Root mean square (RMS) is simple to compute and to interpret, easy to adapt to different kinds of data, and relies on no assumptions. It is therefore a ubiquitous measure of fluctuations in many physical and biological applications. Several data analysis approaches involve RMS, one of which (detrended fluctuation analysis) will be the focus of this chapter.

To compute RMS, just think about each term in reverse order: first, square all data values (squaring eliminates the need to worry about negative values); second, average all squared data values; third, take the square root of the average. The code below shows two ways to compute RMS.

```
step1 = data.^2;
step2 = mean(step1);
step3 = sqrt(step2);
rmsx = sqrt(mean(data.^2));
```

Is there a difference between variables `step3` and `rmsx`? (`rms` is a function that computes RMS, so I'm using `rmsx` as a variable name.) RMS is closely related to variance. The primary difference is that RMS reflects the "raw" fluctuations, while variance reflects fluctuations around the average value.

21.2 Fluctuations in Time Series

One of the reasons why RMS is so widely used is that it can be applied to many different features or dimensions of data. To start with, we will compute RMS from a single channel on each of 99 trials. To make things interesting, I've added noise to one trial. The noise is randomly distributed around zero, making it almost unnoticeable when inspecting the trial average. But the RMS plot clearly reveals the outlier trial. For this reason, RMS can facilitate the initial stages of data cleaning and quality control.

```
rmsx = sqrt(mean(eeg.^2, 1));
subplot(211), plot(mean(eeg,2))
subplot(212), bar(rmsx)
```

Notice that because the variable eeg is a time-by-trials matrix, taking the mean over the first dimension means computing RMS over time separately for each trial. If you compute the mean over the second dimension, you will be computing the RMS over trials separately for each time point (figure 21.1).

21.3 Multichannel RMS

RMS can also be computed over many channels at each point in time, providing a time series of multichannel fluctuations. Increases in RMS can

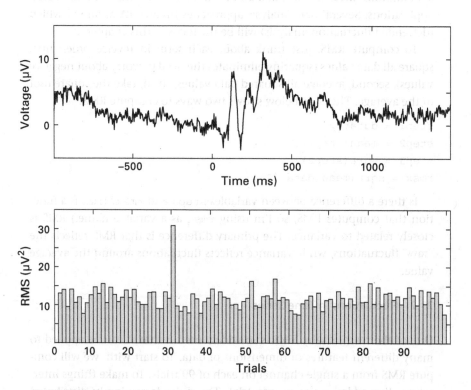

Figure 21.1
The trial-averaged plot of EEG data (top plot) reveals nothing sinister about the data. The single-trial RMS, however, reveals the single trial with excessive noise.

indicate more differentiated activity (although the RMS provides no information about spatial localization), and they can also indicate the presence of artifacts.

The code below will compute RMS over 64 EEG electrodes ("topographic RMS") for each trial separately, and then for the average over all trials. The variable data is a 3D channels-time-trials matrix. Therefore, averaging over the first dimension computes the topographic RMS.

```
rmsxT = squeeze(sqrt(mean(data.^2, 1)));
rmsxA = squeeze(sqrt(mean(mean(data,3).^2, 1)));
```

The second line is tricky because it has two embedded mean functions. Remember when interpreting long lines of code to start from the innermost piece of code and work your way outward, one set of parentheses at a time. The topographic RMS from the trial average produces a vector (one value per time point, where the value is the RMS over all electrodes at that time point), whereas the topographic RMS from all trials produces a time-by-trials matrix (topographic RMS separately per trial). When this matrix is averaged over trials, the difference with the trial-averaged RMS is striking (figure 21.2).

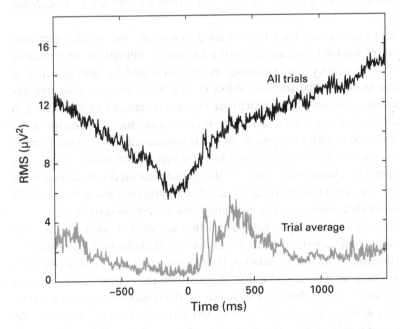

Figure 21.2

Topographic RMS, computed individually on each of 99 trials and then averaged (black line) or computed on the average of all trials (gray line).

These two time courses are so different because a lot of dynamics are lost during trial averaging. The reason why the average of the single-trial RMS time series dips just before time 0 is that time course at each electrode is baseline-subtracted. Thus, by definition, the immediate prestimulus period has a low level of activity.

Among other applications, sudden changes in topographic RMS time series are interpreted to indicate sudden transitions in the state or configuration of brain networks (Lehmann, Pascual-Marqui, and Michel 2009).

21.4 Detrended Fluctuation Analysis

In section 21.2, we computed RMS in a single electrode over the trial period. What if we computed RMS over a shorter time period or over a longer time period? Would the magnitude of the fluctuations change over different timescales? The answer to that question depends on the system that generated the data. Systems that produce pure noise, for example, show slight increases in RMS as the timescales increase. Systems that have autocorrelated behavior, in contrast, show larger increases in RMS as timescales increase. Some systems produce autocorrelated behaviors that look similar regardless of the timescale (figure 21.3). These are referred to as *scale-free* systems.

Scale-free systems have been studied in natural and abstract geometry (think of fractals) and are garnering increasing interest in neuroscience with the discovery that processes in the brain and in human behavior exhibit scale-free-like and fractal-like organization. In turn, scale-free-like organization is taken as evidence that a system is complex and operating in a state of criticality, which facilitates computational memory and increased flexibility to respond to changes in the environment. If you would like to learn more about the theory and key findings of the study of scale-free dynamics in neuroscience, consider the special issue on this topic published in the journal *Frontiers* (http://journal.frontiersin.org/researchtopic/505/scale-free-dynamics-and-critical-phenomena-in-cortical-activity).

The two main analysis approaches that are used to identify scale-free dynamics in time series data are called detrended fluctuation analysis (DFA) and demeaned fluctuation analysis (DMA). We will compute DFA and DMA from a data set recorded in a human, in which the volunteer had to use a mouse trackpad to follow a moving line on the computer screen. Each time point is given a value of +1 (correctly following the line) or –1 (missed the line, thus making a mistake) (data are taken from Cohen 2016). The overall goal of the analysis is to cut the continuous data into epochs of various sizes

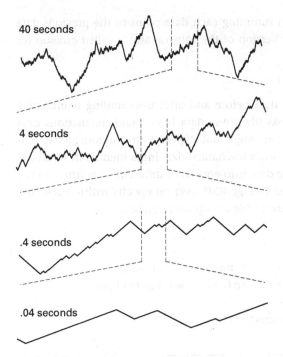

Figure 21.3
Scale-free systems exhibit time series behavior that appears self-similar over different timescales. That is, no matter how much you zoom in or zoom out, the time-varying fluctuations have similar characteristics.

and compute the average RMS across the epochs. Thereafter, a line is fit between the logarithm of the RMS magnitudes and the logarithm of the window lengths. The slope of this line is called the *scaling exponent* and is taken as evidence for a scale-free system if it is higher than the scaling exponent of white noise (0.5). Let's first define our parameters.

```
nScales = 20;
ranges = log10([ 1 400 ]);
scales=ceil(logspace(ranges(1),ranges(2),nScales));
```

The above code means that we will create 20 logarithmically spaced timescales between 1 and 400 seconds. The next step is to prepare the data for the DFA by making it meander like the gait of a drunken college student stumbling home at 4 a.m. (the more formal expression would be transforming the time series into its unbounded form). For a purely random time series, this would be called Brownian motion. It is created by subtracting the global mean and then summing each data point to the previous data point. This is the discrete version of the integral and is called cumsum for cumulative sum.

```
x = cumsum(x-mean(x));
```

Figure 21.4 shows the data before and after unbounding (cumulative sum). Qualitatively, it looks like these data have trend fluctuations over larger time periods that you might not expect from random noise. It is these fluctuations that we want to characterize. Implemented in MATLAB, this means we will cut the data into epochs of different sizes (the scales variable) and compute the average RMS over all epochs within each size. The following code happens inside a loop over scales.

```
% epoch data
N = length(x);
n = floor(N/scales(si)); % number of epochs
epochs=reshape(x(1:n*scales(si)), scales(si),n)';
% detrend
depochs = detrend(epochs')';
```

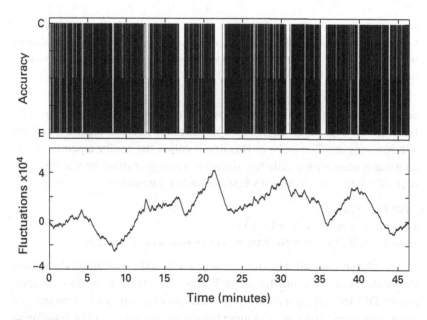

Figure 21.4

Binarized accuracy data (top panel; C means correct, and E means error) are converted to an unbounded form by cumulatively summing all successive time points (bottom panel).

The epoching code should look familiar from section 18.4 of chapter 18. Detrending is a procedure in which a linear trend is fit to the data and then removed (figure 21.5). Fitting the linear trend is done with least squares (see chapter 28), and removing that trend involves subtracting the best-fit line from the data (in other words, taking the residual). The `detrend` function always works column-wise; hence, we need to transpose the epoch matrix and then transpose it back to keep the matrix in the same orientation. Now we are ready to compute RMS for each epoch and then average the RMS values over epochs.

```
rmses(si) = mean(sqrt(mean(depochs.^2,1)));
```

After the code above loops through all timescales, the final step is to compute a least-squares fit of the timescales to the RMS in those timescales. The code below should look a bit familiar from chapter 10; chapter 28 will discuss least-squares fitting in more detail.

```
A = [ones(length(scales),1) log10(scales)' ];
dma = (A'*A) \ (A'*log10(rmses)');
```

The average RMS values are plotted as a function of the scales in figure 21.6. When transformed into logarithms, a straight line seems to be a good fit. A scaling exponent between 0.5 and 1 is often taken as evidence for a scale-free system. A value of 0.8 is in the range of previous studies of scaling

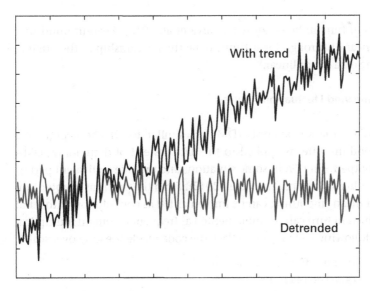

Figure 21.5
Illustration of a time series before and after being detrended.

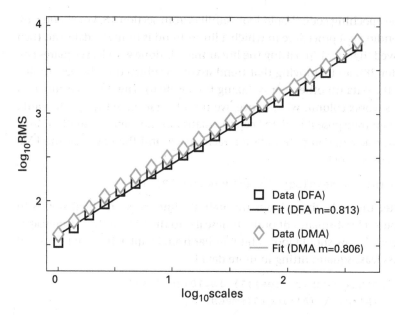

Figure 21.6
This plot shows the logarithm of epoch-averaged fluctuations (y axis) as a function of the logarithm of the epoch length in seconds (x axis). A straight line in this log-log space with a slope greater than 0.5 is often taken as evidence for a scale-free or self-similar system that exhibits long-range time autocorrelations.

exponents in human behavior (e.g., Palva et al. 2013). Keep in mind that this is a linear fit through logarithmic data; the relationship in the original scale of the data is nonlinear.

21.5   Demeaned Fluctuation Analysis

Demeaned fluctuation analysis (DMA) is similar to DFA, except that a mean-smoothing filter is applied to the data instead of detrending. DMA seems to outperform DFA in some situations and provide equivalent performance in other situations (Shao et al. 2012). From the previous chapter, you know that mean smoothing can be implemented as convolution with a flat kernel, which in turn can be implemented as frequency-domain multiplication. The following code would replace the code inside the loop over scales.

```
% create kernel for this scale
nConv = N+scales(si)-1;
kernel = fft(ones(scales(si),1)/scales(si),nConv);
```

```
hfKrn = floor(scales(si)/2)+1;
% mean-smooth as convolution
convres = ifft(fft(x,nConv) .* kernel);
convres = convres(hfKrn:end-hfKrn+1+mod(nConv,2));
```

Notice that the time series is mean-smoothed before epoching the data. This is faster and helps prevent edge effects when smoothing epoched data. In chapter 12, I mentioned that it's best to construct convolution kernels to have an odd number of points. Here you see the reason why. Different smoothing kernels have an odd or an even number of points, and so the size of the convolution wings differ on different iterations. To write code that is robust to this variability, we need to accommodate the parity of the kernel. One mechanism to accomplish this is to add a mod function when clipping the wings. Next, the difference between the smoothed time series and the raw time series (i.e., the residual) is computed.

```
residX = x-convres;
```

Finally, the data are epoched, RMS is computed per epoch, and then the RMSs are averaged over epochs, just like with DFA.

```
n = floor(N/scales(scalei)); % number of epochs
epochs=reshape(residX(1:n*scales(si)),scales(si),n)';
rmses(scalei) = mean(sqrt(mean(epochs.^2,1)));
```

As a final thought, it is interesting to compare DFA with the discrete Fourier transform. Conceptually, they are similar—both are designed to quantify fluctuations over various timescales. The main conceptual difference is that DFA measures the magnitude of any fluctuations at different timescales; Fourier-based analyses measure fluctuations that match a sinusoidal template. With that in mind, you can appreciate why the $1/f$ shape of the Fourier power spectrum is sometimes interpreted as indicating scale-free activity.

21.6 Local and Global Minima and Maxima

Finding local and global extrema admittedly doesn't really fit into the theme of this chapter, but it's an important topic that didn't really fit any better in any other chapter, and I wanted to try to balance the lengths of different chapters.

Global extreme points can be obtained during the functions min and max. You are already familiar with these functions, so let's move on to finding local minima and maxima. If you know the range in which you are

Box 21.1

Name: Simon-Shlomo Poil
Position: CTO and Co-founder
Affiliation: NBT Analytics BV, Amsterdam, Netherlands (www.nbt-analytics
.com)
Photo credit: Private

**When did you start programming in MATLAB, and how long did it take
you to become a "good" programmer (whatever you think "good" means)?**
I was introduced to MATLAB during a university course in 2005. At that
time I already had a lot of programming experience from other languages,
so it did not take much time for me to learn MATLAB. MATLAB is a quite
easy language, so I think a beginner with no prior programming experience
will learn MATLAB within a few weeks. It takes a longer time to become
a "good" programmer, and it depends on how diverse your programming
skills are.

**What do you think are the main advantages and disadvantages of
MATLAB?**
The advantages of MATLAB are the interactive environment, a rich collection
of toolboxes, and the well-written documentation. In the field of neuroscience, the advantage is also the broad selection of toolboxes, such as EEGLAB,
Fieldtrip, SPM, and the Neurophysiological Biomarker Toolbox. The disadvantages of MATLAB are the license costs, and for certain tasks also the speed of
the code.

> **Do you think MATLAB will be the main program/language in your field in the future?**
> Yes. I think MATLAB will still remain a core language for EEG analysis in the near future. At NBT Analytics, we explore alternative languages, such as Python and Julia, mainly to reduce costs and improve performance, but MATLAB will likely remain a core pillar of our analysis of EEGs from clinical trials for quite some time.
>
> **How important are programming skills for scientists?**
> It is essential for most scientists. If you need to analyze a lot of data, it makes your life easier if you can write your own scripts and understand other people's scripts. Programming gives you the freedom to perform the analysis you want to perform.
>
> **Any advice for people who are starting to learn MATLAB?**
> My advice is to find a simple programming task. The only way to learn how to program is by doing it. As a beginner, you should only worry about how to get from A to B in your program, do not spend time on making the code fast and efficient. Just make it work. Start by outlining your program in line-by-line comments, and then write your code. This will help you see and understand the steps going from A to B.

looking for extreme points, you can simply constrain the data that you input into `min` or `max`. Let's sharpen our extrema-hunting skills with a double-sinc function.

```
x = -40:.01:120;
signal = 3*sin(x)./x + sin(x-20)./(x-20);
```

Visually (figure 21.7), it is clear that there is one global maximum, a second large peak, and many other local maxima and minima. To find that second large peak, we can input part of the signal into the `max` function.

```
ridx = dsearchn(x',[10 40]');
[maxLval,maxLidx] = max(signal(ridx(1):ridx(2)));
plot(x(maxLidx+ridx(1)),maxLval)
plot(x(maxLidx+ridx(1)),signal(maxLidx))
```

The first point in `ridx` needed to be added to the index because we want the index of the subvector, not the index of the entire vector. The second `plot` function produces the identical result as the first (why?). A quick glance at the plot suggests our hunt was successful (figure 21.7). However, closer inspection reveals that we missed our mark by one point.

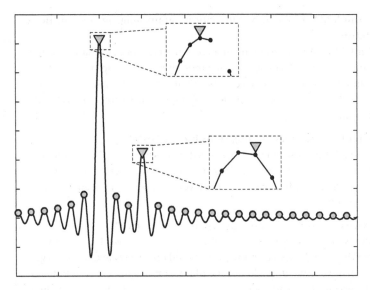

Figure 21.7
Hunting for global and local maxima in a double-sinc function (a search for Poké-
mon was unsuccessful, so I had to resort to local extrema). The global maximum
is accurately identified, but the second maximum is one point off. (That's not my
mistake—it's a challenge for you to find and fix the bug!) Other local maxima are also
indicated.

To appreciate why the result is off, think about what the value of
`maxLidx+ridx(1)` would be if the very first data point in the subvector
were the maximum. When you figure it out, fix it in the online MATLAB
code, and use the zoom tool to make sure your adjustment is correct.

This method of finding local maxima by searching for "global" maxima
within specified subvectors can be useful to identify two maxima (or minima; simply replace `max` with `min` in the above code), but it is not a scalable
solution for finding all local maxima. To find all local maxima, use the
following strange-but-effective code.

```
peeks = find(diff(sign(diff(signal)))<0)+1;
```

I know, it's a really weird piece of code, but basically it finds extreme
points in the derivative of the signal. Try plotting each part of the code
(`plot(signal)`, `plot(diff(signal))`, etc.) step by step to get a feel for
what it does. You can change the less-than sign to a greater-than sign to
get local minima instead of local maxima. This code finds all of the local
peaks in a signal, where a peak is defined as any point that is higher than

both of its neighbors. For this reason, the code works well only for smooth functions; excessive noise will cause each noise spike to be considered a peak.

Two questions: (1) Can you figure out why I called the variable `peeks` instead of `peaks`? (2) What if we want to exclude some peaks, for example the two peaks with the largest values (because those have already been identified earlier)? The code below provides one possible solution; it's your job to figure out what it does and why it works.

```
[~,idx] = sort(signal(peeks));
peeks(idx(end-1:end)) = [];
```

21.7 Exercises

1. The variable `EEG.data` is a 3D matrix (channels by time by trials). Your friend Frances wants to compute RMS by averaging over time, which is the second dimension. Why is the following line of code wrong, and how can you fix it? What is your advice to Frances to avoid confusion in situations like this?

```
rmsf = sqrt(mean(squeeze(EEG.data(30,:,:)).^2,2));
```

2. Frances is back. She's made her life easier by measuring only one electrode and only one trial. But she still has MATLAB troubles. She tried to compute RMS with this line of code, but the result was a complex number. What is her programming bug, and why did it produce a complex RMS?

```
rmsf = sqrt(mean(data));
```

3. Creating fixed-sized epochs means some data are lost at the end of the continuous time series. A mitigating strategy sometimes used in DFA is to run the DFA twice: once forward-epoching (data at the end are lost), and then again backward-epoching (data at the beginning are lost). The two scaling coefficients are then averaged together. Implement this. Are the two scaling exponents very different from each other?

4. The theoretical scaling exponent of random white noise is 0.5. Can you empirically confirm this, and does it depend on how much data you have and how many scales you use? Run simulations for N data points ranging from 1,000 to 1,000,000 and S scales ranging from 1% to 20% of the length of the signal. Do you get a result close to 0.5? How about if the data are not random noise, but a random series of +1 and –1 (what's a good way to create a random series of +1 and –1)?

5. Frances again. She's offered to treat you to dinner if you can write code to implement the following problem in the double-sinc function. She wants code to find the local maximum that follows the first minimum that follows the global maximum. Can you earn your free dinner?

6. It would be nice if the triangles indicating the two maxima in figure 21.7 would be automatically positioned above the data rather than overlaying the data. What's the best way to implement this? Make sure your solution would still work when the function is scaled up or down (e.g., if the sinc function were multiplied by 1,000 or by 0.001).

7. Let's revisit the FWHM of the Gaussian in section 20.4 of chapter 20. Can you compute this width more efficiently using some new tricks you learned in this chapter?

8. In the double-sinc function, find all local maxima, and then remove from the list any peaks with an amplitude greater than 1 or smaller than 0.1. And then remove peaks after $x = 40$. Make a plot like figure 21.7 to confirm that your filter works.

9. I mentioned earlier that local extrema detection is not robust to noise. Add some random noise to the double-sinc and find the local minima. Plot the function with all of the minima on top. How does the plot look? What could be a strategy to find local minima in the presence of noise?

10. It's a bit annoying to modify the convolution code for even-length versus odd-length kernels. Instead, modify the DMA code to force all scales to have odd lengths.

11. The variable hello is a 5-by-1 vector. What are the sizes of the outputs of the following lines of code? Figure out the answer first, then confirm in MATLAB.

```
mean(hello)
mean(hello,1)
mean(hello,2)
```

12. The initial step of computing DFA or DMA is to integrate the signal to create an unbounded form. How can you go from the unbounded form back to the original bounded form? (Hint: What's the opposite of integration, and how do you implement this in MATLAB?)

IV Analyses of Action Potentials

22 Spikes in Full and Sparse Matrices

The brain contains many neurons. No one really knows exactly how many, but an adult human brain contains somewhere around 100 billion. A significant amount of research in neuroscience over the past century has involved sticking small wires (electrodes) into the brain, locating one or a few neurons, and then recording how their activity changes in response to different sensory or cognitive variables, medications, and so on.

The action potentials of individual neurons are fast events, lasting a few milliseconds. They are so fast and so stereotyped that researchers often treat them as binary events. That is, you can imagine a Boolean or logical time series that contains mostly zeros (false—no spike at that time point) and some ones (true—a spike occurred at that time point) (figure 22.1). In this and the next chapter, we will treat action potentials as Boolean events. In chapter 24, we will see how the shape of the time course of the action potential can be used to isolate action potentials from different neurons that are measured by the same electrode.

The term "spike" is colloquially used to refer to action potentials. Most people use those terms interchangeably.

22.1 Spike Times as Full Matrices and as Sparse Vectors

The focus of the analyses in this chapter is on data recorded during tasks with repeated discrete trials; for example, when data were recorded while experimental animals viewed images, experienced somatosensory stimulation, or initiated motor responses. The data we'll use in this chapter were downloaded from crcns.org (https://crcns.org/data-sets/motor-cortex/alm-1/). In the first data set (Li et al. 2015), a mouse was instructed to stick his tongue out repeatedly (perhaps at the experimenters).

Our first goal will be to import the data into MATLAB from the data files provided online. We will put the data into two matrices, one full and one

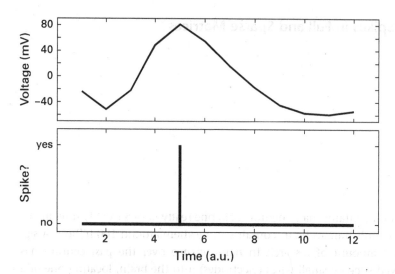

Figure 22.1
A single action potential is a continuous event that unfolds over several milliseconds
(upper plot). In neuroscience data analyses, however, spikes are often treated as
binary events (lower plot).

sparse (see figure 10.8 for review). Viewing this file with a text editor reveals
that the file contains some meta-data followed by spike times, and then
more meta-data, and so on. Each section of meta-data is indicated by hash
marks (#) and separates different trials of data. We don't know *a priori* how
much data there will be, so a while-loop seems appropriate.

```
fid = fopen('03-04-11-aa_sig1_spikes.dat');
spikenum = 1;
trialnum = 0;
APsS = zeros(100,2);
APsM = zeros(1,1001);
```

The spike times will be stored in two matrices: APsS is for "action potentials sparse" and APsM is for "action potentials matrix." APsS will be a sparse
matrix that stores the trial number and time stamp of each spike; APsM will
be a full trials-by-time matrix in which each element of the matrix is 0 (no
spike) or 1 (spike).

Full matrices are larger and take up more space in MATLAB's buffer and
when saved to disk, but are a convenient representation of the data and
allow for easy access. Among other advantages, full matrices make it easy to
(1) compute the trial-averaged time course of spiking activity, (2) extract

the average number of spikes in some time window (e.g., 100–300 milliseconds after stimulus onset), and (3) perform matrix-based computations such as least-squares fitting or principal components analysis. The main advantage of sparse matrices is that they take up less space in the MATLAB buffer. For small amounts of data like what we have here, the matrix representation might be preferable. In practice, however, with very large recordings of dozens or hundreds of neurons over long periods of time, sparse matrices might be preferable. For storing data on your hard disk, sparse matrices are almost always preferable.

We are ready to read in the data from the file. The while-loop will continue until the end of the file; in other words, `while ~feof(fid)`. The code inside the while-loop is separated into four sections. The first two sections are fairly straightforward: read in a line of code, and skip to the next line of code if the current line is empty.

```
d=fgetl(fid);
if isempty(d), continue; end
```

The goal of the next section of the code is to skip through the trial information section. If the line starts with a hash, we can skip forward. But we want special treatment for the last line of the comments, because that is our cue to increment the trial count variable.

```
if d(1)=='#'
    if strcmpi(d(1:5),'# int')
        trialnum = trialnum+1;
    end
    continue;
```

If you use a text editor to look into the data file, you will see that each trial contains information about the stimulus intensity. The full line of text is "intensity (dB): 97.3333" but there is no other line that starts with "int," which means we need to match only the first few characters. If there is no match, the if-then statement is false, `trialnum` is not updated, and the script skips forward to the next line of the file.

The fourth section of the while-loop extracts the spike time and enters it into the matrices. Notice the difference between how the data are entered in the sparse versus the full action potential matrices.

```
else
    spiketime = ceil(sscanf(d,'%g')/1000);
    APsS(spikenum,:) = [trialnum spiketime];
    APsM(trialnum,spiketime) = 1;
    spikenum = spikenum+1;
```

Figure 22.2 shows image representations of the matrices. Typing whos reveals that the sparse representation is about 20% of the size of the full representation. As the matrices get larger, this difference becomes more extreme.

```
>> whos AP*
   Name       Size              Bytes    Class

   APsM       90x1008           725760   double
   APsS       8959x2            143344   double
```

Before moving on, I'd like to share a little MATLAB plotting trick. The top bar of figure 22.2 shows trial count and time stamp, but trial count has

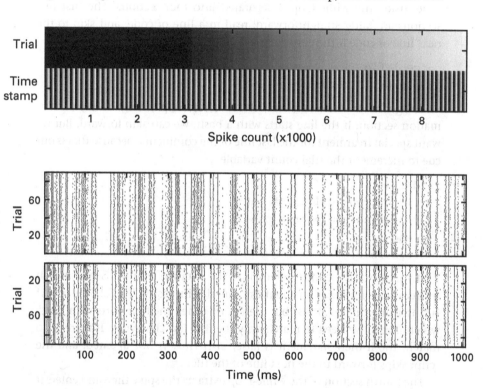

Figure 22.2
Spike times shown in different representations. The top plot shows the trial and peri-stimulus spike times as a 2-by-N matrix. This is the sparse matrix. The middle plot shows those same data in a different format, with each dot indicating that an action potential was measured at each time-trial point. The lower plot shows an image of the full time-by-trials matrix. The point here is that both the sparse and the full matrices contain the same information; they are just stored in different ways.

a range of 1–90 while time stamp has a range of 0–1,008. If you type imagesc(APsS'), the top bar will be solid black. This happens because MATLAB scales the image to the color range of the entire matrix, which means that the maximum trial number is less than 10% of the range of the image. A simple solution is to scale the trial count by a factor of 10 (this should be done only for illustration; you don't want to permanently change the trial numbers). There are two ways to accomplish this; I'll let you guess which is the method of choice among savvy and stylish programmers.

```
% method 1
APsS2 = APsS; % don't overwrite original data
APsS2(:,1) = APsS2(:,1)*10;
imagesc(APsS2')
% method 2
imagesc(bsxfun(@times,APsS',[10 1]'))
```

22.2 Mean Spike Count in Spikes per Second

How many action potentials occurred between 0 and 1 second? This can be answered regardless of whether the data are stored as a matrix or as a vector. I will show how to do it with matrix format here, and you will have the opportunity to repeat the analysis using the sparse format in the exercises.

```
tidx = dsearchn(timevec',[0 1000]');
spikeRates = sum(APsM(:,tidx(1):tidx(2)),2);
```

So far we've computed the *number* of spikes in each trial between 0 and 1,000 milliseconds. To make this finding more comparable across different studies, stimuli, and so forth, the number should be converted to units of spikes per second (often written sp/s). You might be tempted to use the unit hertz instead of spikes per second. Although this seems to be the same as spikes per second, hertz generally connotes rhythmicity. Without checking the temporal pattern of spikes, using hertz might give the wrong impression. That is, although 10 Hz and 10 sp/s both mean that the neuron emitted 10 action potentials within a period of 1 second, "10 Hz" implies that there was one spike roughly each 100 milliseconds, whereas "10 sp/s" simply indicates that a total of 10 spikes were observed somewhere in the time frame of 1 second.

To convert from spike count to spikes per second, simply divide by the total amount of time in seconds (figure 22.3).

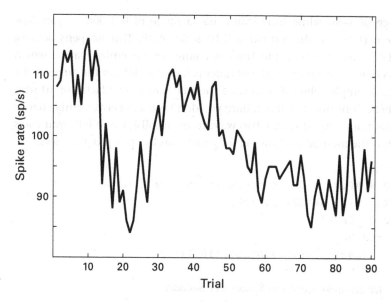

Figure 22.3

This plot shows the spikes per second on each trial, in a time window of 0–1,000 milliseconds.

```
windowTime = timevec(tidx(2)-tidx(1))/1000;
spikeRates = spikeRates / windowTime;
```

Here, I divided by 1,000 because the time vector was in milliseconds. If you have the time vector already specified in seconds, don't divide. Because our goal here was to compute the number of spikes within one second, the divisor is mostly unnecessary (it is 0.9990). But it's good to keep in the practice of making sure the units are computed accurately. Watch what happens to the *y* axis when you change the code to compute the number of spikes in the first 500 milliseconds, with and without the appropriate division (it's not actually in the figure; you're supposed to do this on your own).

22.3 Peri-event Time Spike Histogram

Extracting the average spike count within a time window is useful but provides coarse temporal information. You might want to know how the average spike rate changes over time. The standard way to examine the time course of spiking activity is to make the time windows shorter, and then have many windows over time. The results produce a histogram-like plot of averaged spike rates over time. If we have our time windows be very short (i.e., a single time point), then our task is very easy when the spikes are stored in full matrix form.

```
plot(timevec,sum(APsM))
```

Repeating this analysis with the sparse format data is a bit trickier, which is why I'm leaving it up to you to complete in the exercises. Careful inspection of this simple line of plotting code reveals the same problem identified earlier: The result shows spikes, but not spikes per second. This sum should be divided by the time width of each bin.

```
dt = mean(diff(timevec));
plot(timevec,sum(APsM)./dt)
```

Okay, so in this particular data set, dt is 1, meaning the first plot command was accurate. But we just got lucky in that case, because the data happened to have been stored using units of milliseconds. In practice, you should always divide by the time bin width. In fact, we can simulate a different time window length by down-sampling the spike time series by a factor of 5. We start by defining new bins by using a similar procedure as we used to discretize reaction time data into bins (chapter 14).

```
bins = ceil(timevec/5);
```

You can see that the vector bins comprises 1,1,1,1,1,2,2,2,2,2,3... We will use these numbers to average spike counts across time points with each unique bin number. Results of the fine- and coarse-sampled time courses are shown in figure 22.4.

```
for i=1:length(spikesbins)
    spPerbin(i) = mean(mean(APsM(:,bins==i),2),1);
    timevecBin(i) = mean(timevec(bins==i));
end
% plot
dt = mean(diff(timevecBin))/1000;
plot(timevecBin,spPerbin./dt)
```

You should keep in mind that these are *estimates* of instantaneous firing rates; we cannot accurately measure spikes per second in a window of 5 milliseconds.

22.4 Exercises

1. Smooth the spike time course by convolving it with a Gaussian. Repeat using different widths of the Gaussian. What is the approximate

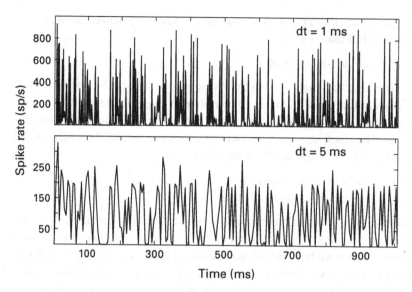

Figure 22.4
Spike time histograms using different window widths.

minimum FWHM (in milliseconds) to have a noticeable effect? What is the FWHM at which the data seem too smooth? When you have these two boundaries, take the FFT of the two Gaussians and plot their power spectra in the same plot. Does this tell you something about the temporal precision of spike time courses? (Keep in mind that your findings from this data set do not necessarily generalize to all spike timing data sets.)

2. Sort the trials in the spiking data set to be in order of the average firing rate from the entire trial. Would you expect figures 22.3 and 22.4 to change with the new sorting?

3. What is the difference between the following two lines of code? Is there one? When would there be a difference?

```
dt = mean(diff(timevec)) / 1000;
dt = mean(timevec(2)-timevec(1)) / 1000;
```

4. Reproduce figure 22.3 using the sparse representation of the spike times.

5. Reproduce figure 22.4 using the sparse representation of the spike times.

23 Spike Timing

A single neuron cannot produce cognition on its own. It is the complex interactions across huge numbers of neurons that—neuroscientists generally believe—give rise to cognition and behavior. This idea motivates investigations of spiking activity beyond how the average spike rate is related to an external stimulus. In this chapter, you will learn about three aspects of action potential timing:

1. Rhythmicity of action potentials within a single neuron (spike rhythmicity)
2. Timing of action potentials relative to action potentials of another neuron (spike-time correlations)
3. Timing of action potentials relative to the local field potential (spike-field coherence)

23.1 Spike Rhythmicity

The goal of spike rhythmicity analyses is to determine whether a neuron emits action potentials with temporal regularity; for example, if the neuron tends to spike every 100 milliseconds. This could occur if the neuron is a pacemaker cell or if it receives strong rhythmic input. To obtain evidence for spike rhythmicity, we want to create a line plot that shows the probability of a neuron emitting an action potential at various points in time relative to other action potentials from that same neuron.

This analysis can be done in one of two ways, depending on whether the spikes are stored in full or sparse matrix format (see the previous chapter). Both methods are fairly straightforward, and both provide great opportunities to sharpen your MATLAB skills.

Spikes as Full Matrices

Here the idea is to loop through spikes and build a new vector that sums the data forward and backward in time around each spike. Our data are structured in a trials-by-time matrix, so we'll also use a loop over trials.

```
win = 50; % in ms
spRhy = zeros(1,win*2+1);
n = 0;
for triali=1:trialnum
    % find spikes on this trial
    sps = find(APsM(triali,:));
    for si=1:length(sps)
        newspikes = APsM(triali,sps(si)-win:sps(si)+win);
        spRhy = spRhy + newspikes;
        n = n+1;
    end % end spike loop
end % end trial loop
spRhy = spRhy./n;
```

Let's work through this code. The first few lines specify a time window of 50 milliseconds. This will be the window in which we collect spike times (forward and backward in time, thus the actual window will be 101 milliseconds). In this case, the data are stored at 1,000 Hz, which means that each time step corresponds to a millisecond. The variable win would need to be adjusted for any other sampling rate.

Next is the loop over trials. In each trial, we identify the spike times in that trial (variable sps), and then loop through those spike times. The data from a window surrounding each spike onset is added to the vector spRhy. Finally, after the loops are finished, spRhy is normalized by dividing by the total number of spikes. This normalization transforms spRhy from count to probability, which is convenient because different neurons have different total spike counts. Obviously, the probability of this neuron spiking at time = 0 is 1 (because time = 0 is defined as the presence of a spike).

If you run the online code for this section, MATLAB will give an error. This happens because ... well, it's up to you to figure out why this is and what the solution is.

Spikes as Sparse Matrices

The general idea is the same as with full matrices, but the programming implementation is slightly different. The following three lines would go in a loop over spikes (si is the looping variable).

```
tempsps = spikes-spikes(si);
tempsps(tempsps<-win | tempsps>win) = [];
spRhy(tempsps+win+1) = spRhy(tempsps+win+1)+1;
```

To understand the first line of code, recall the min-abs construction (e.g., figure 18.5). We are shifting the vector of spike onsets such that the current spike gets a time of zero, and all other spike times become relatively negative (before the spike) or positive (after the spike). In the next line, all relative spike times outside our requested window are obliterated from the `tempsps` vector. And finally, the remaining indices in the `spRhy` vector are incremented by one.

If you look carefully at figure 23.1, you might see that the results of this analysis differed very slightly between the full and sparse matrix representations (you might have to look at the figure in MATLAB to see the difference). Any ideas what might be causing this discrepancy?

23.2 Spike Rhythmicity via the Frequency Domain

As you learned in chapter 12, some time-domain analyses can be done faster and more efficiently in the frequency domain. It turns out that computing spike rhythmicity, like convolution and filtering, is a linear operation and can also be computed via the frequency domain.

More precisely, we can compute spike rhythmicity as an autocorrelation. The procedure is fairly simple and also fairly similar to convolution. Recall

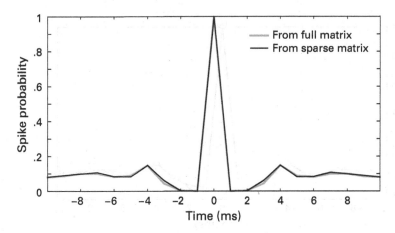

Figure 23.1
This figure shows the probability of action potentials occurring within 10 milliseconds of each action potential.

that in convolution, the power spectra of the data and of the kernel (in chapter 12, this was a Morlet wavelet) are point-wise multiplied, and the inverse Fourier transform is then applied. What's the kernel in an autocorrelation computation?

The answer is the signal. The signal is both the "signal" and the "kernel." Autocorrelation can therefore be performed by multiplying the Fourier transform of the data with itself, and then taking the inverse Fourier transform. In other words, the autocorrelation is the inverse Fourier transform of the power spectrum. Let's see this in action; results are shown in figure 23.2.

```
APsM = bsxfun(@minus,APsM,mean(APsM,2));
spikespow = abs(fft(APsM,[],2)).^2;
spikesifft = ifft(spikespow,[],2);
spRhy = mean(spikesifft) / trialnum;
spRhy = spRhy([end-win+1:end 1:win+1]);
```

There are a few noteworthy differences from convolution as implemented in chapter 12. The length of the FFT here is the length of the signal, meaning we are performing a circular instead of a linear convolution. Second, the inverse time-domain signal is spliced such that the final win+1 points are wrapped around before the first win+1 points. This makes it easier to see the autocorrelation function both forward in time (positive values) and backward in time (negative values). You can also use the function fftshift.

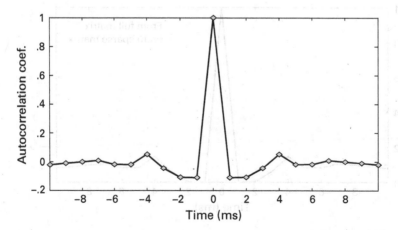

Figure 23.2
Spike autocorrelation function using the same data as plotted in figure 23.1.

If the input time series is real-valued—which is generally always the case when analyzing spiking data—then the result of the inverse Fourier transform will also be real-valued. In digital computers, there are occasional rounding errors, particularly if the values are very small. This can lead to small non-zero imaginary components in the inverse Fourier transform. If you get a complex result from this procedure, you can ignore the imaginary part by using `real(ifft(...))`.

Spike timing analyses must be interpreted cautiously when the data are taken from time windows in which an external stimulus is being presented. Temporal regularities in the stimulus may impose an autocorrelation structure on the data. For example, imagine that you recorded spiking activity from visual cortex while a light flashed every 100 milliseconds. Many neurons will show strong spike rhythmicity at 10 Hz, but this does not necessarily mean that the neurons have intrinsic temporal regularities at 10 Hz; it might simply mean that they spiked with each new sensory input. A similar concern arises when using monitors that redraw the display with each refresh or otherwise have some refresh-related flicker. In this case, an apparent 60 Hz (or whatever is the monitor refresh rate) rhythm in the visual cortex might reflect entrainment to the external stimulus.

23.3 Cross-Neuron Spike-Time Correlations

So far, we've been dealing only with spike times from individual neurons. The next step is to expand this to dealing with pairs of neurons. Now we want to know whether there is a relationship between the timing of action potentials from one neuron and the timing of action potentials from another neuron. This kind of analysis is often used to make inferences about monosynaptic connections among neurons. For example, one neuron reliably spiking 4 milliseconds before another neuron can be taken as evidence for a monosynaptic connection.

As you might imagine, this analysis is fairly similar to the spike rhythmicity analyses shown earlier. The main difference is that we want to identify spikes in neuron A while populating the variable `spRhy` with spikes from neuron B. Figure 23.3 shows results of a pair of neurons.

This analysis can be implemented in the time domain or via frequency-domain multiplication. The main difference from what was shown in the previous section is that we don't want the power spectrum from only one time series; we want the cross-spectral density of two time series. This can be implemented as `f1.*conj(f2)`, where `f1` and `f2` are the Fourier coefficients of each time series. This should sound familiar from spectral

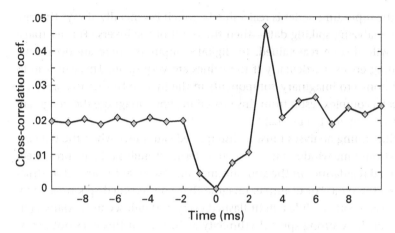

Figure 23.3

A cross-correlation function between a pair of neurons recorded simultaneously. The plot indicates that one neuron was more likely to fire 3 milliseconds after action potentials of the other neuron.

coherence, where the Fourier coefficients of one signal are scaled by the Fourier coefficients of another signal.

23.4 Spike-Field Coherence

The brain operates on multiple spatiotemporal scales. Cross-scale interactions are becoming increasingly studied, in part because it is believed that cross-scale interactions are an important aspect of higher cognitive processes including consciousness (Le Van Quyen 2011). The action potentials of individual neurons reflect one of the smaller spatial scales of brain function (even smaller scales include synapses and ion channels); local field potential (LFP) and EEG reflect larger spatial scales of brain functional organization.

There are only a few methods for quantifying multiscale interactions in the brain that can be linked back to neurophysiology (any data analysis method used in neuroscience should have a neurophysiologic interpretation). One of these cross-scale interaction methods is to test whether the spiking of individual neurons is synchronized to the phase of LFP oscillations. This *spike-field coherence* has been observed in many regions of the brain, perhaps most famously in the rat hippocampus, where the timing of action potentials relative to the ongoing theta oscillation can be used to determine the position of the rat in a maze or a field, and is thought to be

an important aspect of the temporal organization of information processing (Buzsáki and Moser 2013).

There are a few ways to quantify spike-field coherence. Let's start by creating a plot similar to the spike-time correlation plot, except that instead of
averaging together spike time series, we will average together LFP time
series. In fact, this is basically the same thing as an event-related potential
(ERP), one of the classical methods for analyzing EEG data. The difference
is that instead of defining the time = 0 event to be an external stimulus, we
will define time = 0 to be an internal event: the action potential. This
method is therefore sometimes called a spike-triggered average.

The first step is to identify the time indices of the spikes. The data set
we'll work with in this section contains two vectors, one with an LFP
recorded from the rat hippocampus (variable `lfp`) (Mizuseki et al. 2009;
data downloaded from https://crcns.org/data-sets/hc/hc-2/), and one with
all zeros and ones when an action potential was detected in a single neuron
in the hippocampus (variable `spikeTimes`). In this format, finding spike
times is easy:

```
sidx = find(spikeTimes);
```

Next, we compute the average LFP surrounding each spike. If the window size is 200 time points, why will the matrix `sLFP` have `win*2+1` time
points? A second question: What is the purpose of the second line of code
below, and what might happen if that line is omitted?

```
win = 200; % indices, not ms!
sidx(sidx<win | sidx>length(lfp)-win) = [];
sLFP = zeros(length(sidx),win*2+1);
for si=1:length(sidx)
    sLFP(si,:) = lfp(sidx(si)-win:sidx(si)+win);
end
```

Figure 23.4 shows the average of LFP traces surrounding 44,906 spikes.
There seems to be a relationship with the LFP (the null hypothesis, that
there is no relationship between action potential timing and field potential
fluctuations, would produce a flat line plus noise). Is this really reflective of
the data or might this effect be driven by a small number of large outliers
trials ("trials" here refers to spikes)? It is difficult to determine when seeing
only the averaged data. Being able to inspect the single-trial data is the reason why I saved each LFP trace here, in contrast to the code to produce the
spike rhythmicity, in which the vector `spRhy` was repeatedly added to itself,
thus losing the single-trial data.

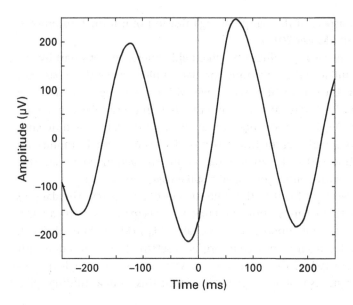

Figure 23.4

A spike-triggered average LFP trace. Each action potential was identified, and the LFP time series was averaged around these spike times.

One way to look at these data more closely is to compare the magnitude of the average response with the magnitude of the non-averaged ("single-trial") data. Inspecting the non-averaged LFP data shows that the maximum y-axis fluctuations tend to be around 1,500 µV, and the peaks of the spike-triggered average are around 200 µV (figure 23.5). In other words, the average response is about 13% of the total response (that's an informal eye-ball-based estimate). I would consider this to be a decent effect size. A very small effect would be if the average response magnitude were less than a percentage point of the non-averaged response magnitude. Of course, the importance and theoretical relevance of an effect does not depend entirely on its effect size, but it is informative to have an idea of how robust different neural phenomena are.

An additional method of inspecting the data is to view all single traces in a 2D (spikes by time) matrix. Figure 23.5 shows this image. It is clear that spikes occurring on the rising slope of the theta oscillation are observed throughout this data set. This picture becomes even clearer after applying a 2D Gaussian smoothing kernel to denoise the image. But you'll have to wait until chapter 27 to learn how to do that.

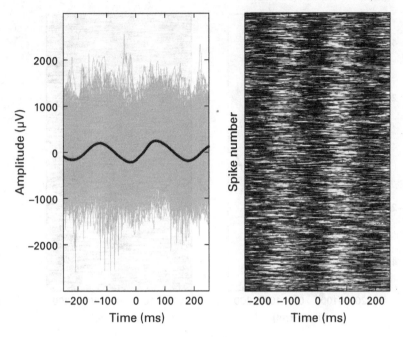

Figure 23.5
Examining all single-trial spike-field data reveals that the averaged effect shown in figure 23.4 is highly reproducible over individual action potentials.

23.5 Frequency-Specific Spike-Field Coherence

Spike-field coherence is usually specific to a frequency range. In fact, computing spike-field coherence using the broadband LFP data as we did above will often yield mixed results. It happened to work well in this example because the rat hippocampus is dominated by very large-amplitude, low-frequency oscillations.

Instead, it is often a good idea to band-pass filter the LFP data before computing spike-field coherence. Let's start by filtering at 8 Hz using Morlet wavelet convolution. The result in figure 23.6 looks quite similar to the nonfiltered result in figure 23.6 but slightly larger (not surprising considering the dominance of theta in this example LFP). For example, the trial-averaged response peak now appears to be over 25% of the magnitude of the single-trial peaks.

Another advantage of having narrow-band data is that the analytic phase is defined and interpretable, and this in turn allows us to use phase-based analyses to quantify the magnitude of spike-field coherence. We'll do

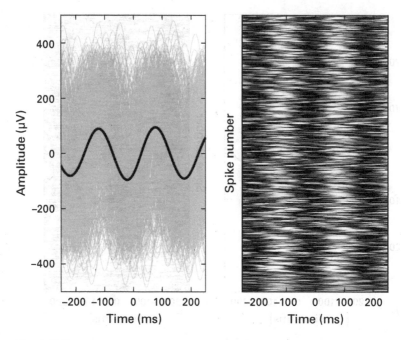

Figure 23.6
Similar to figure 23.5 except the LFP data were filtered around 8 Hz using complex
Morlet wavelet convolution.

this by identifying the LFP phase at which each spike occurs, and computing the non-uniformity of the distribution of phase angles. This is the same
procedure as you learned about in chapter 19 for computing phase clustering over trials. Phase clustering–based spike-field coherence involves two
steps: First, identify the phase values of the LFP at which the spikes occurred;
second, use Euler's formula to find the length of the average vector. In
the code below, variable as is the analytic signal—the result of convolution
between the LFP and a complex Morlet wavelet at 8 Hz.

```
sidx = find(spikeTimes);
angels = angle(as(sidx));
sfc = abs(mean(exp(1i*angels))));
```

This procedure can be repeated for many frequencies to produce a
line plot of spike-field coherence magnitude over frequencies, as shown in
figure 23.7.

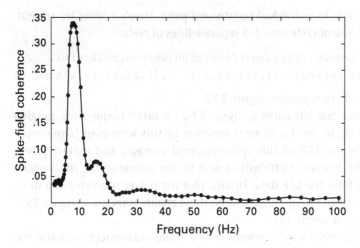

Figure 23.7
This figure shows the frequency specificity of spike-field coherence in the data set used for the previous two figures. Higher values on the y axis indicate stronger spike-field coherence at that frequency.

23.6 Exercises

1. Adapt the spike rhythmicity code by expanding the trials-by-time matrix into one long time-trials vector. Why is it a good idea to remove spikes within N milliseconds (where N is the window for the analysis) of the edges of each epoch, even though the new time series has no "epochs?"

2. To see whether spike rhythmicity changes over time, you could compute rhythmicity in two different time windows. Try this using the data shown in figure 23.1. Which two time windows would you use, and what considerations should you keep in mind (e.g., having a similar number of spikes in each condition to minimize biases)?

3. The results in section 23.1 can be interpreted in milliseconds only because the sampling rate was 1,000 Hz. Modify the code so the window is specified in milliseconds and is then converted into indices.

4. The following line of code is efficient but dense. Imagine that you are writing the Methods section of a paper and need to explain what this line does. Write instructions in full English sentences so that someone else could reproduce this line of code. This exercise practices two skills: (1) explaining code in human language; (2) breaking down a complex

line of code to individual constituent parts. Here's a hint: Try separating the line of code into 3–5 separate lines of code.

```
spikerhyth=mean(ifft(abs(fft(bsxfun(@minus,spikesfull ...
    ,mean(spikesfull,2)),[],2)).^2,[],2),1)/trialnum;
```

5. Write code to reproduce figure 23.7.

6. Reproduce the left panel of figure 23.6 for other frequencies, ranging from 2 Hz to 50 Hz. At each iteration of this loop over frequencies, compute the FFT of this spike-triggered average, and extract power from the Fourier coefficient closest to the frequency of the wavelet used to filter the LFP data. Finally, plot the power spectrum as a function of wavelet frequency. How do these results compare to figure 23.7, and are you surprised?

7. Here is another way to compute spike-field coherence: discretize the LFP phase-angle time series into N bins (e.g., 10 bins), and compute the average number of spikes per bin. Then plot the results as a bar plot of spike count (y axis) as a function of LFP phase (x axis). Try this for a few frequencies and confirm that the results are consistent with figure 23.7.

8. The spike-field coherence value at 8 Hz is around 0.32 (the exact value will depend on parameters). Can this be considered statistically significant? Generate a null hypothesis distribution by randomizing spike times and recomputing phase clustering. Perform 1,000 iterations, plot the null hypothesis distribution and the observed value, and compute a p-value. This method is actually too liberal, because the temporal characteristics of the shuffled action potentials are no longer physiologically plausible. Therefore, also try the "cut-and-shift" procedure, in which the phase-angle time series is cut in a random location on each iteration, and the second half is placed before the first half. How do the two null-hypothesis distributions compare?

9. Instead of the following line of code when computing spike rhythmicity, use the function `fftshift` to obtain the same result.

```
spRhy = spRhy([end-win+1:end 1:win+1]);
```

10. Louis is trying to be meticulous and inspect his single-trial data overlaid on the same plot. But MATLAB takes a really long time to render the figure, in part because he is using an old laptop (from 2014!). He asks you for help. On the basis of his code below, what are (at least) two separate pieces of advice you can give him? (The variable `alldata` contains 60,000 trials and 1,200 time points.)

```
for i=1:size(alldata,1)
    figure(1), hold on
    plot(i+alldata(i,:))
end
```

11. The following line of code was presented in this chapter. Rewrite this line without the "or" operator. (Hint: How can you equate negative and positive values?)

```
tempsps(tempsps<-win | tempsps>win) = [];
```

24 Spike Sorting

If you lower a microelectrode into the brain, that electrode is likely to measure activity from several neurons. Putative action potentials are marked in the data by *threshold-crossings*, when the voltage exceeds some threshold value (often, the threshold is some number of standard deviation units above the mean voltage value). Some data acquisition systems identify these threshold-crossings online and store only data surrounding these exceedances; other data acquisition systems store the continuous data, and threshold-crossings are identified during off-line analyses.

The action potentials from each neuron are fairly homogeneous, which is to say that the waveform shape of all action potentials of a single neuron is very similar. But different neurons have differently shaped action potentials. Furthermore, neurons with different geometric orientations and distances to the electrode tip will have different waveform shapes in the data. This is illustrated in figure 24.1.

The idea of spike sorting is to isolate different neurons in the data according to differentiable statistical characteristics of the action potentials. Spike sorting algorithms have become increasingly sophisticated; for example, by taking advantage of voltage projections onto multiple neighboring electrodes (Rey, Pedreira, and Quian Quiroga 2015). We'll work here with simpler methods that will help you understand the concept of spike sorting and, of course, improve your MATLAB skills.

24.1 Spike Amplitude and Width

What are the statistical characteristics of the action potentials that allow differentiability? Let's start with some simple characteristics to find out. Two basic properties of an action potential are its amplitude and its width. Can we distinguish multiple neurons recorded from the same electrode using these two properties? In theory, yes, if one neuron was fast and high

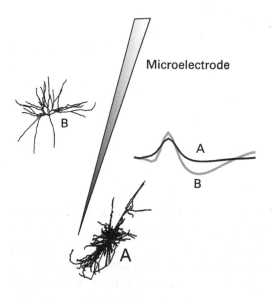

Microelectrode

Figure 24.1
Depending on geometry and distance to a microelectrode, the waveform shapes
of action potentials of different neurons can sometimes be dissociated from single-
electrode recordings. Components analyses and clustering techniques are used to
attribute each action potential to one of several different putative sources.

while another neuron was slow and low. Let's see whether reality is consistent with theory (it rarely is, but we scientists are supposed to be ever-
hopeful that it will be).

We will define spike amplitude as the distance between the peak of the
spike waveform and the preceding dip. And we will define spike width as
the distance in time between the pre-peak dip and the post-peak dip. These
measures must be computed individually for each spike, so we need a loop.
The following code goes inside a loop around spikes. We start by finding
the time indices of the peak, the pre-peak minimum, and the post-peak
minimum. (Variable spikes is a matrix of spikes by time, and variable si
is the looping index.) The data used in this chapter were downloaded from
https://crcns.org/data-sets/vc/pvc-5 (Chu, Chien, and Hung 2014).

```
% find peak
[~,peakidx] = max(spikes(si,:));
% minimum before peak
[~,min1idx] = min(spikes(si,1:peakidx));
```

```
% minimum after peak
[~,min2idx] = min(spikes(si,peakidx:end));
min2idx = min2idx+peakidx-1;
% get premin-peak difference
spikefeat(si,1) = diff(spikes(si,[min1idx peakidx]));
% get min2min
spikefeat(si,2) = min2idx-min1idx;
```

Notice how I used the `diff` function instead of writing out the subtraction. It's a convenient shorthand, but you should always sanity-check it first. The most common mistake to make is having the order backward, which means the difference could be negative when you expected it to be positive.

Now let's see if these two features can differentiate different neurons. We will make a plot of one feature versus the other feature (spike amplitude vs. spike width), with each dot reflecting one action potential. If all action potentials come from one neuron, there will be little spatial structure in the plot. However, if the action potentials recorded by this electrode come from two different neurons, there will be separable clouds of dots, with each cloud corresponding to each neuron.

```
plot(spikefeat(:,1),spikefeat(:,2),'.')
```

The results in figure 24.2 do not look very cloud-like (at least, not like any cloud I've ever seen, not even Arizona desert sunset clouds, which are quite remarkable). Why is it so layered? The problem here is that with only 12 time points, the information is too coarse, and so there is too little variability. This leads to two possible interpretations. One possibility is that all of the spikes in this data set come from one neuron. An alternative possibility is that these two features are simply terrible at differentiating multiple neurons. I have seen the future, and I'm going to place my bet on the second interpretation. In fact, spike amplitude and width are generally not good features to use for spike sorting. We tried them here mainly as an exercise.

24.2 Spike Features via Principal Components Analysis

Amplitude and width alone are insufficient to isolate the two neurons. Perhaps the features that differentiate these neurons are more complex and are related to the waveform shape. Instead of defining *a priori* what features to focus on, perhaps we should try a data-driven approach that allows blind discovery of the important features of the spike waveforms. We will do this

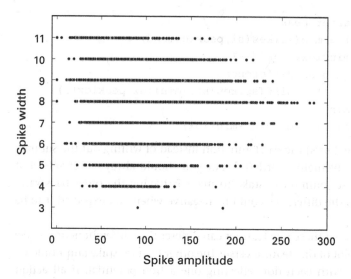

Figure 24.2

Spike width (*y* axis; in arbitrary time steps) and spike amplitude (*x* axis) fail to provide evidence for multiple independent neurons contributing action potentials in this data set.

by applying principal components analysis (PCA) to the waveforms and inspecting whether the projection of each spike onto principal component (PC) space can differentiate two neurons recorded by this electrode (spoiler alert: the answer is yes!).

Recall from chapter 17 that a PCA involves an eigendecomposition of a covariance matrix. And that a covariance matrix can be produced by multiplying a demeaned data matrix by its transform. Before mucking around in MATLAB, let's first think about matrix multiplication and sizes. If we want a time-by-time covariance matrix, how should the spikes matrix be transposed and multiplied? The matrix is natively organized as spikes by time. If we put the matrix transpose on the right (AA^T), we'll end up with a spikes-by-spikes covariance matrix, which is not what we want. Instead, the matrix transpose needs to be on the left (A^TA). If you don't understand why this produces the correctly sized covariance matrix, you might want to review the rules for matrix multiplication in chapter 10.

```
spikescovar = spikes'*spikes;
```

Check the size of `spikescovar`, and also make an image of it. This data set has 12 time points, so the matrix should be 12 by 12. A covariance matrix is symmetric, which appears to be the case from the image (figure 24.3). Did I forget something in the code above? You betcha. In fact, I forgot two important details. First, the `spikes` matrix needs to be mean-subtracted, and second, the covariance matrix should be divided by the number of spikes minus 1. (Readers with a linear algebra background might realize that because MATLAB normalizes the eigenvectors when the input matrix is symmetric, and because we do not use the eigenvalues in computations, the division by $N - 1$ is technically unnecessary; nonetheless, it's good to keep in the habit of doing things correctly.)

Now that we have the correct covariance matrix, we use the MATLAB function `eig` to compute the eigendecomposition and obtain the eigenvalues and eigenvectors. It's easy to forget which output contains the eigenvalues and which contains the eigenvectors, but a quick sanity check via inspecting images of the outputs will reveal which output is which.

Next, we project each neuron onto the first two components. Here again is a good opportunity to think about matrix sizes before writing the code. Per component, we want a single number per spike that encodes how much that spike reflects the properties of that component. In other words, the matrix should be spikes-by-1. The matrix `spikes` has dimensions

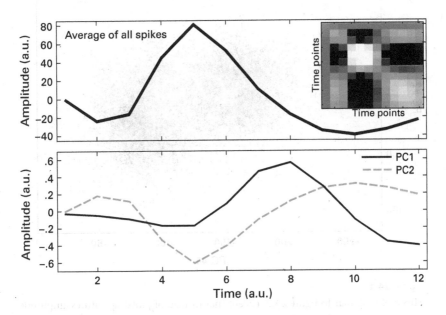

Figure 24.3

This figure shows the average time courses from all spikes (top plot), the time-by-time covariance matrix from which the PCA is performed (inset), and the time courses of the first two components (bottom plot).

spikes-by-time, and the PC weights have dimensions time-by-time (although the values do not represent "time"; the matrix is weight-by-component). This time you got off easy—no transposes required.

```
comp1 = spikes*eigvects(:,end);
comp2 = spikes*eigvects(:,end-1);
```

   Remember that the eigenvectors usually come in ascending order, meaning the final eigenvector is the one with the largest eigenvalue (i.e., the one that accounts for the most amount of variance orthogonal to all other directions of variance). But they are not guaranteed to be sorted, so it's a good idea to check, or to sort them yourself. Now let's generate a plot like figure 24.2, but using the first and second principal components (figure 24.4). And now the story completely changes. It is visually obvious that there are two clouds of dots. This provides evidence that this electrode measured action potentials from two different neurons.

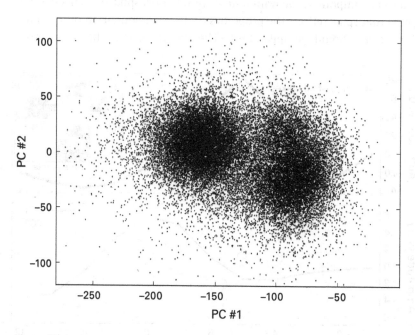

Figure 24.4

This plot is similar to figure 24.2, except that instead of plotting features amplitude and width, we are plotting features PC1 by PC2 (PC, principal component). Separable clusters in this plot is taken as evidence for multiple neurons in the recording.

Box 24.1

Name: Rodrigo Quian Quiroga
Position: Director
Affiliation: Centre for Systems Neuroscience, University of Leicester, Leicester, England
Photo credit: Rodrigo Quian Quiroga

When did you start programming in MATLAB, and how long did it take you to become a "good" programmer (whatever you think "good" means)?
I started using MATLAB about 15 years ago. Good or bad is relative. As with any software language, learning to program is a never-ending learning process. However, I can highlight that for me, a major step change was when I learned how to use the GUI developer. In fact, the possibility to embed some codes within GUIs made a big difference, as these codes were very easy to use for other users, so it boosted their dissemination.

What do you think are the main advantages and disadvantages of MATLAB?
I used to program in Fortran and C before. The main advantage of MATLAB is that you don't have to define variables, compile the code, and so forth. So, in this sense MATLAB is much more user-friendly. Another great advantage is the possibility of developing and using GUIs. The main disadvantage is that it requires specific knowledge in order to optimize the codes (e.g., avoiding for-loops, linearizing the operations) and that it is relatively expensive if one starts adding toolboxes, compared to other options that are free.

Do you think MATLAB will be the main program/language in your field in the future?

It is for me, but I can't say it will stay like this in the future. New generations of scientists may find features of, for example, Java or other programming languages more attractive than MATLAB.

How important are programming skills for scientists?

In my field, neuroscience, it is crucial. People blindly using codes they do not fully understand are constrained to use what somebody else developed and cannot tune the analyses or the development of experiments to their own scientific questions.

Any advice for people who are starting to learn MATLAB?

Get as soon as possible into a specific problem of your own research. It makes no point in doing a course without any practical application. Theoretical knowledge is quickly forgotten if not put into practice. A good way of learning is also to start modifying other people's codes.

24.3   Spike Features via Independent Components Analysis

Before learning how to identify clusters, let's first try an alternative technique to isolating units from this data set of spikes. Instead of principal components analysis, we'll try an independent components analysis (ICA). Recall from chapter 17 that independent and principal components analyses are similar in that they are both blind source separation (BSS) techniques that estimate components of variance based on multichannel data. But independent components analysis separates components according to higher-order statistical criteria and does not require the components to be orthogonal.

As in chapter 17, we'll use the Jade algorithm, which takes the entire time series as input rather than the covariance matrix. If the data matrix is less than full column rank, the function `jader` might return complex outputs. Reduced-rank matrices occur when one or more columns of data provide no unique information. This is the case in the sample data set because the first time point is always set to zero. A column (or a row) of all zeros in a matrix makes the matrix reduced-rank because the zero column is a linear combination of any other column (zero times any column is the zero vector). Thus, we should extract fewer independent components than we have time points.

One effective way to determine the maximum number of components to retrieve is by computing the rank of the data matrix and using the rank as the second input to the function jader.

```
r = rank(spikes);
weights = jader(spikes',r);
icas = weights*spikes';
plot(icas(1,:),icas(2,:),'k.')
```

The output of jader is a time-points-by-components-weight matrix, sorted from the largest component to the smallest (i.e., the opposite order from the MATLAB eig function—always sanity check!). Similar to PCA, the weights are then multiplied by the individual spike waveforms to obtain component scores per spike. And similar to PCA, plotting the scores of the first independent component by those of the second independent component reveals two clusters (figure 24.5). Unlike in figure 17.8, which showed large differences between PCA and ICA, here these two decomposition methods provide comparable results. If anything, the PCA seems to have provided a clearer separation between the two clusters. More generally,

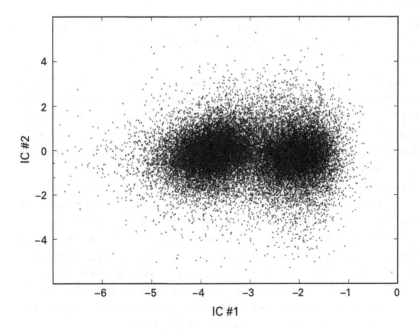

Figure 24.5
Similar to figure 24.4, except independent components analysis was used instead of principal components analysis. IC, independent component.

whether PCA and ICA provide similar results depends on the nature of the covariance patterns in the data.

24.4 Clustering Spikes into Discrete Groups

The interpretation of results like in figures 24.4 and 24.5 is that there are two different neurons measured by this one electrode. Visually, it seems clear that there are two clusters; now it's time to learn how to separate those clusters. We will use a procedure called *k*-means clustering, which tries to separate multidimensional data into *k* clusters. More detail and examples of *k*-means clustering are presented in chapter 31. To run *k*-means clustering, we enter the to-be-clustered data (the component scores), and the number of components to extract.

```
cidx = kmeans([comp1 comp2],2);
```

The variable `cidx` (cluster index) is a vector that contains a numerical label for each spike, categorizing each spike as belonging to group "1" or to group "2." Then we can re-plot the results using different colors for different groups (figure 24.6).

```
plot(comp1(1,cidx==1),comp1(2,cidx==1),'r.'), hold on
plot(comp1(1,cidx==2),comp1(2,cidx==2),'b.')
legend({'unit 1';'unit 2'})
```

You can clearly see the line that divides the two groups. It is a straight line. It's not a perfect separation, but it's also not so bad considering the simplicity of the analysis. In practice, spike-sorting procedures use more dimensions for the classifier and more sophisticated classification algorithms. Nonetheless, professional-grade spike-sorting procedures follow the same basic concepts that you learned here.

Regardless of the fanciness of the analysis, you should always be cautious when interpreting results of spike sorting. The groupings are statistical, imperfect, and probabilistic, and they are open to some amount of subjectivity. It is difficult to demonstrate conclusively that two different spike clusters are produced by two different neurons. For example, if the electrode moves slightly in the brain, the same neuron might have a different geometric orientation relative to the electrode, and therefore might appear to be a different neuron. For this and other reasons, people refer to clustered spikes as "single units" rather than "isolated neurons."

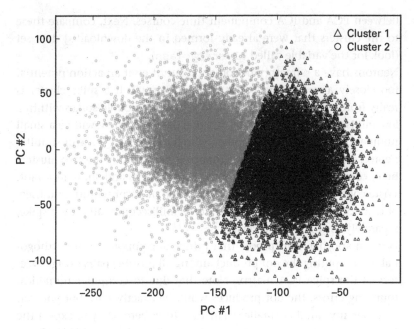

Figure 24.6
The two clusters from the principal components analysis were separated using *k*-means clustering.

24.5 Exercises

1. The following line of code subtracts the mean of the spike time series in the wrong way. How can you tell? (Hint: Sanity check!)

```
spikes = bsxfun(@minus,spikes,mean(spikes,1));
```

2. Re-run the *k*-means clustering that produced figure 24.6 several times. Do the cluster numbers stay the same? Do the boundaries stay the same?

3. Adjust the *k*-means clustering to extract and plot three clusters. Do you agree with the results? What does this tell you about *k*-means clustering and *a priori* knowledge of the number of clusters?

4. Are PCA-based and ICA-based clusters the same? What is the proportion of spikes that overlap? Remember that labeling groups as "1" or "2" is arbitrary; you'll need to devise a method to see if they match even if PCA-derived unit "1" corresponds to ICA-derived unit "2." For the two clusters, plot the average waveform shape for the overlapping PCA and ICA clusters, and, in a separate plot, show the difference

between PCA and ICA component time courses. Next, compare these to the clusters that were already formed in the downloaded data set (look for the variable called `cluster_class`).

5. Neurons have a refractory period and cannot emit an action potential too closely after the preceding action potential. If a spike cluster is really from a single neuron, there should be no double-spikes within a few milliseconds of each other. In large data sets, there will be a small number of such fast double-spikes, which may reflect noise or simultaneous spikes from multiple neighboring neurons. In the two clusters here, compute the interspike-interval histograms and show in a plot. Write code to count the number of spikes that occur within 1 millisecond, 2 milliseconds, and so on, up to 20 milliseconds, after each spike, separately for the two clusters.

6. Perform another "soft proof" that principal components are orthogonal by computing the dot product among all possible pairs of eigenvectors of the spike PCA matrix. Note that due to computer numerical rounding errors, the dot produces won't be exactly zero, but you can consider any number smaller than 10^{-10} to be zero. Do you expect the ICA components to be zero? Confirm your hypothesis in MATLAB.

7. Now that you have two spike clusters, test whether those units interact. Identify action potentials in cluster no. 2 that spike either 3 milliseconds before or 3 milliseconds after spikes in cluster no. 1. Plot their waveform shapes. For comparison, also plot the average waveform shapes of randomly selected spikes.

8. We discovered in section 24.3 that the covariance matrix (variable `spikecov`) has rank 11, although is has dimension 12 by 12. This is the covariance matrix that was computed from data matrix `spikes`. What is the rank of `spikes`? On the basis of this result, can you guess the linear algebra law about the relationship between the rank of A and the rank of $A^{T}A$?

V   Analyses of Images

25　Magnetic Resonance Images

Magnetic resonance imaging (MRI) has revolutionized medical diagnoses since the 1970s and swept in a tide of cognitive neuroscience research starting in the late 1990s. Interesting historical note: It used to be called nuclear magnetic resonance imaging, but too many people were concerned about sticking their head into what they thought might be a small nuclear reactor. It's a good example of smart marketing.

Most scientists who work with MRI data do not program the pre-processing and analyses themselves. The dominant MATLAB toolbox for analyzing MRI-based data is SPM (www.fil.ion.ucl.ac.uk/spm/). Nonetheless, familiarity with importing and working with MRI data in MATLAB is important. This chapter will provide an introduction to importing, plotting, and some basic analyses of MRI data.

25.1　Importing and Plotting MRI Data

Recall from earlier chapters that an image is represented in MATLAB as a 2D matrix of numbers, where the value at each coordinate encodes the brightness of that pixel. MRI data are volumes and therefore stored as 3D matrices. Although 2D color images are also stored as 3D matrices, 3D MRI volumes are colorless, so each dimension in the matrix corresponds to a dimension in physical space (color would therefore require a fourth dimension). Each point in the 3D image is called a voxel (a volumed pixel).

There are several file formats for MRI data; the two most commonly used formats are NIFTI and ANALYZE (NIFTI is becoming the standard). NIFTI-format data (file extension .nii) include "headers" that have basic information about the image (e.g., voxel sizes, 3D rotations, patient information, etc.) and the data matrix. ANALYZE-format data separate the header and the image data into two files (filename.hdr and filename.img). There are several functions that can import NIFTI files into MATLAB. Some functions rely on toolboxes such as SPM. Included in the online MATLAB code is an import function that does not require any additional toolboxes (down-loaded from www.neuro.mcw.edu/~chumphri/matlab/readnifti.m).

```
strMRI = readnifti('MNI152_T1_1mm.nii');
```

The variable `strMRI` ("structural MRI") is a 3D brain scan and is repre-sented as a 3D matrix (confirm this by typing `size(strMRI)`). The image comes with the FSL program (a free, non-MATLAB analysis program for MRI analyses) (Smith et al. 2004), and each voxel measures 1 mm^3 of the brain. Let's see what part of the matrix looks like:

```
strMRI(20:30, 50:60, 90)
```

It's just a bunch of numbers, as you've seen before with images. Those numbers will be more interpretable when converted into grayscale values in an image.

```
imagesc(squeeze(strMRI(60,:,:))')
```

If you are not used to looking at magnetic resonance images, you might think this is some kind of H.R. Giger picture that was colored by a blissfully optimistic 8-year-old girl. Perhaps we can tone it down a bit and flip the image to make it look more recognizable.

```
axis image, axis xy
colormap gray
```

This shows only one slice through a 3D image. I chose to plot the 60th slice of the first dimension more-or-less at random. Notice what we did here—we selected one element in the first dimension and all elements in the second and third dimensions. This gives us a 2D plane, which we then displayed as an image. The MRI slice was transposed to rotate it 90°. This is not necessary but is consistent with the convention of displaying magnetic resonance images (figure 25.1).

Take a few minutes to plot different slices from the same orientation (i.e., picking a different element from the first dimension), and also plot some random slices from different orientations. You can also try changing the colormaps for fun.

25.2 fMRI Data as a Four-Dimensional Volume

The "f" is for *functional* and often is lowercase (fMRI) and sometimes even italicized (*f*MRI), but all four words are equally abbreviated and it seems unfair to "MRI" to make the "f" stand out. But this has become standard

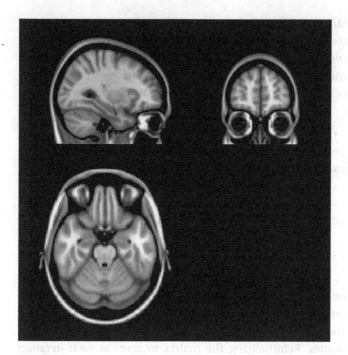

Figure 25.1
MRI data are stored as a 3D volume. This figure illustrates one slice from each dimension of this structural scan.

nomenclature in the literature, and you have to pick your battles (or follow the rule that sentences start with capital letters).

fMRI measures hemodynamic activity, which correlates, albeit in complex ways, with neural spiking and local field potential fluctuations (Singh 2012). fMRI is therefore taken as an indirect measure of brain activity. Each image of the brain is acquired every 1–2 seconds, and the hemodynamic response itself is fairly sluggish—it peaks after 6–8 seconds and decays over the next 8–14 seconds. Standard fMRI measurements have spatial resolutions of around 2–5 mm^3, and innovations in acquisition algorithms and MRI technology are allowing smaller voxel sizes and faster acquisitions, down to the sub-millimeter and sub-second scales (the spatial extent of coverage must typically be sacrificed for smaller voxels or faster acquisitions).

fMRI data can be conveniently stored in MATLAB as four-dimensional (4D) matrices, where the fourth dimension is time. The sample fMRI data used in this chapter come with the SPM analysis toolbox (www.fil.ion.ucl

.ac.uk/spm/data/auditory/). In the experiment, a subject heard spoken sentences for blocks of around 40 seconds, alternating with an equal number of blocks with no stimulus. These data are in ANALYZE format, so each of 96 separate acquisitions (96 time points) are stored in 96 × 2 files. Fortunately, the files are numbered in order, so they can be imported using a for-loop.

```
filz = dir('*img');
for imgi=1:length(filz)
    vol = readnifti(filz(imgi).name);
    % initialize
    if imgi==1
        fmridat = zeros([size(vol) length(filz)]);
    end
    fmridat(:,:,:,imgi) = vol;
end
```

Much of this code you've encountered in previous chapters. It might initially seem strange to initialize the matrix *inside* the loop, but this allows us to initialize the matrix properly without knowing *a priori* the size of the MRI volumes. Reinitializing the matrix to zeros at each iteration would be a mistake, because it would overwrite all the previous data. Hence the if-then statement that ensures this initialization is run only at the first iteration of the loop. The only assumption we make about the data *a priori* is that each image has the same number of elements in three dimensions.

But wait—how do we know that the files were imported in the correct order? Accurate ordering is crucial because we are reconstructing a time series one file at a time, and all subsequent analyses are based on the assumption that the volumes are in the right order. One potential source of error, for example, would be if the files are named as 1, 2, 3 ..., 9, 10, 11; instead of 01, 02, 03 ..., 09, 10, 11. The former sequence is often reordered by computers as 1, 10, 11 ..., 19, 2, 20, 21, and so on. This is useful to know when you are naming sequential data files: if there will be more than nine sequential files, start counting at "01" instead of at "1" (starting at "001" allows for more than 99 sequential files, plus cool things happen after 006).

Rather than blithely trusting that these files are in the correct order, we should check them. One way to do this is by examining the file names by typing {filz.name}' (the transpose prints a column vector, which facilitates visual inspection).

Let's try another, more quantitative method. We'll extract the volume numbers from the file name and store those as a numerical vector. The first step is to figure out how to extract the numbers from the file names. You can see that the volume numbers are always between the underscore and the period (e.g., "fM0023_007.img"), so the goal now is to find the position of the underscore and the period, and take the characters between them. You might be tempted to hard-code this by extracting the 8th to the 10th characters, but we want our method to be more general and work for any files where the volume number is the end of the file name just before the extension.

```
uscore = strfind(filz(1).name,'_');
dotloc = strfind(filz(1).name,'.');
volnum = filz(1).name(uscore+1:dotloc-1);
```

The variable `volnum` is a string, and we can use `sscanf(volnum,'%g')` to convert the string to a number. The three lines above should be placed in the loop over file names, although this doesn't completely solve our problem. The variable `volnum` is a string that contains the image number, but (1) we want this to be numeric, and (2) this variable is overwritten on each iteration of the loop. Adding the following line of code will complete our solution.

```
volnums(imgi) = sscanf(volnum,'%g');
```

After running the loop, you can inspect the vector `volnums`. If the images are imported in order, the volume numbers should always increase by one. To check this, I would inspect the output of `unique(diff(volnums))`. Why would I do this, and what would you expect the output to be if the files were in the incorrect order?

It may seem tedious to spend time checking the file order. But it's better to invest this time early on, rather than reach your final analyses and realize that an innocent but horrible mistake meant the data were imported incorrectly, and all other work needs to be redone.

Anyway, let's get back to the data. Using the same plotting code as we used for the structural MRI image, you'll notice that the fMRI images have a poor spatial resolution compared to the structural image. Because these data were recorded over time, we can plot a time course of activity from a single voxel (figure 25.2).

```
subplot(221)
imagesc(squeeze(fmridat(20,:,:,10))')
subplot(212)
plot(squeeze(fmridat(20,40,40,:)))
```

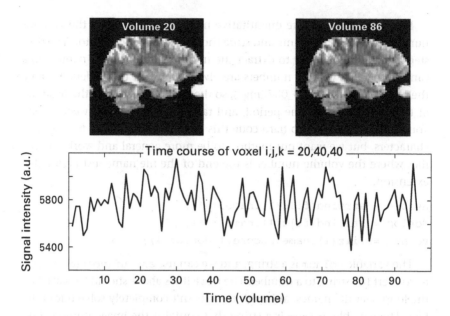

Figure 25.2
The images at the top show two slices of the fMRI volume taken at two different time points; the plot at the bottom shows the time course of the signal changes from one voxel.

25.3 fMRI Statistics and Thresholding

When analyzing fMRI data, the time course of activity at each voxel is treated as the dependent variable with which to perform statistical analyses. We will do a simple *t*-test on time periods when the subject heard speech versus time periods when there was no speech. The SPM manual states that the design had 6 volumes with auditory stimulation, 6 volumes without, and so on. The code below will generate a vector over volumes that indicates whether each volume had ("1") or did not have ("0") auditory stimulation. The first 12 images are discarded for early saturation effects, and so we begin with 18 (12 + 6).

```
onsets = 18:12:96;
timeline = zeros(length(filz),1);
for i=0:5
    timeline(onsets+i) = 1;
end
```

This code works by simultaneously accessing multiple indices into the `timeline` vector. You can inspect the values of `onsets+i` at each iteration. Note that this is accurate only because the timing of the experiment was exactly and repeatedly 6 volumes with and 6 volumes without auditory stimulation. This is often called a *block experiment design*. Nowadays, most researchers use an event-related design, meaning that the experiment events are faster and more frequent, and that deconvolution techniques must be used to isolate overlapping hemodynamic responses. You'll learn in chapter 28 about better data analysis techniques that involve modeling the hemodynamic response and fitting least-squares equations to data. We'll keep it simple here by computing a *t*-test on the average of volumes with versus without stimulation.

```
m0 = mean(fmridat(:,:,:,timeline==0),4);
m1 = mean(fmridat(:,:,:,timeline==1),4);
numerator = m1-m0;
```

The variable `numerator` is the numerator of the *t*-test. The denominator is the variance, and the *t*-test is the ratio between them.

```
v0 = var(fmridat(:,:,:,timeline==0),[],4);
v1 = var(fmridat(:,:,:,timeline==1),[],4);
denominator = sqrt((v0/sum(timeline==0)) ...
              + (v1/sum(timeline==1)));
tmap = numerator ./ denominator;
```

The next step in fMRI analyses is to "threshold" the *t*-statistics map by setting nonsignificant values to zero or to NaN ("not a number"). Plotting the thresholded map on top of the MRI facilitates anatomic localization and interpretability. As a quick-and-dirty threshold, let's reject any voxels with a *t*-value smaller than 2.5 (one-tailed, because we are looking for brain regions that show increased activity during auditory stimulation).

```
tthresh = tmap;
tthresh(tthresh<2.5) = 0;
```

In addition to the pixel-level threshold, it is common to apply a cluster-based threshold. The idea is that because fMRI volumes have some spatial smoothing (particularly after pre-processing), and because many cognitive and perceptual processes recruit distributed neural networks, it is unlikely to observe meaningful activations in only one isolated voxel. Instead, activations are considered interpretable if they are found in clusters of contiguously significant voxels. There are several ways to implement cluster-based thresholding. The simplest way is to remove any clusters that have fewer than k contiguous voxels. This can be done using the MATLAB functions bwconncomp or bwlabeln (both in the image processing toolbox), which will be discussed in more detail in the next chapter. The code below will use bwconncomp to extract all clusters (islands) in the volume, and will set any clusters with fewer than three contiguous voxels to zero. Three voxels is an arbitrary choice (as is $t < 2.5$ above); in practice, more sophisticated techniques are used to determine statistically appropriate thresholds.

```
islands = bwconncomp(tthresh);
islandsizes = cellfun(@length,islands.PixelIdxList);
for ii=1:islands.NumObjects
    if islandsizes(ii)<3
        tthresh(islands.PixelIdxList{ii}) = 0;
    end
end
```

Now we're ready to plot the results. An image of the thresholded statistical result is difficult to interpret on its own (figure 25.3). It would be better to overlay this thresholded map on top of an unthresholded map to

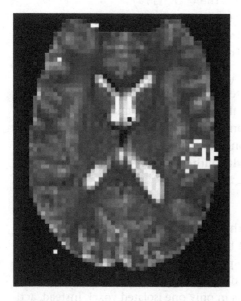

Figure 25.3
A slice of the time-average fMRI volume and, in white, the voxels that survived statistical thresholding in the contrast of auditory stimuli versus quiet rest.

facilitate anatomic localization. The code below will plot a thresholded map on top of an unthresholded map and apply a concatenated colormap to have the unthresholded map in grayscale and the thresholded map in the "hot" colormap. The code is more advanced than the material you learned in chapter 9, but by going through and plotting each line, I hope you'll agree that the complexity is built from simple parts.

```
cdiscr = 64;
% unthresholded image and normalize to [0 64]
img2plot = squeeze(m1(:,:,35))';
img2plot = img2plot-min(img2plot(:));
img2plot = cdiscr * img2plot./max(img2plot(:));
% same for stats map...
stat2plot = squeeze(tthresh(:,:,35))';
stat2plot(stat2plot==0) = NaN; % make invisible
% ... but normalize to [64 128]
stat2plot = stat2plot-min(stat2plot(:));
stat2plot = stat2plot./max(stat2plot(::));
stat2plot = cdiscr+cdiscr*.5 + cdiscr*.5*stat2plot;
% plot and set colormap
h(1) = pcolor(img2plot); hold on
h(2) = pcolor(stat2plot);
colormap([ gray(cdiscr); hot(cdiscr) ]);
set(h,'linestyle','none')
```

As discussed in chapter 1, the analysis here is oversimplified compared to what you would do with real fMRI data. For example, fMRI data are typically smoothed and temporally high-pass filtered, and possibly corrected for head motion. The statistical analysis should include regressors that have been convolved with a hemodynamic response (you'll implement this in exercise 16 of chapter 28), and statistical thresholds should incorporate some correction for multiple comparisons over ~100,000 voxels. All fMRI data analysis packages implement these and other steps.

25.4 Exercises

1. Write code to produce figure 25.1.
2. Write additional viewing code to show the structural MRI scan in 12 linearly spaced slices between the first and the last, in a 3-by-4 subplot matrix. Produce three figures corresponding to the three dimensions.
3. The edges of the structural MRI scan contain no information. Modify the previous exercise to show 12 linearly spaced slices from the slice at

20% and 80% of the way through the volume. Is 20%/80% a good cut-off? Perhaps you can think of better cutoffs based on, for example, the proportion of non-zero pixels in the slices.

4. The structural scan is high resolution. Try down-sampling the image by creating new images that contain skipped voxels (e.g., take every second voxel, or every third voxel, and so on). Compute the percentage of the size of each down-sampled version relative to the original version. Plot the slice closest to 50% on the z-plane for the different down-sampled images with the proportion of decreased size in the title of the image.

5. Using the fMRI data, compute the mean and standard deviation of the signal intensity values over time at each voxel. Then show a few slices from these two maps in a figure. For bonus points, complete this exercise without using the functions mean and std (or var, that would be too easy).

6. Figure 25.2 showed the time course of a single voxel. But one voxel can be noisy, so it is better to average the data together from several neighboring voxels. Write a function that takes three inputs: a 4D fMRI volume, an *ijk* coordinate (note about nomenclature: *xyz* coordinates usually refer to millimeters relative to the anterior commissure; *ijk* coordinates refer to voxel indices along each dimension), and *N* voxels around the coordinate for averaging. The function should generate two plots, one that shows all three slices of the MRI (averaged signal activity across all time points) at the specified *ijk* coordinates, and the time course of activity from that coordinate and the surrounding *N* voxels in all space dimensions. Make sure to include an exception in case one of the coordinates is on the edge of the image. The group of averaged voxels is called a region of interest (ROI). For bonus points, make the ROI be a sphere instead of a cube. To do this, you can specify the radius of a circle along each dimension.

26 Image Segmentation

The purpose of image segmentation is to isolate features of a picture. For example, if you want to share with the Internet a nice picture of your vacation, you might want to segment the picture to cut out your ex-boyfriend standing next to you. This would be an example of manual segmentation. For image analysis, and particularly for images in neuroscience, manual segmentation is often undesirable because it is tedious, time-consuming, and can be too subjective.

Automatic segmentation can be based on low-level image statistics, such as contrast or edges, or based on high-level features like pattern recognition. The sophistication of image segmentation algorithms is steadily increasing because of its role in computer vision and information extraction. Fortunately, most neuroscience applications of image segmentation require only basic segmentation procedures, some of which you will learn in this chapter. Much of the code in this chapter relies on the image processing toolbox in MATLAB.

26.1 Threshold-Based Segmentation

Threshold-based segmentation is often used to identify supra-threshold results in maps of statistical test values. Our first example will be a time-frequency power map, like what we created in chapter 19. The file tfmat. mat includes a time-frequency power matrix (variable tf) and associated *p*-values from a map-wise statistical test (variable p). There is a *p*-value for each time-frequency pixel, and the goal of thresholding here is to find regions in this map that have power values associated with *p*-values smaller than 0.05 (we won't worry about correcting for multiple comparisons).

Our task will be easier if we have a binarized map of the thresholded image. A binarized map is a matrix of the same size as the time-frequency map, but that comprises only zeros and ones to indicate subthreshold (0) and suprathreshold (1) pixels. We can then use this binarized map to draw contours on the unthresholded map that indicate the suprathreshold regions (figure 26.1), as well as to extract important information from those regions (e.g., effect size).

```
threshmap = p<.05;
contourf(timevec,frex,tf,40,'linecolor','none')
hold on
contour(timevec,frex,threshmap,1,'k')
```

The line of code that calls `contourf` produces the contour plot with the colors filled in and with the lines turned off. The line that calls `contour` draws a single contour line. Notice the subtle difference in function names: `contourf` versus `contour`. This picture (bottom left panel of figure 26.1)

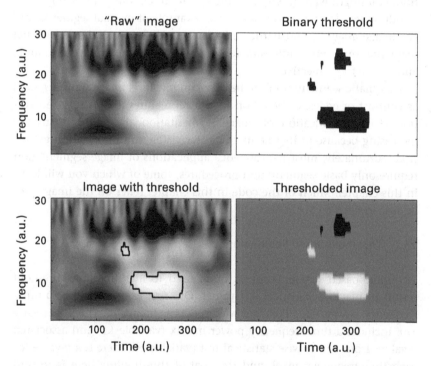

Figure 26.1
The top left panel shows a time-frequency power map in decibel scale. Based on the map of *p*-values, a binarized map (top right panel) was created to indicate significant (black) and nonsignificant (white) pixels. This map can then be used to outline significant regions in the power plot (bottom left panel) or to mask out nonsignificant pixels in the power plot (bottom right panel).

provides only qualitative information for visual inspection. What if you want to quantify these segmented blobs: How many are there? What are their sizes? What are the mean and maximum pixel values inside each blob?

We will use the MATLAB function `bwconncomp` to obtain quantitative information about the segmentation. The same results could be obtained with the function `bwlabel`; you'll learn about that in the next section.

```
islands = bwconncomp(threshmap);
```

The MATLAB function `bwconncomp` identifies contiguous clusters of pixels in an N-dimensional matrix. A *contiguous cluster* means that there is at least one pixel with a non-zero value ("true" in Boolean lingo), surrounded on all sides by pixels with zero values ("false"). Multiple pixels are considered a single cluster if they share a side or a corner (figure 26.2). You can think of clusters as islands in a sea. An optional second input to `bwconncomp` will provide a definition of "neighbor." This input is a number corresponding to the maximum number of possible neighbors. For example, for a 2D image, an input of 4 means that only pixels touching the four sides of a pixel are considered neighbors, while an input of 8 means that pixels touching a corner can be considered neighbors (figure 26.2).

The ouput of `bwconncomp` is a structure that contains several pieces of information about the number and locations of all the clusters. Take a

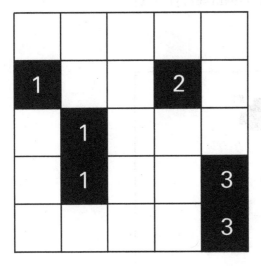

Figure 26.2
Illustration of clusters identified in a 5-by-5 matrix by the function `bwconncomp`. Each cluster is identified by increasing integers.

minute to inspect the variable `islands` and see if you can guess what information is contained in each field.

The output field that is probably the most useful is called `PixelIdx-List`. This is a cell array of linear indices into the matrix, where the indices in each cell tell you which pixels are part of each cluster. For example, if you want to know how many pixels are contained in each cluster, you simply count the number of indices in each cell. You could—if you wanted to—do this in a loop. But your friends might make fun of you. Therefore, in the interest of maintaining your social status, I recommend using the function `cellfun`, which works similarly to the function `bsxfun`.

```
pixelcnts = cellfun(@length,islands.PixelIdxList);
```

The function `cellfun` applies a function (in this example, the function `length`) to each cell contained in the cell array and returns those outputs in the variable `pixelcnts`.

To get an intuitive feel for the output of `bwconncomp`, let's make an image of the clusters. The code below creates a matrix of zeros and sets the value of the pixels in each cluster to a unique integer. We can then view the map as an image to see how `bwconncomp` selects and identifies clusters (figure 26.3).

```
stepmap = zeros(islands.ImageSize);
for i=1:islands.NumObjects
    stepmap(islands.PixelIdxList{i}) = i;
end
imagesc(stepmap)
```

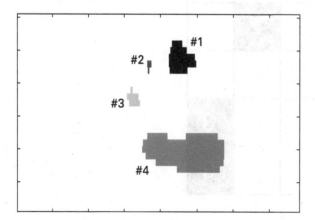

Figure 26.3
This image shows how the output of `bwconncomp` can be used to identify each island, indicated by integers.

26.2 Intensity-Based Segmentation

Let's do another example of image segmentation. Instead of picking a threshold based on a statistical significance value, here we will pick a threshold based on the distribution of pixel value intensities. Our example image will be the structural MRI scan used in the previous chapter. Intensity-based segmentation is often used in MRI and medical image processing more generally.

We start by examining the distribution of pixel intensity values in the image.

```
hist(mri)
```

Oh, that doesn't look very nice. When the function `hist` is given a multidimensional input, it computes and displays histograms for each dimension separately (figure 26.4). We want the distribution of *all* pixels in the image, regardless of which slice they happen to be in. In other words, we need to vectorize the matrix. Of course you know exactly how to do this, but just in case you need a gentle refresher, there are two ways to vectorize a matrix:

```
hist(mri(:))
hist(reshape(mri,1,[]))
```

Now we get a single histogram, but the features of the histogram are small and difficult to observe because they are dominated by a huge number of zeros (figure 26.4). Voxels outside the head have a value of zero, and we can safely exclude these voxels: We brain scientists are generally interested in things that happen inside the head.

```
hist(nonzeros(mri))
```

The function `nonzeros` returns the values of the input without the zeros. This function also vectorizes the output if the input is a matrix, so including "(:)" or the `reshape` function is unnecessary. This is important to remember: the function `nonzeros` will always return a vector regardless of the shape of the input.

The distribution of signal intensity values shows one large peak and a smaller bump to its right. Changing the *x* axis to a logarithmic scale (using the `set` command or by computing the logarithm of the values inside the `hist` function) helps make the positive bump more prominent (bottom panel of figure 26.4). The large peak in the middle of the distribution corresponds to voxels measuring gray matter, and the peak to its right corresponds to voxels measuring white matter (higher intensity).

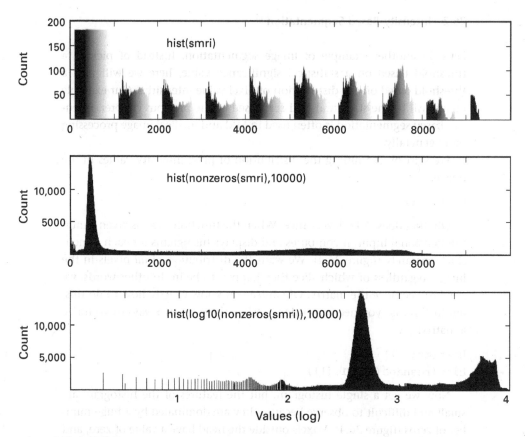

Figure 26.4
Three histograms of the same structural MRI scan; the codes that produced the respective histograms are shown. I'm sure you agree that the bottom panel best shows the features of the pixel intensity distribution.

The goal now is to separate the gray matter from the white matter. We will pick a threshold value based on visual inspection. This threshold will be used to discretize the original image into two separate binarized images, one for the subthreshold voxels (excluding zero) and one for the suprathreshold voxels.

```
% based on visual inspection of the log10 histogram
thresh1 = 10.^[3.0 3.5];
thresh2 = 10.^[3.6 3.8];
smriTh1 = smri>thresh1(1) & smri<thresh1(2);
smriTh2 = smri>thresh2(1) & smri<thresh2(2);
```

Inspecting slices of the resulting thresholded images (figure 26.5) reveals that we have separated gray matter and cerebrospinal fluid (subthreshold) from white matter and bone (suprathreshold) to a reasonable first approximation. Professional-grade MRI analysis toolboxes provide more sophisticated and accurate algorithms that are free from arbitrary user-specified thresholds and that can segment an MRI scan into gray matter, white matter, bone, cerebrospinal fluid, skin, and air cavities. But although the industry-standard algorithms are more complicated, they follow the same conceptual principles that you know now.

26.3 Once More, with Calcium

I'm having fun doing this, so let's do one more example with data from two-photon calcium imaging. This is a technique that allows simultaneous measurement of many neurons from the brain. The data come as a series of pictures captured over time in a mouse. Figure 26.6 shows one image (one

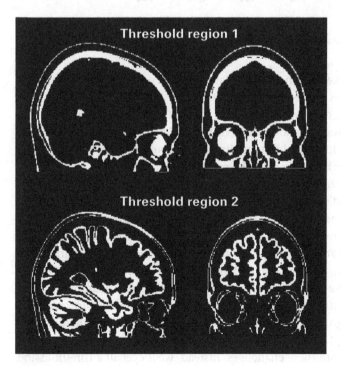

Figure 26.5
Two thresholded images from the structural MRI scan.

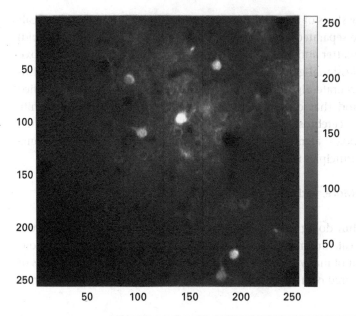

Figure 26.6
One frame of calcium imaging. The bright white spots are active neurons.

time point) taken from sample data that accompanies the Focustack
MATLAB toolbox (Muir and Kampa 2014).

In the image, each white spot is the soma of an active neuron. There are
around six neurons that are clearly visible, and perhaps another dozen that
can be visually identified but are dim (these neurons were inactive at the
time or were farther away from the camera's field-of-view).

To quantify the neural activity in these images, the individual neurons
must be isolated, which means we need to know which pixels correspond
to which neurons. You guessed it—this is an image segmentation problem.
Actually, it's a pretty difficult problem because the image can move slightly
from frame to frame (due to animal movement, blood pressure changes and
breathing, etc.); because different neurons can be superimposed in the
image if they are close together in the z-plane; because neurons are more
difficult to identify when they are not active; and because variations in
local image brightness can mean that a single image-wide threshold may be
inappropriate.

But we'll ignore these challenges. Instead, we'll eyeball a threshold and
see if it's good enough to get us started. Based on visual inspection of the
colorbar, I'll start with a threshold of 120. Any pixels with a value greater

than 120 can be considered pixels of interest. I'll use the function `bwlabel` here so you can see that although the format of the output is slightly different from that of `bwconncomp`, we can achieve the same end result.

```
[islands,numblobs] = bwlabel(im>120);
```

Note the difference between the output of `bwconncomp` and `bwlabel` (also note that there is a function `bwlabeln` for segmenting images with more than two dimensions). The main output of `bwlabel` is a map of the same size as the original image, but with distinct islands given by different numerical labels. In fact, this is the same as the matrix `stepmap` we created earlier in this chapter. We also get the variable `numblobs`, which is the number of islands (how could you get this number from the output of `bwconncomp`?). Either way you do it, there are 82 blobs in this image. That's an important number: 82, the number of clusters in this image. You'll need to remember this number for later.

At first glance, it looks like we're missing the neurons on the left side of the image (figure 26.7). But actually, this is just a color-scaling issue. There are in fact islands on the left, but they are given small numbers and so are difficult to see. Any ideas about how better to show this image?

Perhaps your first thought is to compress the values by plotting the logarithm of the image. But I think the Boolean of the image works best; it gives all blobs equal coloring, and we have no reason to promote certain blobs by highlighting them with different color values.

```
imagesc(islands)
imagesc(log(islands))
imagesc(logical(islands))
```

The segmentation looks okay. One easy way to improve it is by removing islands that have a very small number of pixels. The idea here is that real neurons take up more space than a single pixel, so any suprathreshold individual pixels (i.e., islands with a pixel count of one) can be considered noise.

```
for i=1:numblobs
    if sum(islands(:)==i)<2
        islands(islands==i)=0;
    end
end
```

This code loops through clusters, tests whether the number of pixels in each cluster (more precisely, the number of pixels with a value

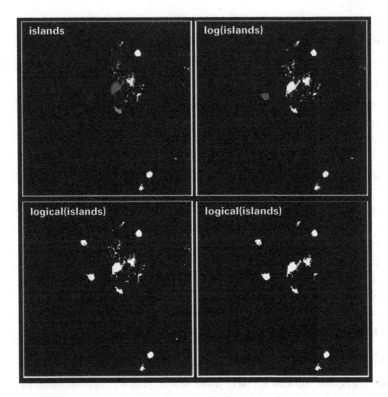

Figure 26.7
Images of segmented versions of the calcium image.

corresponding to the value for each cluster) is less than two, and, if true, sets those pixels to have a value of zero.

I'm not satisfied with the results. There are still too many small clusters. Let's try it again, using 10 contiguous pixels as the cluster. That looks better. Keep running this code, successively increasing the threshold. After a cluster size of 50, the real neurons get removed. So that's good, we now know that 50 is too high. It's good to test boundary limits. Now set the threshold back down to 20 and re-run the code. What happened? Nothing, actually. The pixels were irretrievably deleted from the image, and re-thresholding won't bring them back. You need to re-create the variable islands from the bwlabel function. This is a good opportunity to give some advice that I learned the hard way many times: If you want to test irreversible changes to a variable, make a backup of that variable first, particularly if the variable took a while to create. Here is my strategy:

```
% make backup
islands_o = islands; % "_o" for original
% <lots of testing here>
islands = islands_o; % recover
```

The thresholded image looks better now. There are some overestimations near the center of the image because of the lighter background color. A more sophisticated method would incorporate shape (neural soma tend to be round) as well as local changes in image contrast.

Now we have our putative identified neurons (a more conservative term would be *regions of interest*). Imagine we want to know which regions give the highest signal. This is fairly straightforward: Loop through all clusters and get the average pixel intensity from each cluster. Remember that we want to take the pixel intensity values from the original image, not from the islands variable (islands contains no pixel intensity values; only cluster numbers).

```
for i=1:numblobs
    activity(i) = mean(im(islands==i));
end
bar(activity)
```

Uh-oh. There should be 82 regions, but there are only 10 bars (figure 26.8). Do you see the error? One possible error when getting empty plots after computing the mean is that there are NaNs in the data, because the mean of a vector that includes NaNs is NaN. Try re-running the code using the function nanmean instead of the function mean. This changes nothing. Any other ideas?

Actually, there is no error here. It was a trick question. What happened is that we deleted the smaller islands without updating the numblobs variable or the remaining values in the islands variable. It's an easy source of confusion. You should always try to be cognizant of situations when changing one variable requires changing other variables.

Real two-photon data (and other imaging modalities) comprise many images recorded over time. By extracting the time course of pixel intensity values from each identified neuron, neural activity could be linked to different experimental conditions, medication treatments, and so on. The principle is similar to fMRI, except that different aspects of neural activity are measured (e.g., calcium or voltage) instead of hemodynamic activity.

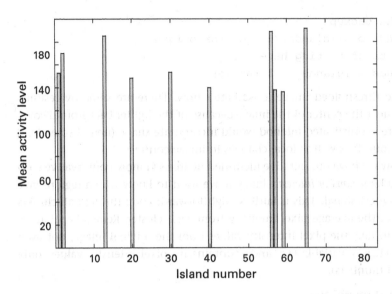

Figure 26.8
This figure shows the average intensity values from the 82 original clusters. Clusters
that were removed because they were too small are shown here as zeros.

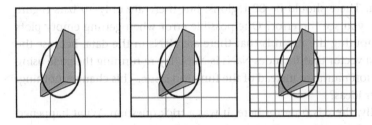

Figure 26.9
We want to segment an image into a grid of regular-sized boxes, like in these images.

26.4 Defining Grids in Images

Our goal in this section will be to define a grid of boxes on a 2D plane using
differently sized boxes, like in figure 26.9. This procedure can be used
to segment images into blocks. For example, image-compression formats
like JPEG work by cutting an image into equally sized blocks and compress-
ing each block separately (in case you are curious, the within-block compression involves a dimensionality reduction procedure, like a Fourier
transform or components analysis; when the compression is too stringent,
you can see the block boundaries).

Box 26.1

Name: Dylan Richard Muir
Position: Postdoctoral Researcher
Affiliation: Biozentrum, University of Basel, Basel, Switzerland
Photo credit: Ayumi Isabelle Spühler

When did you start programming in MATLAB, and how long did it take you to become a "good" programmer (whatever you think "good" means)?
I started programming in MATLAB during my undergraduate studies in electronic engineering. But at that stage I had been programming in other languages (C, C++, Assembler, Basic) for quite some years and already had some training in software engineering and object-oriented programming. So I think I had a bit of a head start!

What do you think are the main advantages and disadvantages of MATLAB?
As a language for learning to write code, especially if you're learning on the job as many neuroscientists are, MATLAB is pretty ideal because it's so forgiving. Because it was designed as a simple scripting language, you don't need to worry about the usual housekeeping of most languages—importing libraries, declaring variables, allocating and freeing memory, and so on. Because all those things are taken care of for you, most of the code you write has a direct effect in manipulating your data. But because MATLAB also supports more

modern language constructs such as classes and objects, you can learn to write more structured code as your experience increases.

Personally, I really like the ability of MATLAB to express vectorized operations so succinctly. It saves a lot of junk overhead code to loop around a matrix, for example. When you're working with experimental data, you usually want to perform an operation on a large chunk of data at once. MATLAB lets you do that very compactly, which makes your code easier to read and understand. That's very valuable when you have to work with someone else's code!

As a platform, it's hard to understate the value of knowing that if someone has MATLAB installed, they are guaranteed to have a fairly complete set of toolboxes that are guaranteed to work. That means that code working on one system will almost surely work identically on another system, even between Mac, Linux, and Windows.

I think MATLAB's main disadvantage is its age, and—paradoxically—it's large user base! The language comes from the days when Fortran seemed like a good idea, and so it suffers under the baggage of a lot of old-fashioned constructs that probably wouldn't have been introduced if it was designed from scratch. But because millions of scientists and engineers rely on MATLAB staying compatible with itself, the language can't change very quickly. That has also left a bunch of small functions laying around with inconsistent calling conventions that are slowly being cleaned up.

Do you think MATLAB will be the main program/language in your field in the future?

MATLAB is under a bit of pressure from newer languages like Python, R, and Julia. But I think MATLAB still has the edge in ease of use and portability of code. And because neuroscientists usually have a background in biology or psychology, and not in computer science, they will continue to benefit from a language with a forgiving learning curve.

How important are programming skills for scientists?

I think it's crucially important for scientists to know how to write code. There is an enormous number of statistical techniques, especially the newer "data science" methods, that are just inaccessible unless you can write your own analysis functions. Even if you're using off-the-shelf functions, you still owe it to yourself—and to your data—to be able to understand how the code works. On top of that, if you're using new experimental techniques that aren't off-the-shelf, many of the systems need a very hands-on approach to produce reliable data. That means you need to be able to tweak the code of the system you're working with.

On top of that, the world today runs on code. Why not allow yourself to peek under the hood?

> **Any advice for people who are starting to learn MATLAB?**
> Learn to use the online documentation efficiently! There is almost always a
> function in MATLAB that will do what you want, but unless you use that func-
> tion every day, you probably won't remember its name.
>
> Learn how to write vectorized code; that's one of the things MATLAB does
> really well.
>
> Start simple, but learn how to write structured code. It will make your life,
> and your colleagues' lives, much easier in the long run. Structured code means
> code reuse—it's much, much better to have a library of code that you can all
> use and improve on together, rather than each PhD student reinventing the
> wheel!
>
> Use a source code repository, especially for code that you're sharing with
> others. This will keep your code consistent across machines and across the lab,
> and let you easily roll your code back to 6 months ago, when you need to work
> out why some analysis that worked then is failing now.

Let's start by creating a 2D grid with N-by-N pixels.

```
n = 300;
[gridX,gridY] = ndgrid(1:n,1:n);
```

The columns and rows of this grid are numbered from one to N (here,
300). We want to down-sample the numbering from 1 to k, where k is our
arbitrary discretization parameter.

```
k = 7;
gridX = ceil(k*gridX./n);
gridY = ceil(k*gridY./n);
```

If this construction looks vaguely familiar, it's because we used something similar when discretizing reaction times into k bins (see chapter 14).
But we don't quite yet have what we need. We now have two matrices that
are numbered 1 through 7 (one matrix with rows, one matrix with columns), but we want one matrix with 49 boxes that each has a unique number. See if you can figure out how to get our final matrix before reading the
next paragraph.

By now, you've either (1) figured it out and are curious if I solved it the
same way, (2) tried but couldn't figure it out, or (3) are lazy and didn't even
try (now is your last chance to stop reading and try it!). A first thought
might be to add the two matrices together, but this will produce a matrix
with many redundant numbers (i.e., multiple boxes will have the same
numerical value). My solution is to scale up one of the matrices, then sum

the two, and then renumber the final matrix. The procedure is graphically illustrated in figure 26.10.

```
tempG = gridX + gridY*(k*1000);
u = unique(tempG);
grids = zeros(n);
for ui=1:length(u)
    grids(tempG==u(ui)) = ui;
end
```

Notice that when gridY is multiplied by k*1000, it will have non-overlapping values with gridX. These two matrices can then be summed with no risk of redundant number values. Technically, that's all we need to do. But it would be nice to have the numeric identifiers be increasing integers. Hence, the rest of the code relabels the elements in the final grid. You should take a few minutes to confirm that this procedure works for a variety of N and k parameters.

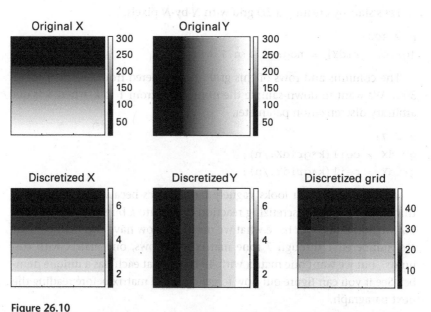

Figure 26.10
A solution to discretizing an image into a regular grid. An extra scaling-and-summing step was necessary to provide each box its own unique value.

26.5   Fractals and Boxes

One of the applications of defining arbitrarily sized grids on images is
to compute fractal dimension. In chapter 21, you learned that fractal-
like processes can be measured in time series data using detrended or
demeaned fluctuation analyses. Conceptually, the idea is to determine
whether the magnitude of fluctuations increases with the size of the time
window used to measure those fluctuations. The analogous procedure with
images is to determine whether the size of an object depends on the size of
the ruler used to measure that object. This was famously characterized by a
paper entitled "How Long Is the Coast of Britain?" by Mandelbrot (1967),
who is the grandfather of fractal geometry. Curiously enough, the length
of the coastline of England appears to get longer as the ruler used to measure it gets smaller. This inverse relationship—particularly a logarithmic
relationship—between the apparent size of an object and the unit of measurement provides evidence for self-similarity, which in turn provides
evidence for fractal geometry. And this is not just some quirky British thing.
There are myriad examples of fractal geometric objects, ranging from clouds
to ferns to volcano eruptions to the dendritic branching of neurons.

Several methods exist for analyzing images to determine whether they
exhibit fractal-like characteristics. One method is to determine whether the
number of boxes required to cover an image depends on the size of the
boxes. We will compute a classic fractal geometric object, and then we will
use variously sized boxes to show that the object appears to get larger as the
boxes that measure the object get smaller. Our fractal object will be the
Sierpinski triangle. You may not know it by name, but you probably recognize it in figure 26.11. Below is code that will produce an approximation of
a Sierpinski triangle (it's imperfect, but with enough dots the approximation is pretty good).

```
N = 10000;
[sx,sy] = deal(zeros(1,N));
for i = 2:N
    k = ceil(rand*3);
    sx(i) = sx(i-1)/2 + (k-1)*.25;
    sy(i) = sy(i-1)/2 + (k==2)*.5;
end
plot(sx,sy,'.')
```

The code above produces compact notation using sparse matrices to
indicate the xy coordinates at which to draw dots. This sparse format is fine

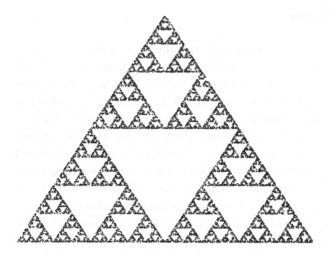

Figure 26.11
The Sierpinski triangle made from dots.

for making a plot, but for our analysis we want a full 2D image to discretize into grids. Your initial guess for how to convert the sparse format to a full matrix format might be to try something like this:

```
siertri = zeros(N);
siertri(x,y) = 1;
```

But this won't work, for two reasons. First, the *xy* coordinates have decimal points, and we're only allowed integer indices into a matrix. Second, the *xy* coordinates are actually all between 0 and 1, so even if we rounded the coordinates to the nearest integer, we'd still get only two unique elements (0 or 1). And zero would give an error because you cannot index the zeroth index in a matrix. The solution is to scale the *xy* coordinates from their initial range of [0 1] to the new range of [1 *N*], and then round up to the nearest integer (rounding down could produce zeros).

```
newX = ceil(sx*N);
newY = ceil(sy*N);
```

Now we're getting somewhere. But we're not quite there yet. The variables sx and sy contain a few zeros at the beginning due to the initialization; those zeros need to be removed. Second, it might be easier to convert these *xy* coordinates to linear indices. There is also a third change we need to make, but we'll get to that later.

```
wherezeros = newX==0 | newY==0;
newX(wherezeros)=[]; newY(wherezeros)=[];
linind = sub2ind([N,N],newY,newX);
siertri(linind) = 1;
```

Technically, this works. But that image is huge (a matrix of size 10,000 by 10,000), and the colored pixels are tiny. I won't show the plot here because it's nearly impossible to see the Sierpinski triangle. It will also take a really long time to run the box-counting analysis. Let's down-sample the image as our third processing step.

```
Nd = 200; % down-sampled N
linind = sub2ind([Nd,Nd],ceil(Nd*sy),ceil(Nd*sx));
siertri(linind) = 1;
```

The down-sampled version looks much better (figure 26.12). Finally, we are ready to apply the box-counting analysis. The idea of this analysis is to create a series of discretized grids like in the previous section and determine

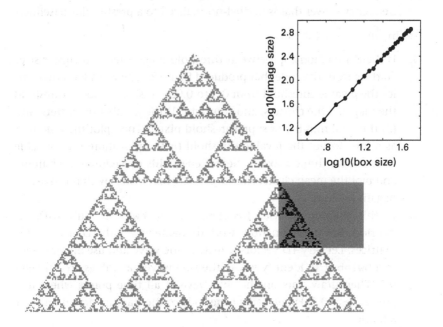

Figure 26.12

The full-matrix version of the Sierpinski triangle. The gray box illustrates one step of the box-counting method. For each discretization size, we determine how many boxes are needed to cover the image. The inset shows the linear relationship in log-log space between the image size and the box size.

whether each box in the grid overlaps with the image. This process is repeated over many different discretizations (box sizes). The results are shown in figure 26.12. I'm sure you are excited to see the code for this, but alas, you'll have to write it yourself in the exercises.

26.6 Exercises

1. The thresholded map in figure 26.1 has colors reversed from what's written in the code (black is 1, white is 0). Reproduce this inverted map by modifying the following line of code (do not simply redefine the color axis). There is one way to do it by adding one character and one way to do it by adding two characters.

```
contourf(threshmap)
```

2. The following line of code does not produce a MATLAB error, but it probably does not produce the intended result. What is the bug, what is the effect, and how you can fix it? (dbmap is a 2D matrix of time-frequency power that is decibel-normalized to a prestimulus baseline.)

```
imthresh = abs(dbmap>3);
```

3. Is there a relationship between threshold magnitude and cluster size? Starting from the code that produced figure 26.1, write a loop that varies the p-value threshold from 0.5 to 0.001. Inside the loop, threshold the map, get the clusters, and compute (1) the number of clusters, and (2) the total number of suprathreshold pixels. Then plot these metrics as a function of the p-value threshold (is the plot more interpretable with a logarithmic x axis?). Next: Repeat this procedure but compute and plot the mean cluster sizes separately for clusters with positive and negative values.

4. In this chapter, we used bwconncomp and bwlabel only with 2D matrices. But these functions work the same way with 1D or 3D or 9D matrices. Let's try 1D vectors. Create a sine wave and use bwconncomp and bwlabel to identify where the sine wave has values greater than 0.5. Then draw stars on the sine wave at all time points where it is greater than 0.5. Finally, adapt the code so the stars are drawn on the y = 0 line.

5. Let's try increasing the sophistication of the size threshold in the calcium imaging example. We assume that the soma of the neurons in this image should be about the same size. Extract the sizes of all clusters, and figure out a way to use that cluster size distribution to

determine a data-driven threshold for removing small clusters that are unlikely to be somas.

6. After eliminating pixels from small clusters, the numbering of the clusters in the calcium image became non-incremental. Write code to renumber the image so the cleaned version has cluster numbers from 1 to N.

7. Did you like the fractal examples? Many naturally occurring objects in nature are thought to have fractal geometric qualities, including snowflakes, trees, and mountains. Find a few pictures of natural objects and implement the box-counting algorithm. It is likely that you will need to threshold the image first.

8. The .PixelIdxList output given by bwconncomp is pretty handy. Re-create this output using the output of bwlabel.

9. The online materials include a picture of a Nissl-stained brain slice (part of the amygdala of a rhesus macaque) downloaded from http://brain-maps.org. The neurons are stained as purple. Import the image into MATLAB and isolate the neurons.

10. The function nonzeros returns a 1D vector, regardless of the size of the input matrix. Write code that will leave the matrix in its original size, but will convert the zeros into NaNs.

11. Write code that can implement the box-counting method, and reproduce the inset line plot in figure 26.12. The online code has some hints to get you started.

12. In section 15.6 of chapter 15, you learned about one method for correction of multiple comparisons using extreme values. Now that you know about threshold-based image segmentation, you can learn multiple comparisons correction based on cluster sizes. Instead of taking the extreme values at each null hypothesis iteration, threshold the null hypothesis image based on a statistical criterion (e.g., $p < 0.05$), and extract the number of pixels in all suprathreshold clusters. Save the number of pixels in the largest cluster. After looping through all iterations, the 95% of the maximum-cluster distribution is the cluster threshold. The original data can be thresholded at $p < 0.05$, and any clusters that are larger than the threshold are kept.

13. The function you should use to add folders the MATLAB path (until MATLAB restarts) is called addpath. There is another function called path. It is easy but dangerous to confuse the two functions. Read the help files for these functions to figure out what the difference is. Then test it (warning: this exercise requires exiting MATLAB, so save any

open scripts first). First type path and you will see the entire MATLAB
path. Next, type path('../'). That seems innocent enough, but you
just replaced the entire MATLAB path with whatever happens to be in
the previous folder. Most MATLAB functions are now inaccessible
(don't worry, it's only temporary). Try, for example, to compute the
mean of a few numbers. Or try to plot some random numbers. MAT-
LAB is really screwed up now. Type path again, and follow MATLAB's
sound advice to close the program and restart. The previous path will
be restored with no permanent damage.

27 Image Smoothing and Sharpening

27.1 Two-Dimensional Mean Filtering

A 2D mean-smoothing filter is the same concept as—and has a very similar implementation to—the 1D running-average filter you learned about in chapter 20. It's a form of low-pass filtering and thus produces an image that is smoother (blurrier) than the original image. Mean filtering is an oft-used denoising method because many sources of noise produce high spatial frequency artifacts.

Let's start with that picture of Saturn that we used several times already (if you are getting bored of this picture, then don't worry; this is the last chapter that uses it). It's not a very noisy image, so let's add some noise.

```
saturn = saturn + randn(size(saturn));
```

Similar to the running-average filter, we will now replace each pixel in the image with the average of the surrounding k pixels, where k is an integer parameter that we can select.

```
k=9;
    for i=k+1:size(saturn,1)
    for j=k+1:size(saturn,2)
        temp = saturn(i-k:i+k,j-k:j+k);
        saturn(i,j) = mean(temp(:));
    end
end
```

The code loops through all rows and columns, extracts a submatrix of data that corresponds to a box around each i,j coordinate, and then computes the average of that submatrix (why do we need the " (:) "?). Unfortunately, there are three problems with that code. Can you figure out what they are before reading the answer below?

The first problem is that the code will produce erroneous behavior at the end of each loop. Consider that the loop ends at index 480, which is the width of the image in pixels. But the code will try to extract pixels up to `i+k = 480+9 = 489`. MATLAB should give an error about "Index exceeds matrix dimensions." But it doesn't because of the next, and more devious, problem. The solution here is to have the loops stop k units before the end of the image, or `size(saturn,1)-k`.

The second problem is that the image of Saturn is a 3D matrix (the third dimension is color), but we are indexing it like a 2D matrix. That's not good. MATLAB won't produce any errors because it will try to use 2D linear indexing into the 3D matrix. But it's definitely going to produce undesirable results because pixels from different color dimensions will intermingle. Remember the important rule about matrix indexing: If there are N dimensions, you must index the matrix using exactly $N-1$ commas. Here we have three dimensions and only one comma ($N-2$). The solution to this problem is either to smooth only one dimension or to add another loop over elements in the third dimension.

The third problem is more insidious and also more difficult to spot. As the center of the filter moves around the image, parts of the image get double-filtered. This happens because each submatrix comprises a combination of unfiltered and filtered data. The solution to this problem is to filter the image into a different variable, like `saturnfilt(i,j,1) = mean(temp(:));`

Actually, there is a fourth issue with that code that isn't a problem per se, but is poor programming. In overwriting the image with a noisier version of itself, we've lost the original data. A better implementation would be to create a new variable and preserve the original. That would allow us to compare the noisy with the original version or compare the results for different levels of noise, different types of filters, and so on. The following line of code is better, although the damage is already done—you'll need to load in the image again.

```
noisySaturn = saturn + randn(size(saturn));
```

Anyway, after the bugs are cleaned, the result looks pretty good (figure 27.1). The noise is cleaned up, although it came at the expense of losing some of the sharpness of the image. Depending on the goal of the analysis, that might not be detrimental. Indeed, neuroscience images are often intentionally smoothed prior to data analysis. The analysis takes a while to run, but you'll see later in this chapter that the frequency-domain implementation of this convolution provides a significant speed benefit.

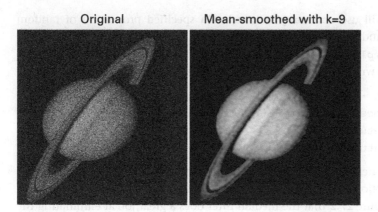

Figure 27.1
Noisy Saturn, and the result of applying a simple mean-smoothing filter.

Before moving on, I encourage you to take a minute to explore the code by changing the *k* parameter to observe the effects on the resulting image. What is a reasonable amount of smoothing before the image becomes really blurry, and what happens at the edges of the image as you increase *k*?

The mean-smoothing filter worked well here because the noise (1) was roughly randomly distributed and (2) had values that were in a similar range as those of the signal.

27.2 Two-Dimensional Median Filter

When the noise is not randomly distributed or when its values are extremely different from those of the image, a mean-based filter will often produce poor results. This was illustrated in chapter 20 for 1D time series, and now we return to the issue. Replication is important in science.

Let's take the Saturn image, but instead of adding random Gaussian-distributed noise, we'll add large-amplitude noise spikes. We can randomly select pixels to replace with noise spikes using the `randsample` function. This function takes two inputs: a list of numbers (the "population") and a number of samples from that population to take. An optional third input allows you to specify whether to draw the numbers with replacement or without replacement (without is the default). Check it out:

```
randsample([1 3 4],2)
randsample(0:.1:1,4)
```

We will use `randsample` to select a specified proportion of random pixels, and then we will replace those pixels with a large number. The `randsample` function is in the statistics toolbox; in the exercises, you'll be asked to write a new function that provides the same output.

```
nPixels = round(.1*numel(saturn));
spikelocs = randsample(1:numel(saturn),nPixels);
noisySaturn = saturn;
noisySaturn(spikelocs) = 123456789;
```

From here, the median filter proceeds similarly as the mean filter, except the function `median` is used instead of the function `mean` (duh). You can see in figure 27.2 that the median filter does a great job at eliminating the speckle noise.

27.3   Gaussian Kernel Smoothing

In chapter 11, you were introduced to the idea of using the Fourier transform to decompose an image into its 2D spectral representation. And in Chapters 19 and 20, you learned how to use the Fourier transform to highlight specific frequency characteristics of a 1D time series. Let's combine these two concepts to low-pass filter (a.k.a. smooth) an image in the frequency domain.

Figure 27.2
Sparse high-amplitude noise is successfully removed with a median filter. On the other hand, the moons are also filtered out, and they are not noise. This is a good illustration of how denoising strategies must always be checked carefully, because removing noise can also mean removing signal.

Remember the spike-field coherence data you saw in chapter 23 (see figure 23.6)? That image comprised several thousand local field potential (LFP) traces surrounding each action potential recorded from a neuron in the rat hippocampus. Single-trial LFP data can be noisy, and our goal in this section is to attenuate this noise by smoothing the time-by-spikes LFP image. Unlike the Saturn picture, which is roughly square-shaped, the spike-field image is asymmetric—it is an order of magnitude taller than it is wide.

A mean-smoothing filter weights all pixels equally. But pixels are not people, so we don't need to treat them as equals. Let's say that neighboring pixels are more important than more distant pixels. We can quantify this weighting-by-distance matrix as a 2D Gaussian. Furthermore, because this image is asymmetric, and because neighboring time points are naturally more strongly autocorrelated than neighboring trials, we might want to smooth more along the y axis (trials) than along the x axis (time). That's easily done; we just need to construct a Gaussian with different widths in each direction. An anisotropic Gaussian is similar to the isotropic Gaussian that you've already seen, except that it decouples the widths along each dimension (figure 27.3).

```
gx = -20:20;
gaus2d = zeros(length(gx));
sx = 10;
sy = 30;
for xi=1:length(gx)
    for yi=1:length(gx)
        xval = (gx(xi)^2)/(2*sx^2);
        yval = (gx(yi)^2)/(2*sy^2);
        gaus2d(xi,yi) = exp(-(xval+yval));
    end
end
```

Now we can perform a 2D convolution between the spike-field image and the 2D Gaussian. Results are shown in figure 27.4. You will need to write the code to create this image.

27.4   Image Filtering in the Frequency Domain

We will now generalize the methods in the previous section to define any arbitrary filter in the frequency domain. Let's start by taking the 2D fast Fourier transform (FFT) of the picture of Saturn. Remember that an easy

Figure 27.3
A few different 2D Gaussians.

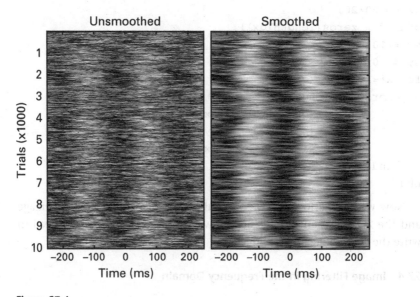

Figure 27.4
Spike-field coherence before and after smoothing. The smoothed version looks very similar to the original, and that's a good thing—data should be smoothed to improve visibility and remove some noise, but you don't want the smoothed result to look qualitatively different from the original result.

mistake to make is to use the function `fft`—that would not produce a MATLAB error, but would also not produce a 2D FFT. For simplicity, we will only use the first color dimension.

```
% get 2D Fourier coefficients
saturnX = fft2(squeeze(saturn(:,:,1)));
```

Before defining the filter in the frequency domain, remember that with a 2D FFT, the low frequencies are at the four corners, while the high frequencies are in the center. The MATLAB function `fftshift` can be used to swap quadrants to put the low-frequency information in the center. There is no mathematical reason to prefer either orientation, but most people find the latter more intuitive. Having low frequencies in the center will also make our filter a bit easier to construct. We'll create a low-pass and a high-pass filter. First, we need to find the indices of the center points of the image, and then we define how many frequencies to use as a cutoff. We'll stick to units of pixels; transforming to a meaningful metric such as cycles-per-degree requires knowing the distance between your monitor and your eyeballs—hopefully it's not too close, otherwise your ophthalmologist might get upset, although your optician would be happy to have a new customer (I admit I had to look up both the spelling and the meaning of those words).

```
% get sizes of image and midpoints
imgdims = size(saturnX);
midX = round(imgdims(2)/2);
midY = round(imgdims(1)/2);
nPix2use = 100;
```

Now we are ready to define our filter. It will be a simple block filter in which frequencies below a threshold (the variable `nPix2use`) will be preserved, while frequencies above this threshold will be eliminated.

```
% create low-pass filter kernel
loPass2d = zeros(imgdims(1:2));
loPass2d(midY-nPix2use:midY+nPix2use, ...
        midX-nPix2use:midX+nPix2use) = 1;
```

Figure 27.5 shows this filter. Now we need to point-wise multiply the filter by the FFT of the image and then take the inverse FFT. You might initially think to have a line of code like this:

```
filtimg = ifft2(saturnX.*loPass2d);
```

Figure 27.5
Effects of band-pass filtering the Saturn image. The top left plot shows the log-power spectrum, and the top right plot shows the filter that will be applied. The bottom left plot shows the result of using the box as a low-pass filter, and the bottom right plot shows the result of inverting the filter to produce a high-pass filtered image.

But this is incorrect. Why is it incorrect? Let's think about how a low-pass filter works. It preserves the Fourier coefficients corresponding to low frequencies while stamping out the Fourier coefficients corresponding to high frequencies. But the low frequencies are at the corners, while our low-pass filter is zeroing everything out except the very highest frequencies at the center. A simple solution is to use the fftshift function on the filter kernel. Try making an image of that to convince yourself it will work. Then point-wise multiply the filter.

```
imagesc(fftshift(loPass2d))
filtimg = ifft2(saturnX.*fftshift(loPass2d));
```

And that's about it (figure 27.5). You can turn the low-pass into a high-pass filter by initializing the filter kernel to ones and setting the pixels around the center to zeros. You should recall from chapter 20 that this filter kernel has sharp edges, and it will therefore produce ripple artifacts (you can see these artifacts when inspecting the moons in the high-pass filtered image in figure 27.5). A better filter would have smoother transitions. This could be accomplished, for example, by defining the filter shape to be Gaussian instead of rectangular. Take a few minutes to play around with the code that produces this figure. Try changing the cutoff values for the filter—how do changes in the filter affect the filtered images? Notice in figure 27.5 that the moons are still visible, meaning the "low-pass" filter cutoff is still fairly high. What's the filter cutoff that will get rid of the moons?

27.5 Exercises

1. Adjust the 2D mean filter code to be able to filter the edges, that is, from 1:k. (This is possible without adding any new lines of code, but it's more important to solve a problem accurately than to minimize the number of lines of code.)

2. Also in the mean-smoothing code, replace the two lines with variable temp to be one line of code that does not require an intermediate variable.

3. If you look in the MATLAB code, you'll see the following line to import the Saturn picture. Why did I convert to double and then divide by 255? What happens if you remove this?

```
saturn = double(imread('saturn.png'))./255;
```

4. Two-dimensional convolution with loops takes a while to run. Add some code to that loop so it prints out progress into the command line. It should report something like

```
15% completed after 8.23 seconds.
```

   But it shouldn't print out this message too often, otherwise it's annoying and will slow the computation down even more. How can you adjust the code so it prints a message whenever the progress reaches another 5%? (When you are considering implementing procedures like this in your analysis code, remember the adage that a watched pot never boils.)

5. In the online code for smoothing, I was lazy and used MATLAB's conv2 function. Replicate the result using convolution via frequency-domain multiplication. Don't forget this is a 2D image, so use fft2.

6. When you do convolution yourself, `dataX.*kernelX` is the same as `kernelX.*dataX`. With `conv2`, does order matter? What does this tell you about using `conv2`?

7. Redo mean smoothing with frequency-domain multiplication. Most of the code you need is already in the code for the Gaussian case; you just need to figure out how to define the kernel of a `mean` function. For a challenge, solve this problem from scratch without looking at the other code.

8. Run the mean-smoothing filter on the noisy version of the Saturn picture that we used to illustrate the median filter. How does the mean filter compare? What is the important lesson here about knowing when to use different kinds of filters, and understanding the nature of the noise in your data?

9. The median filter can take a while to run. You can speed it up by applying the filter only to pixels that have some extreme value. From the spike-noised Saturn picture, run the median filter again but only applied to pixels exceeding a threshold. Use a data-driven method to determine an appropriate threshold. Use the `tic/toc` function pair to determine how much time was saved.

10. Apply the `fft` function to an image instead of `fft2`. What does the power spectrum look like, and what kinds of sanity checks can you think of to check that you used the correct FFT function?

11. Write code to create figure 27.4. Take some time to explore this code. Change the two widths of the Gaussian and see how that affects the result of convolution. When exploring new methods, don't only try reasonable ranges of parameters. Testing extreme or inappropriate parameters gives you an appreciation of how even good methods can produce uninterpretable results. This also helps you learn what to look for when analyses go awry.

12. The `randsample` function is in the statistics toolbox. Based on the description of how it works (in this chapter and in the help file, which you can find online if you don't have this toolbox), write your own function that produces the same functionality.

13. The Fourier coefficients contain information about power and phase. We've worked only with power in this chapter, but the phases are important for images, as you will see in the next few exercises. Write code to separate the phases from the magnitudes of all Fourier coefficients, then reassemble them in the same order. Use the inverse Fourier transform to reconstruct the Saturn image (use only one color dimension to have a 2D image). It's not as simple as separating the real and

imaginary parts, because phase and magnitude both use real and imaginary information. (Hint: A Fourier coefficient can be represented as ae^{ip}, where a is the magnitude and p is the phase.) Make sure your reconstruction is identical to the original image.

14. Now that you know how to extract and recombine the phases, it's time to scramble them. Replace the phase angles from all coefficients with random numbers. What is the appropriate range of random numbers to use? Is the reconstructed image recognizable as Saturn?

15. Now keep all of the original phases, but shift them all by some constant offset, say, $\pi/3$. You will need to write some code to make sure that the angles remain in an expected range (hint: mod).

16. Finally, randomize only some phases while leaving others intact. Using the box-shaped filter from the end of this chapter, randomize only the phases inside that box. Is the reconstructed image recognizable? How about when you randomize only the phases outside that box?

VI Modeling and Model Fitting

28   Linear Methods to Fit Models to Data

Model fitting is an important part of data analysis. For purposes of this and the next chapters, a *model* is a mathematical description of a latent process that might produce or explain empirically measured data. (If you thought that "model fitting" was a behind-the-scenes look at *Victoria's Secret* magazine, then I'm sorry, you have the wrong book.)

Model fitting can be used as a data-reduction technique (by having a small number of parameters that characterize a larger data set); as a hypothesis-testing technique; as a way to relate findings across different types of data, species, and so forth; and as a way to estimate thresholds for perception, memory, and other cognitive processes. The idea of model fitting is to construct a mathematical description of the data, and then determine parameter values that provide the closest match between the model and the data. This chapter will focus on linear methods for estimating those parameters; the next chapter will focus on nonlinear methods.

If you want a good fit between a model and data, you need to have a good model and you need to have good data. Lousy models or excessively noisy data can produce uninterpretable results. Before you start throwing models at your data, think carefully about the data, about the mechanisms and circumstances that produced those data, and about the kinds of models that are most appropriate for those data. Better still: Think carefully about what models you want to fit to your data *before* you even start collecting data. That will help ensure that the data are appropriate (e.g., enough data points to fit the model) and clean.

28.1   Least-Squares Fitting

Least-squares fitting is perhaps the most commonly used data-fitting technique in science. The idea of least-squares fitting is that the best model parameters are those that minimize the squared distances between the observed data points and the model-predicted data points. Those distances (or errors) are squared so that the model can be evaluated regardless of the sign of the errors. Chapter 10 provided an introduction to matrix algebra, including a discussion of the mechanisms of least-squares fitting. Here the focus will be more on the implementation. A quick reminder:

$y = mx + b$ (familiar-looking regression formulation)

$\mathbf{b} = \mathbf{Ax}$ (matrix algebra formulation)

The matrix $\mathbf{A}$ is also called the design matrix, and its columns are called the independent variables. Unfortunately, the notation "$\mathbf{Ax} = \mathbf{b}$" is not universal, although it is the dominant notation in linear algebra. In multiple regression, for example, you are more likely to see the equation $\mathbf{y} = \mathbf{X\beta}$, where $\mathbf{X}$ is the design matrix, β (Greek letter "beta") is the vector of coefficients, and $\mathbf{y}$ is the observed data. For the sake of internal consistency, I'll stick to $\mathbf{Ax} = \mathbf{b}$, but the important part is that although different fields use different notations (a general problem in science and mathematics), the underlying math and the implementation in MATLAB is identical.

As a gentle start, consider one of the easiest least-squares fitting applications, which you probably didn't even realize was a least-squares problem: computing the average of a set of numbers. The model predicts that all data points have the same value, and any deviances from that value can be considered errors. At the implementation level, the model contains a vector of ones (variable A below), and the parameter is a constant that scales all elements in that vector (variable x). The goal of least-squares fitting is to find the best value for x.

```
n = 10;
b = linspace(1,3,n) + rand(n,1);
A = ones(n,1);
x = (A'*A)\A'*b;
```

(If that last line of code looks confusing, you might want to reread chapter 10.) Confirm that the variable x is the same as the output of mean(b). In the parlance of regression, we would say that we are fitting a regression line with some intercept (a "constant" term) and a slope fixed at zero.

Now for the next step in learning least-squares fitting. We'll add a slope, so that we're fitting both a y-intercept and a change in y. How should we set up the new design matrix (variable A)? You might initially think to add a second vector of ones, with the idea that the model should be able to estimate the slope based on the data. But there are two problems with this. First, let's think about our model: We want to test for a *linear change* (the slope), and a vector of all ones cannot capture any changes. Second, MAT-LAB will give a warning about the matrix being singular, which happens because the second column of A is linearly dependent on the first column of A (in fact, it's the same numbers!). So (A'*A) (in math terms, that would be A^TA) is not invertible, and the model cannot be estimated.

On the basis of our extensive knowledge of the data and our *a priori* hypotheses, we can expect a linear increase. This is implemented by having the second column of A be the numbers 1 through *n* (in this case, *n* = 10).

```
A = [ ones(n,1) (1:n)' ];
x = (A'*A)\A'*b;
```

A quick note about this code. When you type 1:n, MATLAB will produce a row vector. We want the matrix A to comprise two column vectors, which is why there is a transpose sign after the n. The parentheses are necessary because otherwise MATLAB would just transpose the variable n, which has no effect because the transpose of a 1 × 1 matrix is the same 1 × 1 matrix.

This works better. I got x=[1.06 0.24] (your values will differ due to the random numbers). These are the parameters that correspond to each column in matrix A: the intercept and the slope. In other words, the coefficients for each independent variable in our design matrix.

Did you notice that the first element of x above is different from the parameter x in the previous example? This is a curious thing. The first column in A hasn't changed, and the data haven't changed. Why did the parameter change? It changed because the parameter for each independent variable (each column in A) depends on the rest of the model. If we would add a third column in A, for example to estimate a quadratic effect, then the parameters of the first two columns would also change, particularly if there were a strong quadratic effect in the data.

How do we know whether the coefficients in x are reasonable; in other words, whether our model is any good? Segue to the next section.

28.2 Evaluating Model Fits

Now you know how to fit a basic model to data. Is the model a *good* fit to the data? There are two ways to evaluate the fit of a model to data—qualitative and quantitative. Both methods involve generating model-predicted data and comparing the predicted data against the real data. In practice, you should use both methods.

To generate the model-predicted data, the design matrix is scaled by its parameters and then by the x-axis values. If you don't have or are unsure what the x-axis values are (this is often the case in toy examples), you can define them as 1 through N. The predicted values are often called $\hat{y}$ (pronounced "why hat?" [insert witty reply here]). In the code below, you'll see that each column of A is scaled (multiplied) by the corresponding element in the coefficients matrix x. The weighted sum of those columns equals the predicted data.

```
yHat = x(1)*A(:,1) + x(2)*A(:,2);
yHat = sum(bsxfun(@times,A,x'),2);
plot(1:n,b,'b-*',A(:,2),yHat,'ro')
```

The two yHat lines produce identical results. The first implementation makes it visually obvious that $\hat{y}$ is computed by summing columns in the design matrix weighted by their corresponding regression coefficients, but this implementation works only when matrix A has exactly two columns. The second implementation is more elegant because it works regardless of the number of columns in matrix A.

As an aside, the process of modeling data by "summing columns of a matrix that are scaled by some coefficient" should sound familiar from chapter 11—the Fourier transform is nothing more than a model of the data in which the *design matrix* contains complex sine waves and the *regression weights* are the complex Fourier coefficients. The main difference is that the Fourier transform is not meant to be a simplified model; it is designed to be a model with zero degrees of freedom and therefore explain 100% of the variance of the data.

Anyway, now that you know how to compute the model-predicted data, you are ready to learn about how to evaluate the goodness of the model. The qualitative method of model evaluation involves plotting the real data and the model-predicted data and comparing them visually. You want to see that the predicted response "looks similar" to the real data. Indeed, if the model were a perfect fit to the data, the circles (actual data) and stars (predicted data) would perfectly align (figure 28.1, top panel).

In practice, there is almost never a perfect fit—if there were, the data are probably boring, and/or the model is too complex. The idea of a model is to have a simplified representation of the real biological or physical process that produced the data; if it's simplified, there should be some leftover variance.

The importance of this visual inspection cannot be understated. It is easy to make a mistake that would produce junk results without

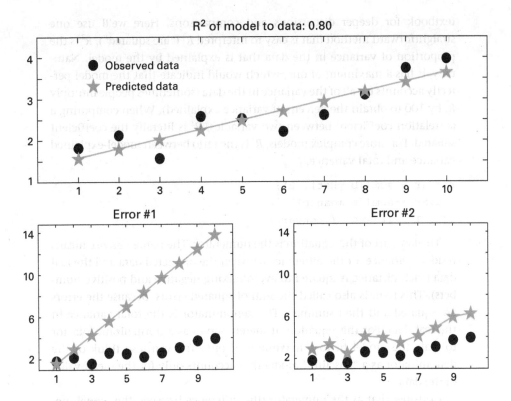

Figure 28.1
Simulated data with a linear trend plus noise (black circles) and the best-fit line generated from a least-squares solution (stars and gray line in top panel). The two bottom panels show the same data but model fits after easy-to-make MATLAB coding errors. These errors are legal operations in MATLAB, so plotting is the best way to ensure accuracy.

producing MATLAB errors. The two most common mistakes to make here are (1) to apply the regression coefficients to the observed y-values of the data instead of to the x-values of the data (remember, the goal here is to predict y from x), and (2) to mix the order of coefficients with respect to their corresponding columns in the design matrix. The bottom panels in figure 28.1 show the results of these mistakes, and the online MATLAB code shows how those mistakes were made. Always check the model fit to the data before doing any further analyses on or interpretation of the model coefficients.

Now for the quantitative method. There are several quantitative methods for evaluating the fit of a model to data, and you can consult a statistics textbook for deeper discussions and comparisons. Here we'll use one straightforward method that is easy to interpret: R^2 ("are squared"). R^2 is the proportion of variance in the data that is explained by the model. Naturally, it has a maximum of one, which would indicate that the model perfectly accounts for all of the variance in the data (sometimes people multiply R^2 by 100 to obtain the percent of variance explained). When computing a correlation coefficient between two variables, R^2 is literally the coefficient squared. For more complex models, R^2 is the ratio between model-explained variance and total variance.

```
resvar = sum((b-yHat).^2);
totvar = sum((b-mean(b)).^2);
r2 = 1 - (resvar / totvar);
```

The key part of this equation is the numerator. The name `resvar` means residual variance, or the difference between the predicted data and the real data (each distance is squared to avoid mixing negative and positive numbers). This term is also called the sum of squared errors, because the errors are squared and then summed. The denominator is the total variance in the real data. For this equation, it merely serves as a normalization factor to scale the model fit to a maximum of one. You can also think of the denominator as a "baseline" model that accounts only for the mean value of the data.

Consider that as the numerator (the difference between the model-predicted data and the actual data) gets smaller, the ratio goes to zero, and R^2 goes to one; conversely, as the difference between the predicted and the observed data increases, the ratio increases, and R^2 goes to zero. If the model were so horrible that its predictions are worse than just predicting the mean, then R^2 would be negative, and you should seriously reconsider the model (or check for a programming mistake).

What are good values of R^2? That depends largely on the kind of data you are modeling, and how complex the system is. When trying to understand something as complicated as the role of personality on financial investments, an R^2 of 0.2 or 0.3 might be a fairly large value. If you have an isolated *in vitro* cell and are testing the relationship between current input and spiking output, you have a very simple system with a small number of biological degrees of freedom, and so an R^2 of 0.9 might be expected. In psychophysics experiments with simple sensory stimuli and analyses based on averaging together thousands of trials, an R^2 of greater than 0.9 might be achievable. So the range of R^2 values that can be considered "good" depends on many factors. You will need to compare the R^2 in your data to R^2 values of other findings in the relevant scientific literature.

One drawback of R^2 is that it increases by adding new independent variables (additional columns of matrix A), even if those columns do not actually improve the model. There are a few ways to ameliorate this concern, such as computing the adjusted R^2 (which scales the R^2 according to the number of degrees of freedom) or by performing formal model comparisons. Consult your trusty statistics textbooks for more information.

28.3 Polynomial Fitting Using `polyfit` and `polyval`

Polynomial models are a class of models in which the design matrix is a series of coefficients of the same term with increasing powers, like x^0, x^1, x^2, and so forth. The first term (x^0) captures the mean offset (intercept), the second term (x^1) captures the linear slope, the third term captures the quadratic effect, and so on. The model in the previous section (intercept and slope) is a first-order polynomial.

You might think that polynomial fitting should be included in the next chapter—the power functions suggest that the models are nonlinear rather than linear. But the model *coefficients* are computed via *linear methods*, even if there are *nonlinear components* in the model. Indeed, polynomial coefficients are typically estimated via least-squares equations.

Let's start by reproducing our results from the previous section using the MATLAB function `polyfit`, which fits a polynomial model to the input data and returns the coefficients.

```
iv = (1:n)'; % iv = independent variable
regcoefs = polyfit(iv,b,1);
```

Note the organization of the inputs to `polyfit`: You don't specify the entire design matrix; you specify only the base vector. You can think of this as the parent independent variable (variable `iv`); each column in the design matrix is created as iv^0 (which is a vector of all ones), iv^1 (itself), iv^2, and so on. The third input is the order of the model. A model of order 1 produces two coefficients, one for the intercept and one for the first (linear) term. It is a bit confusing to have $N + 1$ coefficients from an Nth order model, but that's just how it works. I think the reasoning is that a zero-order model must contain at least an intercept (because $x^0 = 1$), and thus a first-order model would contain an intercept and a slope.

How does the variable `regcoefs` compare with variable x in the previous section? Assuming you didn't recompute the random data, the values

will be identical, albeit in reverse order. This happened because `polyfit`
uses descending order of coefficients, whereas we wrote matrix A to be in
ascending order. This is an important detail, and you should always sanity
check your code to make sure you know the order of the coefficients.
Another important aspect of using the `polyfit` function is that it provided
a sanity check that the manual least-squares fitting was accurate.

   Let's do another example to see how `polyfit` deals with higher-order
models. This example will also illustrate one method for separating signal
from noise that was mentioned in chapter 20. We'll create a signal that
fluctuates slowly and add noise with a broad frequency spectrum. Broadband noise by definition cannot be easily isolated in the frequency domain,
but here we will assume that the noise fluctuations are smaller in magnitude compared to the signal.

```
srate = 1000;
time = 0:1/srate:6;
F = griddedInterpolant(0:6,100*rand(7,1),'spline');
data = F(time) + randn(size(time));
% polynomial fit
polycoefs = polyfit(time,data,7);
```

(If the `griddedInterpolant` line looks confusing, you should review
chapter 13.) Now that we have the polynomial coefficients, we want to
generate the predicted signal based on the coefficients. This we can do
with `polyval`, which generates model-predicted data in the same way
that we created $\hat{y}$ in the previous section (scaling the design matrix by
the coefficients and summing the columns). The results are shown in
figure 28.2.

```
predData = polyval(time,polycoefs);
```

   Now we have our predicted data and our original data. Because the
model captures only relatively slow components in the signal, we can
take this as our time series. Alternatively, if the slow fluctuations are the
noise and the faster fluctuations are the signal, then we can subtract
the polynomial-predicted time series from the original time series and treat
the residual as the signal (figure 28.3).

```
subplot(211)
plot(time,predData)
subplot(212)
plot(time,data-predData)
```

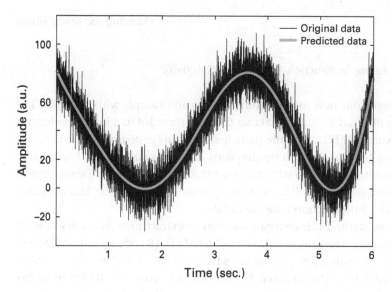

Figure 28.2

A time series that comprises slow-moving components and fast-moving components (black line). A seventh-order polynomial was fit to the data to capture the slow-moving component (gray line) using the MATLAB `polyfit` function.

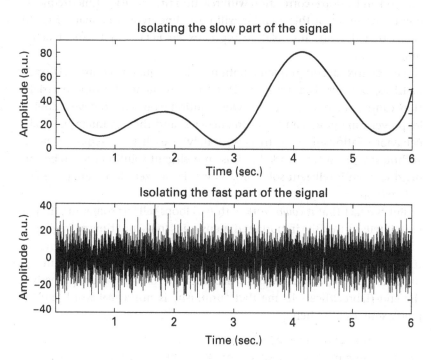

Figure 28.3

Polynomial fitting can also be used as a data-processing technique to separate signal from noise. Note the difference in *y*-axis values between the two plots.

28.4   Example: Reaction Time and EEG Activity

Let's apply our new model-fitting skills to an example with real data. The online material includes a data set that was recorded in a human volunteer who wore an EEG cap while participating in an experiment that involved pressing buttons to indicate decisions about visual stimuli. The button-press latency relative to stimulus onset (a.k.a. *reaction time*) differed on each trial. Our goal here will be to identify some activity in the EEG data that correlates with reaction time over trials.

Before starting this analysis, we need to extract time-frequency activity from the time-domain signal. The online MATLAB code recycles code from chapter 19, with a few missing lines that you can complete based on your knowledge from that chapter. The result of this code is a 3D matrix of frequency by time by trials (for convenience we will run the analysis only on one electrode, but it is easy to expand this analysis to all electrodes). The design matrix comprises a vector of ones and a vector of reaction times, with one line per trial. This means we will estimate an intercept (average activity) and a slope (correlation with reaction time) at each time-frequency point. The result of this analysis will be a time-frequency map of coefficients that describe the relationship between EEG activity and reaction time.

The matrix algebra implementation of least squares requires 1D or 2D matrices, but our data are in a 3D matrix (frequency by time by trials). How can we solve this problem? One solution would be to have a double-loop over time points and over frequencies, and then estimate the brain-behavior relationship over trials separately at each time-frequency point. Technically this would work, but it is an inelegant solution (not to be confused with an intelligent solution). As you know, zero loops are preferable to two loops.

But the fact that it *could* work with two loops tells us something important: It tells us that the important dimension is trials, not time or frequency. This means that the 3D matrix can be reshaped to a 2D time-frequency by trials matrix. It doesn't matter that time points and frequency points get mushed into the same dimension, as long as we unmush them properly at the end (LibreOffice tells me that "unmush" is not a real word, but I'm pretty sure it's okay here).

```
A = [ ones(EEG.trials,1) rts' ];
tf2d = reshape(tf3d,EEG.pnts*nFrex,EEG.trials);
x = (A'*A)\A'*tf2d;
cormat = reshape(x(2,:),EEG.pnts,nFrex); % unmush
```

The variable A is 99 by 2, corresponding to trials by columns in the design matrix (intercept, slope). The variable tf2d is 99-by-19,200. The 19,200 comes from reshaping 640 time points and 30 frequencies into a vector per trial. Before moving forward, think about what the size of variable x will be (the key multiplication that defines this size is A'*tf2d). Figure 28.4 shows images of the design matrix and the data matrix.

By the way, when making this figure, I wanted maximal control over the sizes and positions of the subplots. Rather than using the subplot function, I specified precisely where I wanted the axes to be. The code is below; I'll leave it up to you to figure out what the four numbers mean. (Hint: Try searching the Internet for setting axes positions in MATLAB.)

```
ax1_h = axes;
set(ax1_h,'Position',[.05 .1 .1 .8])
```

Figure 28.5 shows the result. What are these values? They are not correlation coefficients; they are unstandardized regression coefficients, and they are mostly uninterpretable on their own. To see why, run the code again

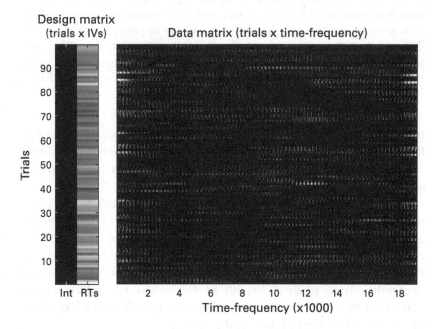

Figure 28.4
Design matrix (left) and data matrix (right) used in the regression of reaction time on EEG activity. Column "Int" is for intercept, a column of ones. RTs, reaction times; IVs, independent variables.

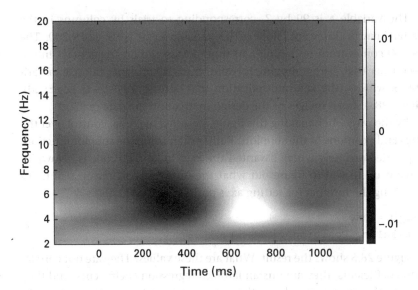

Figure 28.5
Results of the least-squares analysis using the matrices shown in figure 28.4. The
value at each time-frequency point is the cross-trial covariance (non-normalized correlation coefficient) between reaction times and time-frequency power of the EEG
signal. Results indicate that trials with initially smaller (black regions) and then later
larger (white regions) theta-band power were associated with longer reaction times.
The map is more easily visually interpreted on your monitor (assuming you have a
color monitor).

to produce this figure but convert reaction times to seconds instead of milliseconds. The relationship between brain activity and reaction time hasn't
changed, but the results have changed by a few orders of magnitude. That's
awkward.

There are two ways to normalize these data that will facilitate interpretation. One way is to Z-transform all the data (the reaction times and the
power values over trials at each time-frequency point), which will give us
Pearson correlation coefficients (converting covariances to Pearson correlation coefficients was discussed in chapter 16). A second way is to evaluate
the results relative to a null hypothesis distribution, which has an additional advantage of allowing statistical inferences. You'll have the opportunity to try both of these options in the exercises.

28.5 Data Transformations Adjust Distributions

Linear methods to model fitting, and linear models more generally, have several advantages. Compared to nonlinear models, they are easier to implement and tend to be more robust to noise and to reasonable ranges of parameters, and the results are more likely to be easily interpretable. Linear models can be sophisticated and insightful; do not confuse *linear* with *limited*.

One feature of linear models is that they typically work best with data that have well-behaved distributions, such as Gaussian (a.k.a. normal) or uniform. Distributions that contain outliers or have extreme values might result in model parameters that are not representative of the data. Many types of data used in neuroscience and cognitive neuroscience have distributions that are logarithmic, have long positive tails, are circular, or have other nonlinear characteristics. Examples include frequency-band-specific power, spike counts, and reaction times. If you have such data, before rushing off to build complicated nonlinear models, think about whether the data can be transformed to a distribution that is more amenable to linear model fitting.

There are three motivations for transforming data to a normal or uniform distribution. One is for model-fitting procedures in which parameter estimates are influenced by extreme values. Second, and related, is to prepare the data for analyses that assume Gaussian-distributed data, including some methods of classification and clustering. Third is to facilitate direct comparison across variables with otherwise incomparable scales, such as weight and height, or volts and Teslas (this third motivation is more about transformations in general, rather than transforming to a specific distribution).

Figure 28.6 shows two examples of non-normal distributions and the transforms that help them approach normalcy. The online code shows a few more transformations. It is generally a good idea to build a habit of inspecting your data and their distributions using plots and histograms.

Actually, by combining transformations, any distribution can be converted to a normal distribution. Here is the one simple trick you need: First, rank-transform the data; second, scale the ranked values to a range of −1 to +1; third, apply the inverse hyperbolic tangent (MATLAB function atanh), which is also commonly referred to as the Fisher-Z transform. This method works well except when the data are exactly −1 or +1, because the inverse hyperbolic tangent of these numbers is minus and plus infinity,

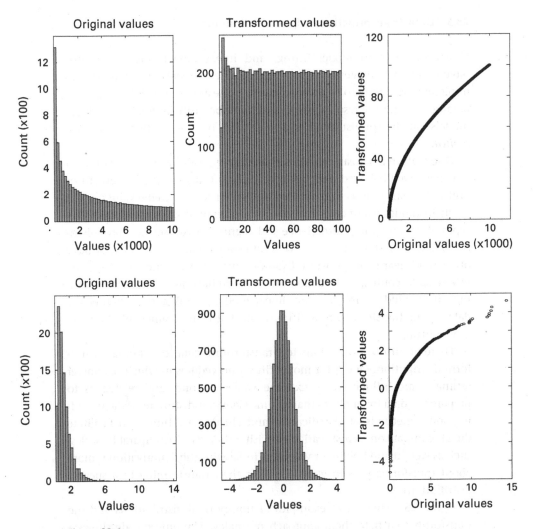

Figure 28.6
Two examples of non-normal distributions (top and bottom left-most panels) and possible transformations to a uniform (top middle and right panels) or Gaussian (bottom middle and right panels) distribution.

respectively. If you use this method to transform data, you'll need to write an exception for these values.

If you apply transformations to your data before analyses, remember to interpret the results correctly. The model parameters reflect the relationship between the independent variables and the *transformed* data, not the data in their original scale.

28.6   Exercises

1. The following line of code does not contain an error, but probably does not produce the intended result. What is the problem and what is the solution? Try to figure it out before evaluating the code in MATLAB.

```
data = 10*round(rand(10,1));
```

2. Create a 3×20 matrix of random integers that range from -10 to $+35$. Sort the rows in this matrix according to the second column, and store the result as a new matrix. Confirm visually that the rows have not been broken up, just swapped around. Next, modify the code so that the sorting is done in descending order.

3. In the second example in this chapter (using least squares to fit a linear slope), the second column of the matrix A was `1:n`. Does this column need to be integers increasing to n? Try re-running the model but change the second column of A to: (1) integers counting down from n to 1; (2) n linearly spaced numbers from 0 to 1; (3) steps of 1,000 starting at 17,000; (4) logarithmically spaced numbers between 10 and 100. Use the same data for all models so the parameters can be directly compared. Inspect the resulting parameters and model-predicted data in each case. What have you learned about constructing the design matrix and interpreted the resulting coefficients?

4. The following lines of code were written by a student who means well but makes a lot of mistakes. Find and fix the errors on each line (if there are errors ...). Imagine the student asks you for advice about how to sanity check the code so he can debug his own code in the future. What is your advice (your answer cannot be "ask someone else")?

```
x = (A*A')\A'*b;
x = (b'*b)\b'*A;
x = (A*A)\A'*b;
x = (A'*A)\A'*b;
x = inv(A'*A)*(A'*b');
```

5. The examples for polynomial fitting involved simulating data that had only zero-order (mean offset) and first-order (linear) characteristics. Create data that have a zero-order and a second-order (quadratic) effect and random noise. Use `polyfit` to estimate polynomial coefficients and `polyval` to obtain the predicted time series. Plot the results to confirm that the model is a reasonable fit to the data. Check that the zero- and second-order coefficients are large and the first-order coefficient is close to zero.

6. There is nothing special about the order of columns in matrix A. Change the column order to confirm that the results are the same. Of course, the order of the coefficients is different. By convention in statistics, the intercept comes first. How do these results compare with the output of `polyval`?

7. Reproduce figure 28.2, but replace the following line of code with the one thereafter. What happens to the result and why?

```
yHat = polyval(polycoefs,time);
yHat = polyval(polycoefs,1:length(time));
```

8. The code to compute `yHat` does only interpolation, which is to say, there are no predicted values outside the measured x-axis variables. Repeat the first example in this chapter, but extrapolate `yHat` to a range of –5 to +15.

9. The color limits in the contour plot in figure 28.5 were set manually. Devise a method to set the color limit based on the data (in this case, variable `cormat`). Make sure the color limits are symmetric.

10. In the reaction time–EEG correlations, the `rts` variable in the design matrix needed to be transposed from a row vector to a column vector. Adjust the code that creates this variable so it does not need to be transposed when creating the design matrix.

11. Convert the time-frequency map of covariance in figure 28.5 to Pearson correlation coefficients. Perform a sanity check on the final result by using the `corr` (statistics toolbox) or `corrcoef` functions to confirm the correlation for a few time-frequency points.

12. Run the reaction time–EEG correlations again, but apply permutation testing to transform the coefficients into statistical z-scores. Consult chapter 15 if you need a refresher on permutation testing for statistical inference. What do you shuffle at each iteration during permutation testing? Generate three plots: one showing the raw coefficients from the least-squares fit, one showing the statistical z-values, and one in which all z-values with an absolute value less than 1.96 are turned to

zero ($z = 1.96$ corresponds to $p < 0.05$, two-tailed and uncorrected for multiple comparisons).

13. When the data and regressors in a linear model are z-normalized, an intercept term is not necessary. Why is this the case? Think about how the intercept accounts for data shifted on the y axis, how we used an intercept to estimate the mean, and what the average data values are before versus after z-scoring. Of course, in real analyses it is best practice always to include an intercept even if the data are normalized.

14. The reaction time–EEG correlations involved a design matrix with two variables (intercept and reaction time). Redo the analysis using a third independent variable corresponding to beta-band power. The value of the third column on each trial should be power from 14 to 20 Hz and from –600 to –200 milliseconds from that trial. How do the results look? Of course, the correlation in the "seed" window should not be interpreted. This is one method of analyzing brain functional interactions (in this specific case, the results suggest that trials with stronger prestimulus beta-band power also had stronger poststimulus theta-band power).

15. Generate random data that contain a linear trend. Next, remove the trend line using the least-squares methods shown here. Check your results against the output of the function detrend. In general, you may have noticed that we spend a lot of time in this book replicating outputs of MATLAB functions. In addition to allowing sanity checks while learning to program, the point is to realize that most functions are created to simplify work flow, not because those procedures are necessarily incredibly long, complex, or difficult.

16. (This question is related to chapter 25.) Hemodynamic activity does not suddenly increase and decrease, like the boxcar-shaped regressor we used in the t-test. FMRI analyses involve generating a canonical hemodynamic response and convolving the design matrix (the independent variable) with that canonical hemodynamic response (the kernel). Search the Internet to find a function called spm_hrf.m and use it to generate a canonical hemodynamic response function. Then convolve the regressor (variable timeline in the code in chapter 25) with the hemodynamic response function kernel using techniques you learned in chapter 12. Then create a design matrix that includes an intercept and the regressor, and fit the model to each voxel's time course. How do these results compare with the t-test we implemented in chapter 25?

29 Nonlinear Methods to Fit Models to Data

Linear least-squares model fitting is a powerful approach that is suitable in many situations. But there comes a time in life when linear models are just too … linear. Sometimes you need to express yourself in more creative and nonlinear ways. MATLAB is here for you.

Nonlinear model fitting often falls into the category of *optimization*, which is a big and important field in mathematics and engineering. This chapter does little justice to the rich and highly developed field of optimization. But hopefully you'll learn enough in this chapter to be comfortable fitting basic models to data and can use this as a springboard for getting deeper into the subject.

There are two steps to nonlinear fitting in MATLAB: Define the model or function that should be fit to the data, and then use one of many MATLAB functions to find parameters that optimize the model fit to the data. We'll mostly use the functions `fminsearch` (function minimization search) and `fminbnd` (function minimization, bounded). Using other optimization functions such as `fmincon` or `lsqnonneg` is generally a matter of minor adjustments.

29.1 Nonlinear Model Fitting with `fminsearch`

Before learning about nonlinear model fitting, you need to learn about function handles. You need to learn about function handles because the first input to `fminsearch` is a function—not the output of a function, which you've seen many times, but a function itself. But MATLAB does not understand a function being directly inputted to another function. Therefore, we need to use a little trick: Instead of inputting a function, we will input something called a *function handle*. A function handle is a pointer to a function, similar to h in `h=plot(x,y);` or to `fid` in `fid=fopen('data.dat');`.

To illustrate how function handles work, we'll create a simple quadratic function. A quadratic function is defined as $ax^2 + bx + c$, where a, b, and c are scalar parameters, $a \neq 0$, and x is the input number, vector, or matrix. In the simplified function below, $b = c = 0$.

```
funch = @(t) t.^2;
```

It's a funny-looking notation, but what we've just done is create a function that takes its input and returns the squared version of that input. The @ symbol tells MATLAB that we want a function handle (called funch), the (t) is the input into the function, and the rest of the line after the space is the function itself. By typing whos funch you'll learn that the class is function_handle. Now we can move to the second step, which involves inputting this function handle into the function fminsearch. The fmin* functions work by repeatedly calling the pointed function with different parameters, until some criterion has been reached. The criterion is generally defined as the objective minimization function not getting any smaller. I know it's probably still confusing; an example should help.

```
[xval,funcval] = fminsearch(funch,-2);
```

This line of code passes the function (via its handle) into fminsearch and starts with an x-axis value of –2. That's obviously not the minimum of the function, and the point here is for fminsearch to find the value (starting from –2) that minimizes t^2. Of course, you know that the answer needs to be $t = 0$, and the output xval does not disappoint (figure 29.1). The second output, funcval, tells you the value of the function; in this case,

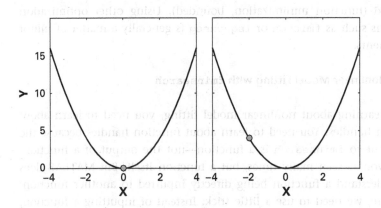

Figure 29.1
Results of function minimization. The left plot illustrates the minimum obtained from fminsearch; the right plot illustrates the bounded minimum between –30 and –2 using fminbnd.

both y and t are zero (due to computer rounding errors, you might get something like 10^{-25}). To differentiate the two outputs, try adding a constant to the function.

Take a minute to play around with this code. Test it with different starting values, and alter the function by changing the quadratic (e.g., add an offset or a linear term), and so on.

Let's now adapt the code to learn about fminbnd. This function finds the minimum within specified boundaries.

```
[xval,funcval] = fminbnd(funch,-30,-2);
```

Now the "minimum" is located at $t = -2$ (funcval = 4). This is not the minimum of the function—it is the minimum of the function *within the specified boundaries*; in other words, between –30 and –2 (figure 29.1). This would be a good time to remind yourself about the importance of visually inspecting and sanity checking your results; without any inspection or critical consideration, you might think that –2 is the minimum of t^2.

29.2 Nonlinear Model Fitting: Piece-wise Regression

Now let's see a better example of the power of nonlinear model fitting using a model that you might need to apply in real research: a piece-wise linear regression. A piece-wise regression is used when there is a sudden transition in the relationship between two variables. The key nonlinearities in piece-wise regressions are the "breakpoints," the x-axis positions that define the boundaries between the linear functions. We will use fminsearch to find the optimal breakpoint of a piece-wise linear function. For simplicity, we'll stick to one breakpoint and two linear pieces.

For this example, we will generate a so-called triangular distribution of data (you'll never guess why they call it that). It's a useful distribution to illustrate piece-wise regression. You can already have a peak at figure 29.2 to see where we're going with this example.

```
% parameters
a=.2; c=.6; b=.9;
% generate random data and then modulate
x = rand(1,10000);
y(x<c) = a + sqrt(x(x<c).*(b-a).*(c-a));
y(x>c) = b—sqrt((1-x(x>c)).*(b-a).*(b-c));
% plot distribution
hist(y,100)
```

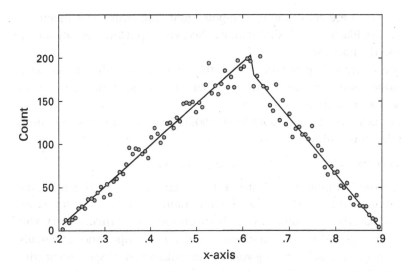

Figure 29.2
A triangular distribution can be modeled as a piece-wise linear regression. This figure
shows the distribution and the two linear pieces that were each fit using least squares.
The nonlinear fitting procedure was used to find the optimal breakpoint. This figure
is actually the final frame of a movie that shows how `fminsearch` tries different
breakpoint values in search of the best value.

A triangular distribution takes three inputs that correspond to the bottom, middle, and top points of the distribution. They must be between zero
and one. The two lines that modulate the variables are a MATLAB implementation of the mathematical definition of the triangular-distributed random numbers. Don't inspect the code for too long, because you'll be asked
to reproduce it (without cheating!) in the exercises.

Actually, we don't care about the variables x and y. What we need for
our model fitting is the shape of the distribution.

```
[y,x] = hist(y,100);
```

What are the sizes of x and y? What were the sizes of variables x and y
before running that line of code? That's right, we overwrote the previous
variables. This is legal behavior but it's bad programming because (1) we've
lost the original data, and (2) it increases the possibility of confusion.

Before we can get to the fun part of using `fminsearch`, we need to
define a function that we want to minimize (well, this part is also fun, just
not quite as much fun as using `fminsearch`). This function will be more
complex than that in the previous section, so we should put this new

function into a separate file. The implementation of a piece-wise linear regression is not so difficult—start with some arbitrary breakpoint (call it variable bpoint), and use standard linear least-squares fitting on either side to fit the two pieces. The two variables will be called x and y (this is inside a function, so we're not going to overwrite the previous x and y variables).

```
% first piece
x1 = [ones(bpoint,1) x(1:bpoint)];
y1 = y(1:bpoint);
b1 = (x1'*x1)\(x1'*y1);
% second piece
x2 = [ones(length(x)-bpoint,1) x(bpoint+1:end)];
y2 = y(bpoint+1:end);
b2 = (x2'*x2)\(x2'*y2);
```

This is basically the same as the least-squares fitting you learned about in the previous chapter. The only difference is that the data are cut into two pieces, and linear models are fit to each of them separately.

But we're not finished yet. Functions like fminsearch search for parameters that minimize some error objective. What is the objective that we want to minimize? What is always the minimization objective in model fitting? Of course, it's the fit of the model to the data, which we can quantify as R^2. Well, almost. R^2 is defined as 1 minus a ratio, and we want R^2 to be as high as possible. But MATLAB wants to minimize something. If we drop the "1 minus" bit, smaller values indicate a better model fit, and we can use this as our minimization objective. To compute this ratio, we need to compare the predicted values against the observed values.

```
yHat = [b1(1)+b1(2)*x1(:,2); b2(1)+b2(2)*x2(:,2)];
sse = sum((yHat-y).^2) / sum(y.^2);
```

If you compare the variable sse to how it was defined in the previous chapter, you'll notice two differences, one in the numerator and one in the denominator. The difference in the numerator doesn't matter (why not?). The difference in the denominator doesn't really matter either for our purposes here. In fact, we don't even need a denominator at all in this function, and it's your job to figure out why.

We're getting close, but there is one more issue. The breakpoint is meaningful only if the x-axis values are sorted, because the breakpoint is an index. To understand why this is the case, consider what happens in the following situations when the breakpoint is the third index.

```
x = [ 1 2 3 4 5 ];
x = [ 3 4 1 2 5 ];
```

In the first case, 1 and 2 are before the breakpoint, while in the second case, 3 and 4 are before the breakpoint. The conclusion is that the variables need to be sorted. The variables x and y cannot be independently sorted, because the relationship between them needs to be preserved. Instead, we sort x and then apply the same sorting indices to y.

```
[~,i]=sort(x);
x=x(i); y=y(i);
```

Finally, we are ready to move forward. The online MATLAB code has this entire function in a file called fit2segLinear.m. In the first example of this chapter, we wrote an in-line function and entered the function handle into fminsearch; now we have a function file, but the solution is the same: Use a function handle.

```
[~,initB] = min(abs(x-.5));
funch = @(initB)fit2segLinear(initB,x,y);
[optBreakPoint,sse,exitflag,fmininfo] = ...
          fminsearch(funch,initB);
```

There are a few important things to discuss in this code. First, note the initB variable, which specifies the initial estimate of the breakpoint. Parameters in optimization procedures always need some starting value, and you will see later in this chapter that good starting values are important. We will initialize the breakpoint to be the center of the data distribution, and we use the min-abs construction because we want initB to be an index, not a data value.

The first output of fminsearch is the parameter that provided the best fit, then the value of the objective function (in this case, the sum of squared errors), then the "exit flag" (1 if everything was okay; 0 if it exited without thinking it did a job well done), then some meta-information about how many iterations it went through, and so on.

The output sse is pretty small (0.0053 when I ran it; random noise will cause slight differences but it should always be close to zero), which means the R^2 is pretty big. So the model was a good fit to the data. The breakpoint was 60.2 (again, it will be slightly different each time you run it). We need to be careful when interpreting this output. It is not the actual breakpoint. Indeed, the x-axis values range from 0.2 to 0.9. Instead, this is the index into the variable x at which the breakpoint is optimally placed.

"But Mike," I imagine you interjecting, "indices must be integers. I know because I've gotten 10,000 MATLAB errors about this!" Yes, gentle reader, you are correct. However, MATLAB doesn't know that this is supposed to be an index; it just knows that this is a parameter of a model. MATLAB will happily try to give it non-integer values, which is why line 10 in the function fit2segLinear rounds the index.

You might now be thinking that this couldn't get any more fun. But it can and will. To get a better idea of what's going on inside `fminsearch`, we can plot the data and the fit to the data each time `fminsearch` calls the function fit2segLinear. That function contains a commented line of code that plots the data. Uncomment this line (and save the file) and re-run `fminsearch`. You will watch the function searching the parameter space for the optimal breakpoint. Don't feel embarrassed to watch it over and over again (figure 29.2).

In fact, you should always do this kind of plotting when writing new functions or when using existing functions for the first time. This is the same as the qualitative model-fit inspection discussed in the previous chapter, and it is an excellent way to determine whether the parameters of the model are sensible and interpretable. For example, try running the model fitting with `initB` set to the index corresponding to an *x* value of 0.1. On the other hand, not all models and data sets can be easily visually depicted like in this example. You might need to get creative to figure out how to show the data and the fit to the data.

29.3   Nonlinear Model Fitting: Gaussian Function

Let's try another example using another function that is often used to fit neuroscience data: a Gaussian. A Gaussian function has three parameters—width, peak amplitude, and center (figure 29.3). We will create a noisy Gaussian and see how well `fminsearch` can recover the original parameters. The code below constructs the noisy Gaussian.

```
peak = 4;
fwhm = 1;
cent = 3;
nois = .5;
x = -10:.1:10;
gaus = peak*exp(-(x-cent).^2 / (2*fwhm^2));
gaus = gaus + nois*randn(size(gaus));
```

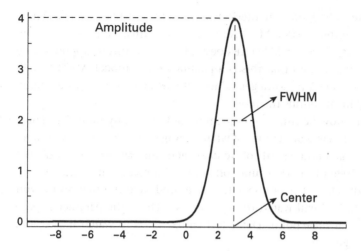

Figure 29.3
A Gaussian function is defined by three parameters.

Defining the objective function is fairly simple: It's the Gaussian with parameters that are defined by three inputs (for simplicity, the three parameters can be inputted as a three-element vector). One important difference from the piece-wise linear fit example is that here we want to keep the original data to evaluate relative to the model-predicted data, rather than inputting the distribution of the data (although the code could be adapted to fit the distribution rather than the raw data). In fact, we'll input the original data into a Gaussian-fitting function.

```
InitParms = [ 2 2 -2 ];
funch = @(initParms) fitGaussian(initParms,x,gaus);
[outparams,sse] = fminsearch(funch,initParms);
```

Otherwise, the code is overall fairly similar to that used in the piece-wise linear regression example, except of course the piece-wise regression is replaced with a Gaussian. Figure 29.4 shows the final result. It looks like a good fit, don't you think? If you run the code with the plotting line in the fitGaussian function uncommented, you'll be able to watch the result of each iteration as fminsearch tries to find the minimum in the 3D parameter space.

Nonlinear search procedures can get tripped up or caught in a local minimum. In general, you should try to make the models be as simple as possible, and try to make sure the data are as clean as possible. If there are many parameters in the model, try to specify some of the parameters *a*

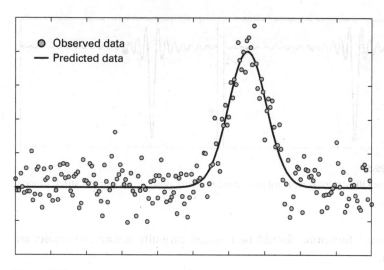

Figure 29.4

This figure is similar to figure 29.2 except a Gaussian replaced the piece-wise linear regression.

priori to decrease the total search space of the model. Nonlinear optimization functions are powerful when used correctly but can produce misleading or uninterpretable results when used sloppily. Always check the model fits to the data, and test your models using a range of different parameter starting values.

29.4 Nonlinear Model Fitting: Caught in Local Minima

Are you having fun with these models? I am. Let's squeeze in another quickie, this time to illustrate how optimization functions can fail because of local minima. A negative sinc function provides a good illustration, because it has one global minimum and many local minima.

```
funch = @(x) -sin(x)./x;
[xval,funcval] = fminsearch(funch,0);
```

Figure 29.5 shows that we successfully found the global minimum. But that's because we gave it a starting value close to the global minimum. Try running the code again using a starting value of 20. The model gets caught at the nearest local minimum and then, thinking it has accomplished a job well done, happily quits. Finding the global minimum in a sea of local minima is a difficult problem in optimization. This is another illustration of

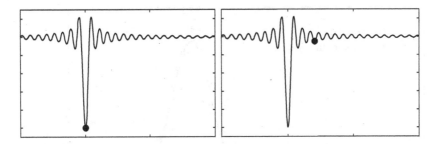

Figure 29.5
This figure illustrates how nonlinear model-fitting procedures can get stuck in local minima.

how `fmin*` functions should be checked carefully before the results are trusted.

29.5 Discretizing and Binning Data

Often in model fitting, it is the *distribution* of values, not the values themselves, that you want to model. You already saw this in the piece-wise regression example—in fact, we destroyed the raw data and replaced those variables with their discretized distributions.

Another example: If your experiment involves detecting weak sensory stimuli in the presence of noise, the raw data from each trial are binary (seen or not seen). This is not what you use for estimating the model, for example, of a sigmoid function. Instead, the model is fit to the average accuracy from trials that contain a specific level of noise. In other words, n trials are binned into r groups, and trials are averaged within each group. We could also say that the data are discretized into r bins.

Sometimes it's easy to discretize trials. If your experiment explicitly manipulates sensory noise in 10 steps, then discretization simply involves averaging the data within each noise level. This section will show you how to discretize data when there are no *a priori* defined bins. This would happen, for example, if you want to bin trials according to reaction time, spike rates, or oscillation power.

One way to discretize data is via the outputs of the `hist` or `histogram` functions. You've already seen these functions being used to produce a histogram plot. But they can also provide outputs that allow you to reproduce that plot (figure 29.6).

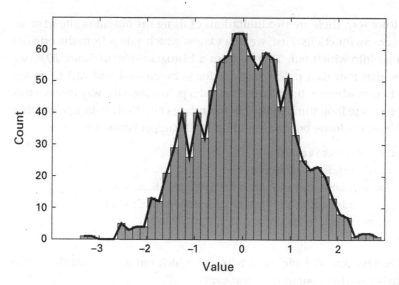

Figure 29.6
The outputs of the `hist` or `histogram` functions can be used to discretize the data according to equally spaced bins on the *x* axis.

```
x = randn(1000,1); hold on
hist(x,40)
[yy,xx] = hist(x,40); % doesn't plot anything!
plot(xx,yy,'r','linew',3)
```

MATLAB recently introduced an updated version of the `hist` function, called `histogram`. The idea is the same; the outputs are slightly different. Actually, there is only one output, which is a structure that contains useful information about the distribution, including the raw data. One feature of this output, which is sometimes useful and sometimes a bit annoying, is that it doesn't return the bin values themselves; it returns the upper and lower edges of each bin. That's why there are $N + 1$ bin edges for N bins. Thus, to reconstruct the histogram accurately, you will need to compute the bin centers, which are the averages of the bin edges.

```
hdata = histogram(x,40);
xvals = (hdata.BinEdges(1:end-1) + ...
         hdata.BinEdges(2:end))/2;
plot(xvals,hdata.Values)
```

If you are just getting into MATLAB, I recommend using `histogram` over `hist`. If you are a rigid old dinosaur, stubbornly stuck in your 1990s habits, then by all means continue using the `hist` function.

Either way, there are two limitations of using the output of the hist or histogram functions. First, we don't know which values from the original data go into which bin. That is, from a histogram like in figure 29.6, we know that four data points had a value between –3.4 and –3.12, but we don't know where in the matrix those data points are. One way to solve this problem is to loop through the bin boundaries and find all data points that are above the lower boundary and below the upper boundary.

```
xidx = zeros(size(x));
for bini=1:hdata.NumBins
    ix = hdata.Data > hdata.BinEdges(bini) & ...
         hdata.Data < hdata.BinEdges(bini+1);
    xidx(ix) = bini;
end
```

The resulting variable xidx tells you which entries in variable x were placed into which bin in the histogram.

The second limitation of the hist (and histogram) function is that it discretizes according to equal spacing of the range of the distribution, not according to an equal number of data points per bin. This is, of course, the entire purpose of making a histogram, but there are situations in which equal points per bin is preferred over equal bin spacing. The solution to discretizing data into N equally sized bins is to use the procedure described in section 14.5 of chapter 14. As a brief reminder, discretizing data into some number of equally sized bins can be implemented as follows.

```
N = length(data);
nbins = 8;
temp = tiedrank(data)/N;
discr = ceil(temp*nbins);
```

29.6 Exercises

1. Adding code to the minimization function to plot results is an excellent way to sanity check the process, but it also increases computation time. Redo the piece-wise and Gaussian examples using tic/toc to time how long it takes to find the minimum with the plotting line uncommented. Run each procedure 100 times. Then repeat with the plotting line commented. Make a bar plot of the average computation times for the four tests, along with error bars showing the standard deviations.

2. Look on the Internet for the formula for generating random variables with a triangular distribution (Wikipedia, for example, should have a relevant entry). Without looking at the code for this chapter, implement this formula in MATLAB, in two ways. First, loop through elements in the vector of random numbers; second, do it without a loop. Do you also get this existential feeling of wholesome goodness by solving problems without loops?

3. Here is another formula that you should convert into MATLAB code. A reasonable range of variable t would be -5 to $+5$. Plot this function.

$$|(2 \bmod t) - 0.66|\sin(2\pi10t) - 0.08\sin(2\pi10(t-1))/(t-1)$$

4. Would you have any difficulties using `fminsearch` to find the global minimum of the function in the previous exercise? How about the global maximum? How might you find the maximum without `fminsearch`?

5. In section 29.2, you learned how to sort multiple variables according to the order of a single variable. This task is easier to accomplish if the variables are columns in a matrix. Go back to that example, put variables `x` and `y` as column vectors in a matrix, and then figure out how to use the function `sortrows` to sort both columns according to values of the first column (variable `x`). Are you sure the result is correct? Maybe it's a good idea to test `sortrows` using a small matrix where you can visually confirm the answer.

6. In the triangular-distribution example, perform optimization by "brute force." This means you should write a loop over all possible x values, use each x value as a breakpoint and fit the piece-wise linear regression, and save the R^2 for each breakpoint. The best breakpoint parameter is the one with the highest R^2. Confirm that you get the same (or very similar) answer as with `fminsearch`. Then run a time test on your brute-force code and compare that against using `fminsearch`. This brute-force method is okay for functions with one parameter. What would be involved in applying the brute-force method in the Gaussian example, which has three parameters?

7. We used the function `fminsearch` in the piece-wise linear regression. Get it to work with `fmincon` and `fminbnd`. The inputs are slightly different, so you might need to look up the help files or look online. In which situations would you prefer the different `fmin*` functions?

8. You saw that proper initialization of `initB` is important. Knowing that this is a triangular distribution, devise and implement an algorithm to make a good initial guess for the breakpoint parameter. Then try

changing the "c" parameter in the code that produces the triangular distribution to confirm that your algorithm works well.

9. How robust is the Gaussian parameter search to noise? Re-run the Gaussian fitting procedure over different noise levels, ranging from 0 to 4 (these are the numbers that scale the output of the randn function when creating the data). Use at least 10 steps of noise. For each level of noise, compute the squared differences between the parameters you specified and the parameters returned by fminsearch. Repeat these procedures 50 times to be able to take the median over noise levels (you are likely to get some horrible fits, so median is preferred over mean). Plot the differences as a function of noise level.

10. Following up on the previous exercise, fix the noise level to 0.5, and vary the distance of the initialization of the Gaussian center relative to the real center, from –6 to 3 (assuming the actual center is 3). Again, compute the squared differences between specified and fitted parameters. What is a good way to plot the results? What have you learned about the importance of having good starting points?

11. Weibull functions are occasionally used in psychophysics for fitting behavioral data. Find the formula for the Weibull function online and implement it in MATLAB. Try different parameters until you find a curve that you find visually pleasing. Next, come up with six data points that are near that curve (don't be too exact—you should be able to fit the function to the data). Write a function to fit a Weibull function and return the two parameters based on the six data points that you simulated.

12. Re-run the Weibull model fitting in the previous exercise, but first interpolate the data to 100 points. Does up-sampling improve the fit?

13. Zoom into the sinc-minimization in figure 29.5. There is not a perfect match. Did something go wrong in the plotting code? (Obviously the answer is yes.) Find and fix the bug in the code.

14. Sigmoid functions are also common in psychophysics and in neuroscience. A sigmoid is an S-shaped function that, among other things, can squeeze any arbitrary number to range between –1 and +1. You'll see this in practice in the next chapter. The formula is $v/(1 + e^{-s(x-m)})$, where e is the natural exponent, m is the x-value of the function midpoint, s is the steepness of the function (how quickly it goes from the bottom value to the top value), and v is the maximum value of the sigmoid. Implement this function in MATLAB in a way that allows you to specify the m, s, and v parameters and produce a plot of the input x values

by the function output. Spend a few minutes changing the parameters to see the effects on the resulting plot.

15. Here are 6 data points: [0 0.05 0.3 0.7 0.9 1] (these are the y-axis values; the x-axis values are 1:6); fit these to a sigmoid and determine the parameters. Then do it again adding +1 to all the data values. Plot the data and the fitted sigmoid on the same graph. Make sure to select the starting parameters carefully.

Computational modeling in neuroscience and psychology is used to test specific hypotheses about theoretical predictions, to control and manipulate parameters in simulations that are not possible to control in the real world, and to help interpret or explore empirical data.

Like the topics covered in every chapter, computational modeling in neuroscience is a big and fascinating field, and this chapter can provide only a Lilliputian taste of all the possibilities. Encyclopedic books have been written about the mathematical details and programming implementations of computational modeling in neuroscience and cognitive science (Jaeger and Jung 2015). Still, basic models like the ones introduced here can be informative and are widely used in the literature.

30.1 Integrate-and-Fire Neurons

Integrate-and-fire (IAF) neurons are the oldest and perhaps most fundamental conceptualization of the behavior of a neuron. The idea is simple (figure 30.1): A neuron receives input, and it emits an action potential when the sum of those inputs exceeds a threshold. Inputs can be excitatory (brings the neuron closer to its threshold) or inhibitory (further away from its threshold). If you would like to learn more about the biology and biophysics of IAF neurons, you might start with Burkitt (2006).

We will create a model neuron and give it some input. I think Reza is a good name for our simulated IAF neuron. We start by initializing Reza's parameters. Reza will have his resting, spike threshold, and reset values set at –70, –50, and –75 millivolts (mV), respectively. The membrane resistance is 10 megaohms (MΩ), and the decay time constant is 10 milliseconds.

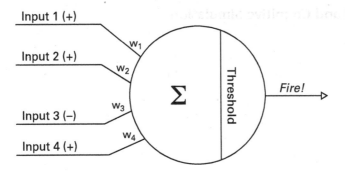

Figure 30.1
Depiction of the integrate-and-fire neuron model. The model neuron receives excitatory and inhibitory inputs; when the weighted sum of those inputs exceeds a threshold, the neuron emits an action potential, which is used, among other purposes, as an input to other neurons.

```
volt_rest = -70; % resting potential (mV)
volt_thresh = -50; % action potential thresh. (mV)
volt_reset = -75; % post-spike reset voltage
% membrane parameters
R_m = 10; % neuron membrane resistance (MOhm)
tau = 10; % time constant of decay (ms)
```

Reza needs stimulation. This can be implemented by defining a time series vector of inputs comprising zeros (no input) and non-zero values when we want input. Another possibility would be to define onset/offset times for stimulation, but defining an entire vector gives us the freedom to create a time series of arbitrarily time-varying inputs, which you'll have the opportunity to do in the exercises. The simulation will last for 1 second, and we will simulate at 10 kHz.

```
srate = 10000; % sampling rate in Hz
sim_dur = 1; % stimulus duration in seconds
time = 0:1/srate:sim_dur - 1/srate;
input = zeros(1,length(time));
input(dsearchn(time',.3):dsearchn(time',.7)) = 3;
```

Next we need to initialize two results vectors: Reza's membrane voltage over time, and a list of spike times. You'll learn later why it's convenient to have these two quantities stored separately.

```
neuronV = volt_rest + zeros(size(timevec));
spiketimes = [];
```

The voltage vector is initialized as the resting membrane potential, whereas `spiketimes` is initialized as an empty vector. We don't know *a priori* how many spikes there will be, so this is a good initialization approach.

Next, we loop over time points. The loop goes from 1 to the number of time points minus 1. You'll see in the simulation that each time point involves updating the membrane potential at the subsequent time point; that's why the simulation stops one time point before the end. The first thing to do inside the loop is to check whether Reza has exceeded his firing threshold. If he has, the membrane potential is reset and the `spiketimes` variable is updated. (Variable `ti` is the looping variable over time points.)

```
if neuronV(ti) > volt_thresh
    neuronV(ti) = volt_reset;
    spiketimes = cat(1,spiketimes,ti);
end
```

The next step inside the loop over time points is to update the membrane potential, which is separated into two lines for convenience. The first line describes the total input, which is the existing membrane potential plus any external input. The second line defines how the membrane potential is updated based on the decay rate and the current input.

```
r_i = volt_rest + input(ti)*R_m;
neuronV(ti+1) = r_i + (neuronV(ti) - r_i) ...
                    * exp(-1000/srate/tau);
```

Here you see that the membrane potential is updated into the subsequent time point. Try running the code again, looping until the total number of time points. The simulation will successfully complete, but the script will crash when plotting. It's an easy issue to fix.

If you compare this code to other implementations of IAF neurons, you might see a variable Δt or `dt` in the exponential, instead of `1000/srate`. Had I used a variable called `dt`, what would its value be, given the sampling rate of 10 kHz?

And that's it for our simulation! It probably wasn't as complicated as you thought it would be. Of course, the simulations get much hairier when you incorporate detailed biophysical mechanisms, ion channels, and realistic morphology. The last thing to do is to plot Reza's membrane potential over time. Here is where we use the `spiketimes` variable. The standard IAF neuron does not actually model the action potential; instead, it just resets the

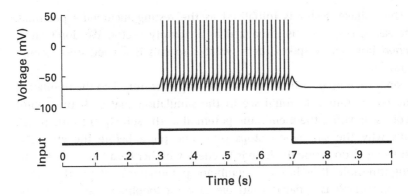

Figure 30.2
The membrane potential of a simulated integrate-and-fire neuron (top panel) in response to DC current input (bottom panel).

membrane potential. But Reza is a bit of a show-off, so for visualization, we will set the membrane potential value at the time of the spikes to be more visible. The following code will change the membrane potential to 40 mV, and now the results can be seen (figure 30.2).

```
neuronV(neuronV==volt_reset) = 40;
```

30.2   From Neuron to Networks

Neurons are social animals. That's why you never see them in isolation in the wild. Let's give Reza some friends before he gets sad. We'll create a network of 100 IAF neurons, 80 of which will be excitatory and 20 of which will be inhibitory. An excitatory neuron makes the postsynaptic cell more likely to spike (depolarization; think of glutamatergic pyramidal cells), while an inhibitory neuron makes the postsynaptic cell less likely to spike (hyperpolarization; think of GABAergic interneuron cells).

Before starting with the code, let's think about how to do this. It's easy to create 100 neurons; that just involves making a 100-by-time matrix of membrane potentials instead of the 1-by-time vector that we made for Reza. But these would all be independent neurons. They need to talk to each other. In the brain, this is done through chemical and electrical synapses; in our simulation, we will set the input to each neuron to be the spikes of other neurons (as in figure 30.1). Spikes from excitatory cells will bring each neuron closer to its threshold, and spikes from inhibitory cells will bring each neuron further away from its threshold. For simplicity, we will wire our network to be all-to-all, meaning that each neuron is connected to each other neuron (with 100 neurons, how many total connections are there?).

Because networks are more complicated than individual neurons, it's best to take this one step at a time. The first step will be to simulate 100 identical unconnected neurons. We should then see exactly the same results 100 times (there is no noise in this simulation). This provides a sanity check: If something goes wrong with 100 independent neurons, we definitely cannot interpret any results when those neurons are wired together.

What needs to be changed from the single-neuron simulation? For starters, we need to initialize the number of neurons we want, and we also need to track the membrane potentials for all neurons.

```
N_exc = 80; % excitatory neurons
N_inh = 20; % inhibitory neurons
```

... a few lines later ...

```
neuronV = v_i + zeros(N_exc+N_inh,length(time));
```

Inside the loop over time points, we need to modify how threshold exceedances are computed. This can be done with a loop over all neurons or it can be done with no loop using logical indexing. You can guess which implementation is preferred.

```
spikedNeurons = neuronV(:,ti) > volt_thresh;
neuronV(spikedNeurons,ti) = volt_reset;
```

The most likely mistake to make here is forgetting that neuronV is a 2D matrix instead of a 1D vector. Indeed, in the lines thereafter, you will need to change neuronV(timei) from the single-neuron script to neuronV(:,timei). You should now have 100 identical neurons. Plot the membrane voltage potentials from a few randomly selected neurons, and confirm that they are all perfect Reza replicas. If you see any deviations, check the code for errors.

Assuming your code passed the sanity check, it's time to connect these neurons into a network. We can use this network as an opportunity to test how well activity propagates through the network. We will program the simulation to apply external input only to half of the neurons. To accomplish this, the input needs to change from a scalar that is applied equally to all neurons to a vector that allows individualized inputs.

```
r_i = volt_rest + input(ti)*R_m;
n2stm = round(N_exc+N_inh)/2;
r_i = [r_i*ones(1,n2stm) volt_rest*ones(1,n2stm)]';
```

Notice what happens in this code. After computing the new input (variable r_i), we specify that the first half of the neurons get that input, while the second half of the neurons get rest-level stimulation. You might initially have thought to provide an input of zero, but zero is actually not the same as no input. In fact, zero-level stimulation is an excitatory drive, because the resting potential is negative. Don't believe me, try it yourself.

And now we're ready to connect the neurons. Each neuron receives input from all spiking neurons, with spiking from excitatory cells being added and spiking from inhibitory cells being subtracted. The inputs are scaled to make the model work well; the numbers "25" and "10" have no direct biological interpretation (although in publications, these numbers would be justified as simulating synaptic strength).

```
r_i = r_i + 25*sum(spikedNeurons(1:N_exc)) ...
           - 10*sum(spikedNeurons(N_exc+1:end));
```

And it works! The neurons that receive no external input are still spiking, because they receive spiking input from neurons that are receiving external input (figure 30.3). And you can see that you get some interesting network patterns that were not apparent when Reza was there just by himself. For example, the nonstimulated neurons spike with the network, but the timing is delayed because the input is indirect and the membrane potential must build up to threshold. It's impressive that even this very simple network with very simple neurons can produce interesting and complex behaviors. You can try modifying a few parameters and adding a bit of noise to gain an appreciation of the complexity of this simple network.

That's the end of the integrate-and-fire section. Wave bye-bye to Reza.

30.3 Izhikevich Neurons

If you would like to work more with IAF neurons, you might be interested in studying and using the *Izhikevich neurons* (Izhikevich 2003). They are only slightly more complicated than the basic IAF neuron introduced here, but they are more flexible because different parameter settings can produce a large range of neural behaviors. The Izhikevich neurons are controlled by four parameters (named *a*, *b*, *c*, and *d*). These are abstracted parameters that

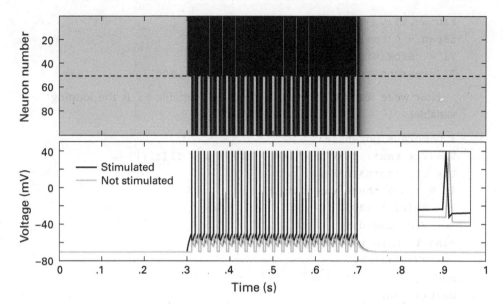

Figure 30.3
Results of a simulation of a network of integrate-and-fire neurons. Only part of the network was stimulated, but the nonstimulated neurons also participated in network behavior. The upper plot shows the activity of all neurons (each row is a neuron). The lower plot illustrates the membrane potential time course of two neurons. The inset plot shows a zoomed version of one network spike.

do not individually map onto specific biophysical processes, but interact in a dynamical model to reproduce behaviors of different classes of neurons (bursting, fast spiking, adapting, etc.). Let's have a look at one neuron. We start with initializations of the four dynamics parameters.

```
a = .03;
b = .25;
c = -60;
d = 4;
```

Next, we initialize the timing-related parameters. The simulation will run for 1 second at a sampling rate of 4,000 Hz (specified in the code as the duration in milliseconds of each time step). We also need to define the neuron's input, which is variable T1. Before looking at the output, figure out when the neuron will receive stimulation based on the fourth line of code below.

```
tau = .25;
tspan = 0:tau:1000;
T1 = zeros(size(tspan));
T1(dsearchn(tspan',200):dsearchn(tspan',800)) = 1;
```

Now we're ready for the loop over time (variable `ti` is the looping variable).

```
% membrane potential
V = V + tau*(.04*V^2 + 5*V + 140 - u + T1(ti));
u = u + tau*a*(b*V-u);
if V > 30% there was a spike
    VV(ti+1)=30;
    V=c; u=u+d;
else % there was no spike
    VV(ti+1)=V;
end
uu(ti+1)=u;
```

The first two lines control the updating of the membrane potential and the membrane recovery. Note that the input (`T1`) is added per time point to the membrane potential. Thereafter, the membrane potential is examined. If the voltage is above threshold, the membrane potential and recovery variables are reset. After the simulation finishes, we plot the results (figure 30.4).

```
plot(tspan,VV(1:end-1),tspan,T1-88);
```

Note that because the membrane potential is updated one time step into the future, the vector of the neuron's activity is longer than the time vector. Hence the `1:end-1`. Spend some time comparing the overall features of the IAF and the Izhikevich neurons. Although there are several differences, there are also many similarities in how the simulations are implemented in MATLAB.

30.4 Rescorla-Wagner

Rescorla-Wagner–style models (Rescorla and Wagner 1972) are designed to help understand the mechanisms that might support how agents (animals, robots, etc.) can learn to select the best actions on the basis of rewarding or punishing feedback. For an animal, this might translate to learning which trees yield more fruit; for a robot, this might mean learning how much pressure to apply when shaking hands with a human. These types of models

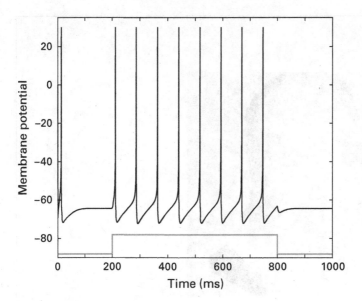

Figure 30.4

Simulation of a single Izhikevich neuron. The membrane potential is drawn in black, and the DC input to the neuron is drawn in gray (the scaling of the input is arbitrary to facilitate visual inspection).

have been around for several decades and are still being used because of the success they have in providing simple but robust descriptions of foraging and reward-seeking behavior.

The model works by selecting one of several actions on the basis of a combination of chance probability and learned preferences for each action. If a selected action produces a reward, the preference for that action increases. In the famous words of Thorndike (the "law of effect"): "Responses that produce a satisfying effect in a particular situation become more likely to occur again in that situation, and responses that produce a discomforting effect become less likely to occur again in that situation" (Thorndike 1898). Translated into twenty-first century English: We do things that we like and we avoid things that we don't like.

In the context of modeling, learned preferences for actions are called *weights* or *values*. I'll use the term weights here to avoid confusion when referring to the numerical values of different variables and equations.

Our goal in this section is to learn one basic implementation of a Rescorla-Wagner–type model. The model will be presented with a few different choices, and, based on trial-and-error feedback, will learn to select

Box 30.1

Name: Eugene M. Izhikevich
Position: President, Chairman, and CEO
Affiliation: Brain Corporation, San Diego, California
Photo credit: Brain Corporation

When did you start programming in MATLAB, and how long did it take you to become a "good" programmer (whatever you think "good" means)?
I started using MATLAB at the end of the past century, right after getting my PhD, and switched completely to MATLAB and C around the year 2000.

What do you think are the main advantages and disadvantages of MATLAB?
The main advantage of MATLAB is the simplicity of use, the ability of implementing parallel programming, and the support by MathWorks. I found the help pages of MATLAB the most useful and intuitive. In fact, I have never seen anything done better than MATLAB help.

The main disadvantage of MATLAB is the awkwardness with which objects are implemented. In my view, this is why it is losing to Python in popularity.

Do you think MATLAB will be the main program/language in your field in the future?
MATLAB was the main language for more than 10 years, starting in the mid-nineties of the past century. Unfortunately for MATLAB, Python became the fast follower, borrowing the best mathematical concepts from MATLAB and replacing it as the main programming language in scientific computing.

> **How important are programming skills for scientists?**
> Programming skills are more important than math skills. One can always simulate an equation that one cannot solve analytically. More and more scientific disciplines rely on simulations and data analysis, and if a scientist cannot implement his or her own simulations, nobody would help, especially in the early career.
>
> **Any advice for people who are starting to learn MATLAB?**
> Read more of the code that is presented on MATLAB help pages—this will teach you good programming skills, and your code will look beautiful.

the best option. First, some initializations. We'll have 100 trials, a learning rate of 0.3, and two actions that reward at 70% and 20% (i.e., whenever option "A" is selected, there is a 70% chance of getting rewarding feedback). The learning rate refers to how strongly the model uses feedback to adjust the weights; after the model is implemented, you'll have the opportunity to see the effects of different learning rates on learning behavior. Finally, we initialize the weights for the two actions to be 50% (no preference).

```
nTrials = 100;
lrate = .3;
rewProbs = [.7 .2];
w = .5+zeros(nTrials+1,2);
```

The weights are held in the variable w. The vector is nTrials+1 long in order to save the weights separately for each trial. Saving the weights for each trial is not necessary for the modeling, but it allows us to watch the model learn over time. Weights get updated into the following trial, similar to the membrane voltage of the IAF neuron. Here, I decided to loop through all trials and compute an extra set of weights at the end.

Inside the loop over trials (variable ti is the looping variable), we first compute the probability of choosing action 1 over action 2. Do you think I'm missing something by not explicitly computing the probability of action 2?

```
pPickAct1(ti) = exp(w(ti,1)) / sum(exp(w(ti,:)));
```

The exponentials in the function (exp) are used to convert arbitrarily valued numbers into probabilities. This is called the softmax rule. If the two weights are equal, the probability is 0.5; relatively larger versus smaller weights cause the probabilities to go toward 1 and 0, respectively.

Let's take a minute to explore this code snippet to understand the soft-max rule better. The code below will create a 2D space of probabilities resulting from applying the softmax rule to the numbers –3 to +3 (imagine these numbers are weights for two actions).

```
[v1,v2] = deal(-3:.1:3);
for vi=1:length(v2)
    p(:,vi) = exp(v1) ./ (exp(v1)+exp(v2(vi)));
end
imagesc(p)
```

You can see in the plot (on your screen; it's not shown here) that action 1 is most likely to be picked when its weight is high and the weight of action 2 is low. Vice versa for action 2. When the weights of both actions are equal, the model has no preference; it simply picks by chance. Notice also how the image values changed: v1 and v2 range from –3 to +3, while the probability map ranges from 0 to 1.

Getting back to the model, we now have the probability of picking an action on the basis of the relative weights between the two actions. To pick the action, that probability value is compared against a random number drawn from a uniform distribution. If the random number is less than the action probability, action 1 is chosen; otherwise, action 2 is chosen. To make sense out of this implementation, consider the extremes: What happens when the probability value is 0, 0.5, and 1?

```
action(ti) = 1 + (pPickAct1(ti)<rand);
```

That line of code does two operations at once. First, the probability of action 1 is compared against a random number drawn from a uniform distribution; second, the outcome of that comparison, which is a Boolean true or false result, is added to one, and that is the action our model takes. Take a minute to understand this line. Again, consider what happens when the probability of picking action 1 is 0, 0.5, and 1.

After picking an action, we see whether the model gets rewarded. The reward is given according to a similar algorithm as the action choice: A random number is compared against the reward probability for the selected action, and if the random number is less than the indicated probability, a reward is given. As a human, you probably consider a reward to be choco-late, sex, a scientific publication, or something else equivalent. For comput-ers, the number "1" is a reward and the number "0" is a punishment. And that's why humans and computers will never understand each other.

```
reward = rand < rewProbs(action(ti));
```

Next up is the prediction error. The prediction error is what the model uses to learn its action preferences; in other words, to adjust the weights. The prediction error is defined as the difference between the reward and the weight (remember that the weight is the model's expectation of the reward). When the reward is larger than the weight, this is a positive prediction error (the outcome was better than expected); when the reward is smaller than the weight, this is a negative prediction error (the outcome was worse than expected). When the weight and the reward are equal, there is no prediction error because the model has predicted the state of the world perfectly; hence, no learning is required.

```
rewpred(ti) = reward-w(ti,action(ti));
```

This prediction error is also called a delta (inspired by it being a difference variable) and is often indicated in equations by its Greek character (δ). Rescorla-Wagner–style models became popular in neuroscience in the mid-1990s with the discovery that dopamine-producing neurons in the midbrain produce patterns of activity that closely resemble this delta function (Schultz, Dayan, and Montague 1997). These kinds of models continue to be widely used in neuroscience studies of reinforcement learning.

Finally, we update the weights according to the prediction error, scaled by the learning rate.

```
w(ti+1,action(ti)) = w(ti,action(ti)) + ...
                     lrate*rewpred(ti);
```

You can see that we're updating the weight for the selected action in the following trial. But what about the weight for the nonselected action? We don't want to forget that one, otherwise the nonselected weight will default to zero (or whatever value was used during initialization). One method is simply to carry over the weight without changing it (this is implemented below); another option would be to discount the weight of the nonselected action.

How can we efficiently implement this weight carryover? Consider that the value of action is either 1 or 2. We can access the other action by subtracting 3 from that value: $3 - 1 = 2$ and $3 - 2 = 1$. (This is a good trick to know; you'll probably use it fairly often.)

```
w(ti+1,3-action(ti)) = w(ti,3-action(ti));
```

Figure 30.5 shows the results of this model. Clearly, it learns to prefer the action that rewarded more often.

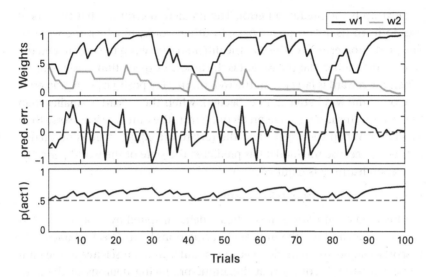

Figure 30.5
Results of the reinforcement learning model. The top plot shows the weights (action preferences) for the two actions over 100 trials, the middle plot shows the prediction error, and the bottom plot shows the probability of selecting action "1," which is computed on the basis of the difference in the weights. Note that as the weights for the two actions diverge, the probability increases.

30.5 Exercises

1. Reza was given DC input. Adjust the code that creates figure 30.2 to provide oscillatory input. Test several different frequencies. Given the amount of time that is simulated, what is the slowest frequency that is sensible to use?

2. Imagine that instead of `neuronV(neuronV==volt_reset)=40;` I wrote `neuronV(spiketimes)=40;` Would that have produced a different plot? Why or why not? First answer the question, and then test it in MATLAB.

3. Let's pretend that the Rescorla-Wagner–style model (call it "Seymour") will need to learn which of two trees has more rubber (they're rubber trees and Seymour is some kind of weird animal or maybe an alien that eats rubber). Both trees provide nourishing rubber, but neither provides food all the time. Tree 1 has rubber 65% of the time, while tree 2 has rubber 45% of the time. Modify the code from section 30.3 as little as possible to get this simulation to work. Figure out a decent criteria for

learning (e.g., greater than 60% tree 1 choices over a sliding window of the preceding 10 trials). How many trials does it take Seymour to learn to select the better action? Randomness might make a single simulation nonrepresentative, so you should average the results over many simulations.

4. Continuing from the previous exercise, modify the learning rate (parameter `lrate`) to determine the impact on Seymour's behavior. Notice that both extremes are bad: At low learning rates, Seymour learns too slowly to have any practical use, whereas at high learning rates, Seymour's behavior is erratic and overly sensitive to every minor outcome (like a teenager on whatever social-network website is popular when you are reading this).

5. Also from exercise 3: Plot Seymour's tree selection on each trial as a function of time. It's a choppy plot, because the choice is binary. Run the simulation 100 times and average the choices across trials. Is the plot smoother? Make the plot even smoother by applying a mean-smoothing filter with a kernel of five trials. Do you get different results if the filter is applied on each trial before averaging, or after the trials have been averaged together?

6. The softmax function is sometimes endowed with a beta parameter, sometimes called a "temperature" parameter: $e^{w/\beta} / \Sigma e^{w/\beta}$. In other words, the weights are divided by a parameter β. What is the value of this parameter in the code? (Hint: It's hard-coded.) Modify the code to include and soft-code this parameter. Then run the code again trying different values. Reasonable values range from 1 to 2. Try ridiculous values as well, just to see what happens.

7. The formulas for creating Izhikevich neurons are not very complicated. Look them up online (e.g., in Izhikevich 2003) and try to translate the formulas into MATLAB code without looking back at the code in this chapter.

8. Using the Izhikevich code for a single neuron, test how the neuron responds to oscillatory inputs of different amplitudes. Using two loops (three including the loop over simulation time), stimulate the neuron using sine waves that vary from 1 Hz to 60 Hz and amplitudes that vary from 1 to 30. You'll need to add an offset to the sine wave input to avoid negative values. For each 1-second simulation, count the number of action potentials and show the results in a 2D plot of stimulation frequency by amplitude. Do the results look different when using different neuron parameters?

31 Classification and Clustering

Data can often be grouped together into clusters. You might, for example, try to use multivariate data to determine whether a heterogeneous group of patients can be classified into smaller, more homogeneous groups. Or perhaps you want to determine a statistical effect size by classifying the experimental condition based on characteristics of the data. Classification is a major topic in image processing and other computer science applications, where it is also called machine learning. In the fMRI literature, researchers often use classification techniques to map spatial patterns of variance across voxels to conditions, a technique commonly referred to as multivariate (or multivoxel) pattern analysis, or MVPA. In this chapter, you will be introduced to three methods for classification: backpropagation learning, *k*-means clustering, and support vector machines (SVMs).

31.1 Neural Networks with Backpropagation Learning

Backpropagation is an important algorithm in neural network modeling of learning. It was popularized in the 1980s and has regained support recently with the realization that backpropagation-based algorithms can be used to train large neural networks to recognize features of images (now called "deep learning" because of the myriad hidden layers). The purpose of backpropagation learning is to form mappings between a set of input patterns (e.g., neural spiking patterns, or pictures of animals) and a set of output categories (e.g., experimental conditions, or cat vs. dog), where the set of input patterns is much larger than the number of output categories. With backpropagation learning, we know the ground truth and can tell the model when it is correct and incorrect. This is how the model learns. Therefore, this form of learning is called supervised learning. Unsupervised learning would mean that the model doesn't know what the correct answer is. Blind source separation methods (PCA, ICA, etc.) are examples of unsupervised learning.

In this section, we will implement a simple three-layer network that can learn a nonlinear input-output mapping. The basic structure is illustrated in figure 31.1.

According to figure 31.1 (and many other more reputable sources), backpropagation learning involves two steps: the forward sweep, in which the inputs are processed and sent through the hidden layer to the output layer; and the backward sweep, in which the outputs are compared against the true result, and the error is sent backward to modify the weights.

Let's start by setting up the model and the input patterns. We'll have our model solve the famous "exclusive or" (XOR) problem: Given two inputs that can independently be on or off, respond TRUE when both are on or off, and respond FALSE when only one is on (figure 31.1). It's a simple but nonlinear problem (nonlinear because there is no linear mapping between the input pattern and the correct answer).

```
inputs = [ 1 0 1; 0 1 1; 1 1 1; 0 0 1 ];
output = [ 0; 0; 1; 1 ];
```

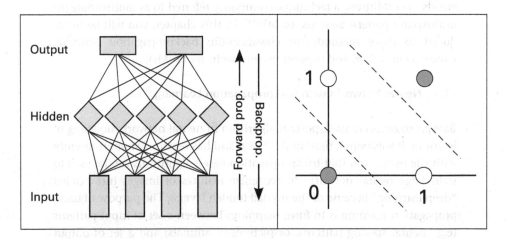

Figure 31.1
This figure illustrates the basic architecture of a three-layer model with backpropagation (left side) and a visual representation of the nonlinear "exclusive or" (XOR) problem (right side). The two white dots are in the same category, and the two dark-gray dots are in a different category. There is no single line that defines category boundaries, which is why this problem is nonlinear.

The inputs are listed in a 4-by-3 matrix. The first two elements of each row specify the patterns—think of 1 being "on" and 0 being "off." Backpropagation models often have an additional "bias" term in the input layer, which can provide an offset or shift of the mapping between the input and the hidden layer, analogous to how an intercept term in a linear model allows the best-fit line to cross the y axis at a non-zero value. This is the third column of all ones. The output vector will act as the teacher to give the model feedback about its accuracy.

Next, we initialize some variables and other housekeeping items.

```
nInputsNodes = size(inputs,2);
nHiddenNodes = 3;
nOutputNodes = 1;
% random initial weights
weights_i2h = randn(nInputsNodes,nHiddenNodes);
weights_h2o = randn(nHiddenNodes,nOutputNodes);
l_rate = .2;
```

The learning rate parameter (variable l_rate) controls how fast the weights are adjusted: too low and the model never learns; too high and the model becomes unstable. It is similar to the learning rate of the Rescorla-Wagner–like model you saw in the previous chapter. I found, through trial-and-error testing, that a learning rate of 0.2 is pretty good for this model and for this problem. More sophisticated backpropagation models adjust their learning rate over time.

Now for the important stuff. We need two nested loops: one while-loop in which the model keeps adjusting its weights until it learns or it runs out of time, and one loop over the four input patterns within each iteration of the while-loop. Remember from chapter 7 that while-loops are generally preferred when you don't know the number of iterations (in this case, we stop when learning is finished), and for-loops are preferred when you do know the number of iterations (in this case, four input patterns). As a precaution to avoid getting stuck in an infinite loop, we program in a toggle that breaks out of the while-loop if the model never learns the correct input-output mapping. That's why we have a variable called max_iter, and exceeding this number of iterations will flip the toggle. Then we can run a new simulation. If at first you don't succeed, try again.

Now let's check out what happens inside the for-loop. The forward sweep of the model involves computing the model's response to the input pattern. This is computed in two mini-steps: multiply the input pattern by the weights from the input layer to the hidden layer, and then multiply that result by the weights from the input layer to the output layer.

In between, the response is passed through an *activation function*, which converts any arbitrary number to a value between –1 and +1 (or sometimes between 0 and 1). Here we use a sigmoid as the activation function. The sigmoid function was introduced in exercise 14 of chapter 29.

```
hdLayerResp = inputs(ini,:) * weights_i2h;
hdLayerResp = 2./(1+exp(-hdLayerResp'*2))-1;
```

The variable `ini` is the looping index around the four input patterns. The variable `weights_h2o` contains the weights from the hidden layer to the output layer (the reference to the molecular structure of water is inconsequential). And `weights_i2h` contains the weights from the input layer to the hidden layer. Those weights get adjusted later.

```
otLayerResp = hdLayerResp' * weights_h2o;
otLayerResp = 2./(1+exp(-otLayerResp'*2))-1;
```

This completes the forward sweep: the inputs are multiplied by one set of weights and passed into the hidden layer, and then the hidden layer activations are multiplied by another set of weights and passed into the output layer. The next step is to compute the prediction error as the difference between the model's output pattern (which in turn reflects how the model transformed the input pattern) and the correct output for that pattern.

```
predError(ini) = otLayerResp-output(ini);
```

The concept and implementation of this prediction error should remind you of the prediction error used in the Rescorla-Wagner–style model in the previous chapter. In fact, prediction errors have been the backbone of learning algorithms and theories across many domains of science (computer science, ecology, economics, neuroscience, etc.) since at least the mid-twentieth century.

Now for the second part of the backpropagation algorithm: sending the error back from the output layer to the hidden layer to the input layer. As the errors are backpropagated, the weights are adjusted. For each set of weights, we compute a "delta"—the amount that the weights should be changed—and then simply subtract that delta from the weights. This is where the learning rate comes into play, just like with Rescorla-Wagner models. Below is the code that uses the prediction error to adjust the weights from the hidden layer to the output layer.

```
delta = l_rate * predError(ini) * hdLayerResp;
weights_h2o = weights_h2o-delta;
```

The delta to adjust the weights from the input layer to the hidden layer are slightly more involved, because the error at both the output layer and the hidden layer are incorporated.

```
bp = weights_h2o .* (1-hdLayerResp.^2)*inputs(ini,:);
delta = l_rate * predError(ini) * bp;
weights_i2h = weights_i2h-delta';
```

And that's about all there is to the model. Finally, we need to check whether to go through another iteration or to break out of the loop. Remember that we break out of the loop either when learning is successful or when we've exceeded the maximum number of iterations. I set both of these thresholds arbitrarily based on trial-and-error guessing.

```
iteration = iteration+1;
totalError(iteration) = sum(predError.^2);
if totalError(iteration)<.01 || iteration>max_iter
    toggle=false;
end
```

Run the model a bunch of times and watch the output. First of all, it's fun to watch, isn't it? (If you don't know what I'm talking about, run it on your computer rather than just looking at figure 31.2.) Second, notice that the model doesn't always solve the problem. It depends on the random initial weights. It can get stuck in one configuration that is sort-of good (it gets two outputs correct) but not optimal (two outputs incorrect). This is called a local minimum, which you also saw in the previous chapter. Advanced backpropagation algorithms have methods to bump out of local minima, but this is an opportunity to reiterate a point made in previous chapters: Never blindly trust model fits (particularly from nonlinear models), even with fancy algorithms. Always carefully and critically inspect the results.

Third, notice how the weights matrices—particularly those related to the hidden layer—change each time you run the simulation. There are many correct solutions to the same problem, and it is difficult or perhaps impossible to interpret the exact pattern of weights. This is not a limitation when the goal is to find solutions to nonlinear problems, but it can be an important consideration when trying to apply these kinds of models to understand how the brain works.

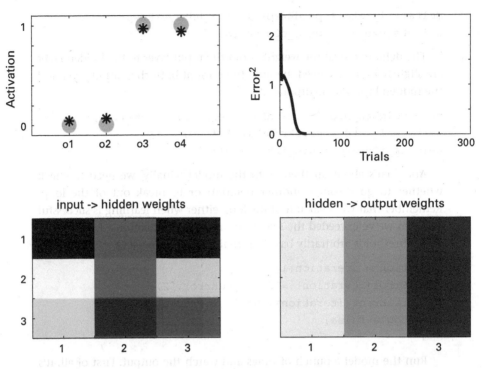

Figure 31.2
Results of backpropagation learning. The stars (model output unit activations) are in
the circles (the correct answers), and the error goes to zero after around 40 iterations.
The bottom panels show the layer-to-layer weights. This is the final frame of a movie
that updates as the model learns to solve the task.

Fourth, notice that the plotting really slows down the computations. Try
moving all the plotting code from inside the loop to outside the loop. It will
plot only the final result, but the code will run much faster, from many
seconds to a few tens of milliseconds. Chapter 33 will show you a few tricks
to speed up the plotting, although in general, plotting inside a loop will
always slow things down.

31.2 *K*-means Clustering

The idea of *k*-means clustering is to separate multivariate data into *k* groups.
An example problem is shown in figure 31.3. Visually, it is clear that the
data are organized into three groups. Quantitatively, however, separating
the data into three groups is a nonlinear problem, because you cannot draw

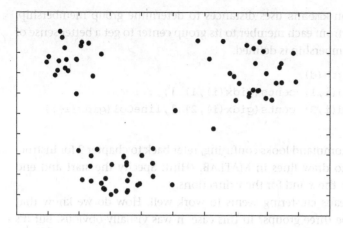

Figure 31.3
Clearly, there are three groups of dots here. *K*-means clustering is one algorithm to label each dot as being part of one group.

a single line on the plane that will separate these groups. *K*-means clustering is one solution to this separation problem. We can use MATLAB's kmeans function.

```
[gidx,cents,sumdist,distances] = kmeans(d,3);
```

The variable d is a 75-by-2 matrix that contains the *x* and *y* values of each of 75 data points, and the input "3" instructs MATLAB to extract three clusters. The number of clusters is a parameter and—no big surprise here—it's typically called the *k* parameter.

The first output variable gidx associates each member of d with a group category (1, 2, or 3; more generally, integers from 1 to *k*). Let's replot the data using different colors for the different groups.

```
lineCol = 'rbk';
hold on
for i=1:3
    plot(d(gidx==i,1),d(gidx==i,2),[lineCol(i) 'o'])
end
```

Notice how I used the variable lineCol to assign a different color to each group. The second output of the kmeans function (cents) identifies the *x* and *y* coordinates of the center of each group.

```
plot(cents(:,1),cents(:,2),'ko')
```

The function `kmeans` uses distances to determine group membership. Let's plot lines from each member to its group center to get a better sense of how group membership is defined.

```
for i=1:length(d)
    plot([ d(i,1) cents(gidx(i),1) ], ...
         [ d(i,2) cents(gidx(i),2) ],lineCol(gidx(i)))
end
```

If this plot command looks confusing, refer back to chapter 9 for instruction on how to draw lines in MATLAB. (Hint: Specify the start and end coordinates for the x and for the y directions.)

So far, k-means clustering seems to work well. How do we know that there should be three groups? In this case, it was visually obvious. But it's not always so clear, particularly for multidimensional data sets that might be difficult to visualize.

Let's see what happens when we extract four clusters. Change the k parameter in the `kmeans` function to 4, and re-run the k-means clustering and plotting. What happens? Well, first of all, you get an error in MATLAB. Before reading the next paragraph, try to find and fix the error.

The source of the error was that the `lineCol` variable contained exactly three letters and the code asked MATLAB to plot the color corresponding to the fourth element. Adding a few extra letters into that variable solves the problem. After fixing this bug, did you get a plot with groups that look like figure 31.4? Possibly, or possibly not. If you re-run the code multiple times, you'll get different groupings. That's not a good thing. The groups should be stable each time you re-run the function (`kmeans` clustering starts with random weights, which is why the results can differ each time). If you run `kmeans` on the same data and get different results, it suggests that you have the wrong number of groups. The exercises will explore this in more detail.

How does k-means clustering work? There are different k-means clustering algorithms, but generally, data centroids are initially placed randomly in the space of the data. Then, each data point is assigned to the closest centroid, and the distances between all data points and all centroids is computed. Centroid locations are then moved in a direction proportional to the average distances of the data points. This process continues until some stopping criteria, for example when the centroids take only insignificant steps at each iteration.

K-means clustering works well in simple situations like what was shown here. However, there are situations—such as noisy data, multivariate data,

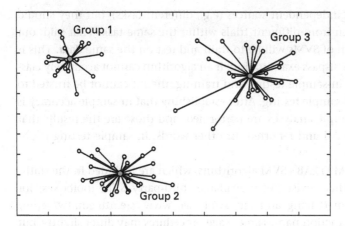

Figure 31.4

K-means clustering successfully grouped the data points into three different categories. This is a fairly easy problem, and *k*-means clustering is very likely to produce a sensible and replicable result in situations like this.

or unclear separations between clusters—when *k*-means clustering will provide disappointing results, for example when repeated calls to the function produce qualitatively different clusters. Expectation-maximization and support vector machines are among the more flexible clustering algorithms (there are many more). This is not to say that you should avoid using *k*-means; just be aware that *k*-means clustering might not produce optimal results for complicated datasets.

31.3   Support Vector Machines

SVMs are a class of algorithms used for categorization in machine learning and artificial intelligence. The idea of an SVM is to define boundary lines or planes through a feature space. These boundaries are defined by distances away from features grouped into different categories (the *support vectors*). Linear SVM algorithms are often used in neuroscience, particularly in the MVPA approach that is increasingly being applied in the fMRI and M/EEG literatures. Shall we do an SVM analysis? I'm glad you agree. In this analysis, we will use EEG data to predict whether a human volunteer subject pressed a button with his right hand or with his left hand.

The general approach of SVMs is to determine model parameters based on a training set, and then apply that model with those parameters to new test data that the model has not seen. Ideally, the training and testing data are taken from independent sources (e.g., different tasks), but they should at least be taken from different trials within the same task. That said, our first exploration of SVMs will be to train and test on the same data. This is an important first-pass examination: If an algorithm cannot accurately classify data using in-sample testing and training, then it cannot be trusted to perform out-of-sample testing. After establishing that in-sample accuracy is high, out-of-sample analyses are performed, and those are the results that can be interpreted and reported. In other words, in-sample testing is our sanity check.

We will use MATLAB's SVM algorithms, which are included in the statistics toolbox. There exist freely available third-party SVM toolboxes; for example, libsvm (Chang and Lin 2011; see www.csie.ntu.edu.tw/~cjlin/libsvm/). The function names and usage procedures may differ slightly, but the overall operation is the same.

The code is fairly straightforward, because MATLAB does most of the hard work. We compute the SVM model using the function `fitcsvm`, which takes a minimum of two inputs: the data and the category labels. The data must be a matrix of observations (in our case, trials) by features (sometimes called predictors; in our case, these are EEG channels), and the category labels are in a vector, one per exemplar (trial). Our data come in two 3D matrices, one for left-hand presses and one for right-hand presses. Each matrix is channels by time by trials. To get a feel for the code, let's start by fitting one time point.

```
t = 200;
data = squeeze(cat(3,l_eeg(:,t,:),r_eeg(:,t,:)))';
trueLabels = [ones(size(l_eeg,3),1); ...
              2*ones(size(r_eeg,3),1)];
svmModel = fitcsvm(data,trueLabels);
```

The first line concatenates the EEG data from both matrices, all trials, at the 200th time point. The first input into the `cat` function is "3" because we want to concatenate on the third dimension, which is trials. The `fitcsvm` function expects the data to be a 2D observations-by-features matrix, so we need to squeeze out the singleton dimension and then transpose the matrix. We have 64 channels and 200 trials; if the data were not transposed, MATLAB would think we have 64 independent observations of 200 features, when we actually have 200 independent observations of 64 features. Matrix orientation is a major source of confusion and errors, and you should always double-check that your matrices are in the order that a given function expects. Part of why this gets confusing is that different functions have different preferences for data organization.

The MATLAB function `fitcsvm` can take several additional inputs that allow you to specify additional constraints and algorithms of fitting. The standard configuration works fairly well in this example, so we'll keep it simple. If you want to use an SVM for real data analysis, it's a good idea to spend some time optimizing the procedure beyond the default settings.

Now that the model has been computed, the next step is to feed in the test data and determine whether the model can accurately predict the category.

```
catlabel = predict(svmModel,data);
accu = mean(catlabel==trueLabels);
```

The MATLAB function `predict` takes the model and test data as inputs, and it outputs the predicted category labels for each trial. Those predictions will be 1 or 2, because that's what I specified in the function `fitcsvm`. You can also use strings or Booleans as labels.

For the model to be accurate, the labels should be "1" for the first 100 trials and "2" for the second 100 trials. Unlike `kmeans` clustering, with an SVM we need to know which observations (trials) came from which category. SVMs are another example of supervised learning.

These lines of code are embedded in a loop that tests each time point separately and stores the accuracy in a vector. Figure 31.5 shows the resulting accuracy over time. There are two conditions, so 50% is chance level. We hit 100% accuracy. That might initially seem really amazing, but keep in mind that this is in-sample testing. Because the same data were used for training and testing, there is a bias from overfitting. Still, this result tells us three important things: (1) the conditions are definitely separable and categorizable; (2) the accuracy is just above chance level up to around 300 milliseconds (we would expect chance-level performance, and the slight increase here is likely due to overfitting); and (3) peak condition differentiation occurs at around 400–500 milliseconds, so we should expect a similar time course in the real analysis. These findings help guide our expectations about the real test with out-of-sample testing. For example, if we found high prediction accuracy before 100 milliseconds and chance-level accuracy around 400–600 milliseconds, then something would be very strange. Most likely a mistake somewhere.

Now that we have established that the conditions are categorizable in the overfitting situation, we can do the analysis in a more appropriate way. This means using different trials for training and for testing. There are

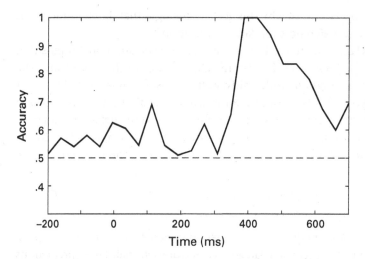

Figure 31.5

Results of in-sample SVM classification, with the dashed line indicating chance-level performance. Although these results should not be interpreted on their own (because in-sample classification is biased by overfitting), they provide an important sanity check that the code works and that the data are categorizable.

several ways to approach such cross-validations; we'll use a leave-one-out procedure, in which the model is trained on $N - 1$ trials and then tries to predict the remaining trial. This is repeated for all trials, and we take the overall accuracy to be the average accuracy from all tests (all trials).

The code needs some adjustments, one of which is an additional loop over trials. This is a good opportunity to think about what code can go inside versus outside a loop. In general, you should always think about whether code can be moved outside a loop for speed and clarity. In each iteration, we want to separate the trial-N data from the data from all other trials. One solution is to make a copy of all the data and then delete the data from one trial:

```
traindata = data;
traindata(triali,:) = [];
```

The variable `traindata` now has 199 trials instead of 200. We use the same approach to remove the label from one trial in the `trueLabels` variable.

```
templabels = trueLabels;
templabels(triali) = [];
```

Testing the model now looks slightly different. We provide to the pre-dict function the missing trial, and our accuracy vector is now a matrix in which we store the accuracy separately for each time point and each trial. In the plotting, we can then average over trials (each trial will have a value of 0 or 1—false or true).

```
svmModel = fitcsvm(traindata,templabels);
catLabel = predict(svmModel,data(triali,:));
accu(ti,triali) = catLabel==trueLabels(triali);
```

The results of this test are shown in figure 31.6. Not surprisingly, the performance is lower than the in-sample test, but it appears to be above chance level from around 350–600 milliseconds (that's a qualitative interpretation—no statistics were performed). If you would like to learn more about using MVPA in neuroscience data, you can start with the papers by Norman et al. (2006) and King and Dehaene (2014).

31.4 Exercises

1. Run *k*-means clustering on the 2D data many times using *k* = 3. Does the grouping change? How about which group is given which integer

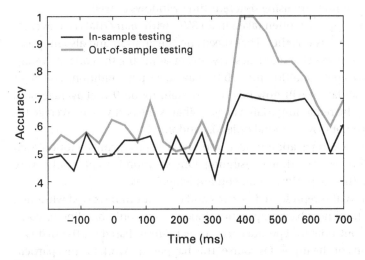

Figure 31.6
Here you see the difference between in-sample testing and out-of-sample testing. In-sample testing provides an important sanity check, but only out-of-sample testing can be interpreted.

label? What does this tell you about the stability of groups versus the stability of group labels?

2.   Run *k*-means clustering on the 2D data for 2–8 groups, 100 times per *k*. Devise a method to determine whether the groupings are stable (e.g., you might compute the average distances to the nearest vs. all other centroids). Plot this stability measure as a function of *k*. This exercise shows that *k*-means clustering can be a powerful technique but can also go awry when suboptimal parameters are used.

3.   Can you replace these two lines with one line? You should do this by extracting $N - 1$ trials from data, rather than deleting one trial in `traindata`. Make sure your code works for any trial, not only the first or last trial (that would be too easy).

```
traindata = data;
traindata(triali,:) = [];
```

4.   One interesting use of an SVM is to train the model on data from one time point, and then use that model to predict data from other time points. This could be done as a full time-by-time matrix, but that can be time-consuming. Here, train the model using the average data from 400 to 550 milliseconds, and then apply that model to predict each other time point. Plot the results and compare against the same-time-point training. Try using different time windows as well.

5.   Another approach often taken in an SVM with many trials is to average blocks of trials together. That speeds up the analysis and also increases signal-to-noise characteristics. Try this by adjusting the code to average 100 trials per condition into 10 trial-averages per condition. The $N - 1$ cross-validation will now be done by training on 9 trial-averages and testing the remaining trial-average. What is the best way to average trials together—by temporal order or randomly?

6.   In the backpropagation model, you saw that producing a plot on each iteration was very time-consuming. One solution is to plot only every 5th or 10th step. How could you implement this?

7.   When XOR is represented as zeros and ones, it can be solved without a fancy nonlinear backpropagation model, using only one line of code. Figure out how to reproduce the correct output based on the first two columns of the input. Of course, this happens to work for this particular instantiation of XOR; you shouldn't think that all problems solvable with backpropagation can be solved with a single line of code.

8.   Another option for an activation function is the hyperbolic tangent (MATLAB function `tanh`). Your mission here is to use a sigmoid

function (see the formulation in exercise 14 of chapter 29) to obtain the exact results as the following code. You'll need to figure out how to set the sigmoid parameters; depending on your level of math background, you can try this either analytically or empirically.

```
x = -3:.1:3;
af = tanh(x);
```

9. In the backpropagation model, I guess you played around a bit with the learning rate. Let's make this more quantitative. Put the model into a function that takes a learning rate as input and outputs whether the model successfully learned or stopped because it exceeded its maximum iteration limit (you might want to turn off plotting). Then run the model 50 times using 30 different learning rate parameters between 0 and 1 (thus, 1,500 simulations). Plot average successes as a function of learning rate.

10. In the backpropagation model, the value for `max_iter` (300) was selected arbitrarily. Is it possible that the model would have learned the solution if it had more iterations? Run an experiment in which you determine the number of successful learning solutions as a function of the maximum number of iterations. Run 100 simulations (don't forget to reinitialize the weights for each simulation!) for each of many different iterations. You can select the range and number of iterations to use. Make a plot of the proportion of successes (you'll need to define "success" in this model) as a function of the number of iterations.

VII User Interfaces and Movies

32 Graphical User Interfaces

A graphical user interface (GUI; pronounced "gooey") is a useful tool for interacting with and viewing data. GUIs are particularly useful for visualizing multidimensional data sets. Thanks to MATLAB's GUI-creating utility (cleverly called GUIDE), GUIs are fairly simple to design. The construction and use of simple GUIs is underutilized in neuroscience data analysis.

32.1 Basic GUIs

Let's start with a few simple GUIs that involve very little programming. The first will be a dialog box that informs the user of some event (figure 32.1).

```
h=msgbox('Press if you like chocolate','ChocTitle');
```

Obviously, you should press the OK button in the message box. These kinds of dialog boxes are useful for providing important information; for example, if an existing file is about to be overwritten or if there is an error or a warning. However, MATLAB will not pause when a message box opens—when called in a script or function, the message box will open, and

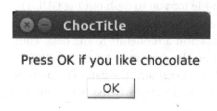

Figure 32.1
Press OK!

then MATLAB will continue to the next line of code (see exercise 1). The variable h is a handle that allows you to control visual and functional features of the message box. It works the same way as handles for axes and figures, as you learned in chapter 9.

Other simple interfaces allow the user to provide input. This is often useful when you want to specify a parameter or input a file name, but it is not a scalable method for providing large amounts of information.

```
s=inputdlg('How much do you like chocolate (0-9)?');
```

The output of this function will be a cell that contains the text the user entered. The text is given as a string, so you might need to convert it to a number. Another function will allow you to select a file from the computer.

```
[fname,fdir] = uigetfile('*.txt','Pick a file');
```

The '*.txt' is the search filter (MATLAB always additionally provides an option for all files), and the 'Pick a file' is printed in the top bar of the dialog box. Note that this function does not import any data into MATLAB, nor does it open any files; it simply provides string outputs of the file that was selected and the full path to that file.

There are several other small user interfaces similar to these three. And they are all just as easy to use. You can type help msgbox to see a list of other similar functions.

32.2 Getting to Know GUIDE

Cute little interfaces like msgbox can be useful. But real MATLAB programmers use GUIDE to make their own GUIs. And that is what you will do—because this is chapter 32 and you are now a real MATLAB programmer.

GUIDE is the MATLAB GUI that helps you build and customize your own GUIs. You use GUIDE to create and adjust the layout of your GUI, and GUIDE then creates a figure and an m-file for you to flesh out the GUI. The GUI we will create here will generate a set of correlated data based on some user-specified parameters, plot the data and a model fit to the data, compute a principal components analysis on the data and plot the results in a separate axis, and create an output file that contains the data. The purpose of this GUI (other than teaching you how to create GUIs) is to help you gain a more intuitive feel for how PCA works.

Type guide in the MATLAB command, and then create a new blank GUI. You'll see a window like figure 32.2. In our GUI we will need two axes

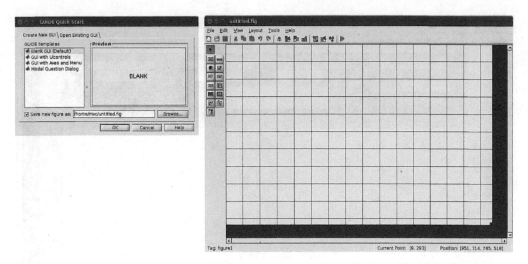

Figure 32.2
This is what you see when you start GUIDE (by typing guide) and create a new GUI.

for viewing data, an input region for specifying five parameters (the number of data pairs, the correlation between them, x-axis and y-axis offsets, and an option to remove the mean of the signal), and an output region for specifying options for saving data. A nearly completed GUI is available in the online code, but I recommend creating your own GUI from scratch as a learning experience.

You can see in figure 32.3 how I laid out the GUI. Set up yours to be similar, including the names of the components. When creating this layout, I used the following buttons on the left side of GUIDE: Axes, Panel, Edit Text, Static Text, Push Button. Note that you can use the solid panel to group similar objects. You can also set the background color to purple if you want.

After creating and placing an object inside the GUI, double-click on it to see a list of properties of that object. Some properties are purely aesthetic, like font size and color, while others are important for programming. The "Title" or "String" properties define the text that you see in the GUI. You may need to click outside the just-edited field for MATLAB to recognize and update the changes.

Don't forget to save! I called my GUI "PCA_GUI." The first time you save, you'll notice that MATLAB automatically generates an m-file with the same name and opens it in the Editor window. This file will contain all of

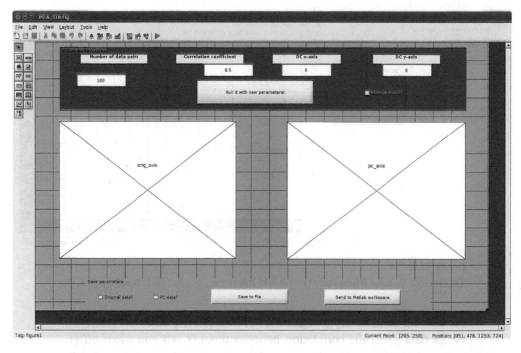

Figure 32.3
After adding axes, buttons, dialog boxes, and other features, your GUI skeleton might look something like this.

the actual code you need to breathe life into the GUI. We'll get to this part soon. The GUI window is called PCA_GUI.fig.

One very important object property that you should always customize is called "Tag." These are the names of handles that identify objects and their properties and are important for programming. You should give these tags meaningful names, particularly if you have multiple similar objects. For example, "origdata_axis" and "PC_axis" are much more meaningful than "axis1" and "axis2." Similarly, input boxes should be called "dataCorr" or something similar, rather than, for example, "input3."

When your new GUI looks something like figure 32.3, move on to the next section.

32.3   Writing Code in GUI Functions

GUIDE produces a skeleton; it's up to you to add muscle and skin. That's what we'll do in this section. First, take a minute to look through the m-file that GUIDE created when you saved the GUI. You will notice a repeating structure: Each object that you created in the GUI has two associated functions, one called `<objectname>_Callback` and one called `<object-name>_CreateFcn` (recall from chapter 6 that multiple functions can be embedded into a single file). They contain a few lines of comments, and the `<objectname>_CreateFcn` may contain a few lines of code that sets the background color if you are using a Windows computer and if the GUI and default figure color is set to white. Those lines are purely decoration, and you can delete them if they bother you.

The `_CreateFcn` function is called when you first open the GUI, and the `_Callback` function is called when you take action on the corresponding object, such as clicking or entering data. Let's see how this works. Find the tag name associated with the button at the top of the GUI that is called "Run it with new parameters!" in figure 32.3. I named the tag "update_data." Now find the function called `update_data_Callback` (or `whatever_Callback`). Inside that function—that is, anywhere before the next function is defined—type `msgbox('You pressed the button!')`. Save the m-file.

Open the GUI. You can do this either by typing `PCA_GUI` in the MATLAB command or by pressing the green right-arrow button in GUIDE. You should see the figure open. Now press the "Run it!" button, and the message box should appear. You can keep pressing the button, and more and more message boxes will pop up. Now you see how the GUI and the script work together—when the user interacts with the GUI, MATLAB finds the corresponding function (`<tagname>_Callback`) in the m-file and runs all the code in that function. This is the first important concept of MATLAB GUI programming.

The second important concept in GUI programming is that each object in the GUI has a set of states that can be queried and changed using the handles to those objects. You can see from the m-file that MATLAB defines each function with three input variables, of which `handles` is the most important, and probably the only input variable you will use in practice. This variable is a structure that contains handles to all objects in the GUI. For example, `handles.nDataPairs` is the handle for the input box with the tag name "nDataPairs." The input variable `hObject` is the handle to the current object and is therefore redundant because it is the same thing as one of the fields in the `handles` structure. I recommend against using `hObject` because of the potential for confusion—this same variable name points to different objects depending on where it is in the file, whereas the variable `handles` is exactly the same in every subfunction. Finally, the

Box 32.1

Name: Vladimir Litvak
Position: Senior Lecturer
Affiliation: Wellcome Trust Centre for Neuroimaging, UCL Institute of Neurology, London, England
Photo credit: Ashwani Jha

When did you start programming in MATLAB, and how long did it take you to become a "good" programmer (whatever you think "good" means)?
I studied computer science as an undergraduate and had to learn MATLAB during my second year as it was used for several courses I took at the time. As I was intensively doing all kinds of programming at the time, I had to get good quite quickly, probably within weeks, but on the other hand MATLAB is really vast, and even now after 20 years of continually using it, I think I am only familiar with a tiny fraction of all the functions.

What do you think are the main advantages and disadvantages of MATLAB?

The main advantage is that there is a lot of functionality already implemented in the built-in functions, MathWorks toolboxes, and also other toolboxes available online. So only rarely does one need to implement something low-level, and then it's mostly about combining existing building blocks in the right way, which for me is appealing as I don't like low-level stuff. Also the fact that most of the code is not compiled, so when an error happens it's easy to get to the bottom of it quickly. The main disadvantage is that MATLAB is not free, which for some people creates difficulties (e.g., they don't have a particular toolbox or don't have enough licenses to run a distributed computation). Also, we are all at the mercy of MathWorks, which is now at the point where they make changes in the software just for the sake of making changes and not necessarily improving things.

Do you think MATLAB will be the main program/language in your field in the future?

Yes, if only because years have already been spent on creating the code base that we have now, and no one in the world has the manpower to re-implement it all on a different platform. It might be that other platforms (e.g., Python) will gain popularity in parallel, which I see as a good thing.

How important are programming skills for scientists?

Basic programming skills are probably important for just about everybody in the modern world. Even people who don't program would benefit from being able to better understand the logic of people who created the software they are using. As a child I was always interested in biology, but when I took an internship in a neuroscience lab after high school, the colleagues I worked with told me in very strong terms that with a biology background alone I won't get very far and convinced me to also take computer science as a second major. Although this transition was not easy, I am very grateful for this advice that really changed my life and career and made it possible for me to be involved in research that I find very exciting also as a biologist.

Any advice for people who are starting to learn MATLAB?

Try to find a piece of code written by someone else that is similar to what you need and modify it using your basic knowledge. Don't be afraid of the red error messages. You won't break anything by making mistakes, and with time you'll get better in figuring out what these messages mean as they can be quite cryptic.

variable eventdata mostly contains information that is available through the handles variable, and also changes its values depending on where in the file it is.

The best way to edit these functions is to step into them. Put a little red ball to the left of the line that contains the msgbox (see figure 6.1), and click the "Run it!" button on the GUI. Now you are inside the update_data_Callback function (see section 6.10 of chapter 6 if you need a reminder about stepping into functions). You can access information about all of the GUI objects using the get function, the same way you would access information about an axis or figure. For example:

```
get(handles.nDataPairs)
get(handles.nDataPairs,'String')
```

The first command will return all of the properties of the nDataPairs object. The data entered into the field is called the 'String' property of the object (for other types of objects, the user-specified piece might be called 'Value'). You can also use the set command here. Try, for example, setting the property backgroundColor to have a value of [1 0 1].

When plotting inside a GUI, MATLAB has no idea which axis to use for plotting (well, that's not entirely true; MATLAB will always plot to the active axis, but it is really unlikely that the active axis happens to be the one in which you want to plot). You should always call a specific axis before any plotting commands. For example:

```
cla(handles.orig_axis)
plot(handles.orig_axis,data)
```

Save often! It takes only a few seconds to save, and these few seconds can save you minutes to hours of time spent recoding. GUIs can crash for various reasons, and they will crash more often than MATLAB will. For example, if you delete an object and a stale link to that object is called, the GUI might crash. If the crash is particularly problematic, the .fig file itself might become corrupted. Keep backup copies just in case or keep your files on a system that preserves backups and old versions, like github or Dropbox. One way to minimize the risk of crashes and corrupted GUIs is to avoid changing low-level parameters like object names. And if you do change object names or make other modifications, do these through GUIDE instead of through modifying the associated m-file.

32.4 Exercises

1. The `msgbox` does not pause MATLAB to wait for input. Read the help file for this function to learn how to make MATLAB pause until the OK button is pressed. Implement this in the first example of this chapter so people are forced to profess their fondness for chocolate before continuing to use MATLAB.

2. Although my GUI is really awesome, it needs some help. Will you help make the GUI better?

   a. For one thing, the text fields on the top are misaligned. Visual clarity is important in GUI design. Open the GUI in `guide` and adjust these objects. Look for the Align button on the top to align and distribute different objects. You might also consider changing the colors of different objects.

   b. The *Save to file* and *Send to MATLAB workspace* buttons in my GUI do nothing. Write code in the GUI m-file to save the variables `data` and `pc` (depending on what was checked). You will need to find a way to access these variables from inside the *save* subfunction.

   c. Start the GUI, set the number of data pairs to zero, and then run the analysis. It crashes. Implement a method to prevent this from happening. There should be a minimum of 10 pairs of data points.

   d. Draw principal component vectors in the axis like in figure 17.2.

3. Write code in your own GUI so that it reproduces the functionality of my GUI. You will need to write code to generate random data, compute a PCA, perform linear least-squares fitting, and plot the results. You should run my GUI to confirm that your version reproduces the core functionality, but I'm trusting you not to copy my code.

4. A useful tool for inputting data or points on an axis is a function called `ginput`. Read the help documents for this function to open a blank figure, collect 10 points of user input, and store those input points in two variables for the *x* and *y* coordinates. Update the plot after each click so the dots appear and remain on the graph.

5. Use `ginput` to generate 30 points of data that are grouped into two regions of space, and then run *k*-means clustering on those data. Put the code into a function (perhaps called "test_kmeans"). The function should take one input—the number of data points to generate using

ginput—and it should produce a plot like figure 31.4. This will allow you to test *k*-means clustering using an easy case (clearly separate points), a "medium" case (sort-of separable points but some points in the middle), and a hard case (inseparable clusters, or maybe just one cluster).

6. Design a GUI that will import an MRI file and display three views, like figure 25.1. Incorporate ginput into the GUI so that when a user clicks on an MRI slice, the other two views update according to the *xyz* coordinates that were clicked.

33 Movies

Everyone likes movies. They are fun to watch and can also be powerful methods for illustrating time-varying data, multidimensional data, or complicated methods. In this chapter, you will learn the basics of creating animations in MATLAB. Much of this involves learning different ways to update data inside axes. In this sense, this chapter is an extension of chapter 9. If you enjoy programming simple animations and would like to combine that with learning about the mathematical descriptions of physical laws that govern motion and other interactions, you might consider the book *Nature of Code* (Shiffman 2012; see http://natureofcode.com).

33.1 Waving Lines

To make a movie, we need to think like a director. Before filming, we need the cast of characters, we need the set, and we need the story line. (A big explosion somewhere would help, but unfortunately, we're working on a tight budget.)

One of the difficult concepts of time-frequency decomposition is the link between the complex representation of the analytic signal in a polar plane and the power and phase values that are extracted from that analytic signal. Let's see if we can make a movie that will help clarify this link.

The characters in our movie are an analytic signal (variable as), a power time series, and the projection of the analytic signal onto the real axis (the band-pass filtered signal). The set will be a figure with three panels: top panel for the polar representation, middle panel for the power time series, and bottom panel for the band-pass filtered result. And the story line is: analytic signal meets Pythagorean theorem, analytic signal then calls out longingly to the real axis, things change over time, and they all live happily ever after. (Okay, maybe not an Academy Award winner, but don't tell me you haven't seen movies with worse plots.)

You already know much of the important parts of the code (time-frequency analysis, plot handles, for-loops); we just need to put them together in a new way. The characters are simulated from cosine and sine waves with randomly varying amplitudes (see the online MATLAB code).

Let's set up the scenery.

```
subplot(311)
as_h = plot(as,'k','linew',2);
set(gca,'xlim',[min(real(as)) max(real(as))],...
        'ylim',[min(imag(as)) max(imag(as))])
subplot(312)
bp_h = plot(timevec,real(as),'k','linew',2);
set(gca,'xlim',timevec([1 end]),...
        'ylim',[min(real(as)) max(real(as)) ])
subplot(313)
pw_h = plot(timevec,abs(as).^2,'k','linew',2);
set(gca,'xlim',timevec([1 end]),...
        'ylim',[min(abs(as)) max(abs(as)) ])
```

The most important thing to notice here is the plot handles; one for the complex analytic signal (variable as_h), one for the band-pass filtered signal (projection onto the real axis; variable bp_h), and one for the power (squared distance from the analytic signal to the origin; variable pw_h). Recall that when you plot a complex vector, MATLAB will plot it as the real part by the imaginary part (see also chapter 19).

Now it's time for the movie. We want to loop over time points and update the plots at each time point. You can see below a skipping variable (variable vidspeed) that controls the update rate. If your computer is graphics card challenged or if you like movies involving high-speed car chases, you can set the skipping variable higher to speed up the animation by plotting fewer frames.

```
vidspeed = 2;
for ti=1:vidspeed:length(timevec)
```

Inside the for-loop we want to update the axes. Here is the important part, and the difference between making animations and normal plotting: We do not *replot* the data; instead, we use the set function to *update* the data inside the axis. This is much faster and looks much nicer than when drawing a new figure. This is why having handles is necessary.

```
set(bp_h,'xdata',timevec(max(1,ti-timelag):ti))
set(bp_h,'YData',abs(as(max(1,ti-timelag):ti)).^2)
```

This is the same `set` command that you use to change the *x*-axis ticks, the color axis limits, and so on. The only difference is that we're setting the data, rather than an aesthetic property of the axis. Because these two lines both call the same handle, they could have been combined into one line. The properties are not case-sensitive, as you can see from calling `xdata` but `YData`. Notice that we're not plotting the entire time series from the beginning; we're trailing the time series by `timelag` points. This is not necessary but (1) shows you how to plot a limited amount of data and (2) facilitates interpretation when comparing with the complex data shown in the polar plot. Question: What is the purpose of `max(1,ti-timelag)`, and what would happen without the `max` function?

The power time series data are updated in the same way.

```
set(pw_h,'XData',timevec(max(1,ti-timelag):ti),...
        'YData',abs(as(max(1,ti-timelag):ti)).^2)
```

So far, the band-pass filtered signal and the power time series plots were updated in the same way: The *x* axis is the time vector, and the *y* axis is the data vector. The complex vector needs to be updated a bit differently. Before reading the solution in the next paragraph, inspect figure 33.1 and try to figure out how the polar plot should be updated. (Hint: Think about how MATLAB plots complex numbers in the complex number plane.)

The solution is to update the `XData` property using the real part of the data and the `YData` property using the imaginary part of the data. Time is not explicitly incorporated here; time is spinning around the polar axis. If this were a 3D plot, we could also update time using the `ZData` property.

```
set(as_h,'XData',real(as(max(1,ti-timelag):ti)),...
        'YData',imag(as(max(1,ti-timelag):ti)))
```

Now the code is set. Run the loop. How does the movie look? I guess you are pretty disappointed, because nothing happened. Actually, MATLAB was dutifully updating the plots in the background, but the figure was not updated until after the loop ended. You need to instruct MATLAB to refresh the figure, and—depending on your computer speed and graphics card— you might want to slow things down a bit so you can see the refreshes. Use the function `drawnow` to force MATLAB to update the figure or use the function `pause` to have MATLAB wait 100 milliseconds between each iteration of the loop.

Now you should see the plot updating. If you were a bit confused when going through chapter 19, hopefully this movie helps clarify the

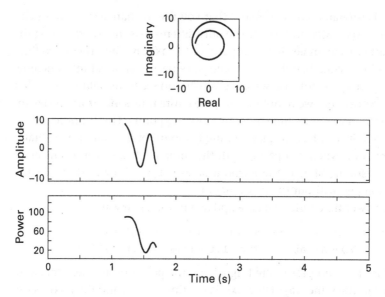

Figure 33.1
One frame of a movie that shows a complex analytic signal (top panel), the band-pass filtered signal (the projection of the analytic signal onto the real axis; middle panel), and the power time series (the distance on the complex plane from the origin; bottom panel).

relationship between the analytic signal and the power and filtered time series. And if you are totally confused about what this plot means, try reading chapter 19.

By the way, one of the advantages of setting up the axis and then updating only the handle rather than redrawing the axis is that other axis parameters need to be set only once, before the loop starts. That is, without setting up plot handles the way we did it here, you would need to re-run `axis off`, `axis square`, `set(...)` and so on at each iteration inside the loop. This makes the code slow and messy, which in turn increases the risk of careless mistakes or forgotten hard-coded parameters.

33.2 Moving Gabor Patches

Let's extend the line movie to an image movie. The concept is the same—define new parameters on each frame of the movie and use handles to update the data in the images. There will be three learning objectives here: learning how to make a 2D Gabor patch, extending the line handles to image handles, and learning how to use MATLAB's `movie` and `frame` functions to create a movie file that is independent of MATLAB. Let's first learn about a Gabor patch.

A Gabor patch (also sometimes called a 2D Gabor filter because it is used to decompose images into different spatial frequencies) is very similar to a Morlet wavelet, but is 2D. To create a Gabor patch, we need a 2D Gaussian and a 2D sine wave, and then we simply point-wise multiply them together. In the 1D case, we used a *vector* of points to define the Gaussian and the sine wave. Now we're in 2D, so we need a *grid* of points, which we can obtain using the `ndgrid` function that was introduced in chapter 13.

```
lims = [-31 31];
[x,y] = ndgrid(lims(1):lims(2),lims(1):lims(2));
```

The variable `lims` (limits) defines the size of the image in pixels. Without looking at the size of variables `x` and `y`, what do you think will be the size of our Gabor patch? After guessing, confirm your hypothesis by typing `whos x y`. To remind yourself of what `x` and `y` look like, type `imagesc(x)` and `imagesc(y)`. Next, we create the Gaussian, the sine wave, and their point-wise multiplication.

```
width = 10;
gaus2d = exp(-(x.^2 + y.^2) / (2*width^2));
sine2d = sin(2*pi*.05*x);
gabor2d = sine2d .* gaus2d;
```

The 2D Gaussian formula should look similar to the 1D Gaussian formula, except that we have two dimensions. The three functions are shown in figure 33.2.

That's the basic Gabor patch. There are several parameters that define how the Gabor patch looks. The Gaussian part has one parameter: the width. Recalling the discussion of the width of the Gaussian in chapter 12 and looking at figure 33.2, what do you think will happen to the Gabor patch when the width of the Gaussian is increased or decreased? After you've generated your hypothesis, test it by changing the width parameter and re-creating the plot. Was your intuition correct?

The 2D sine wave has several parameters.

• *The frequency of the sine wave*, which in the code above is hard-coded to 0.05. This is an arbitrary number, but it could be converted to degrees visual angle (dva) if you would measure the size of the Gabor patch on the monitor and the distance between the monitor and your eye. Try changing this

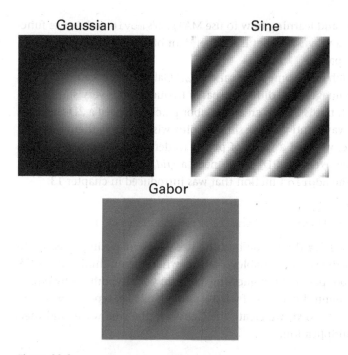

Figure 33.2

A Gaussian and a sine meet to form a Gabor patch.

parameter. With the size of the image we've chosen (variable lims), what are reasonable lower and upper frequency bounds?

• *The phase of the sine wave,* which in the code above is implicitly set to 0. What is the range of parameters that you should test (i.e., what is the smallest and largest value that makes sense to use)? Try some different values. If you don't know how to set the phase in the code above, consult chapter 11, where you learned about 1D sine waves.

• *The amplitude of the sine wave,* which is set to 1. We won't change the amplitude here. Technically, the Gaussian also has an amplitude parameter implicitly set to 1. Let's also not worry about that one.

• *The rotation of the sine wave,* which in the code above is also implicitly set to 0. Imagine that the 2D sine wave were on the face of an analog clock. As time ticks on and your life slips by, the 2D sine wave spins around its center. The trick to rotating a 2D sine wave is to rotate the x and y grid points. This we do by scaling the x and y grid points by the standard 2D rotation matrix. If the angle of rotation in radians is variable th, the code would look like this:

```
xp = x*cos(th) + y*sin(th);
yp = y*cos(th) - x*sin(th);
```

I called the new variables xp and yp because the typical math terms would be x' and y' (x-prime and y-prime; I'll let you guess what to call the rotated variable optimus). To get a feel for the effect of this rotation, make an image of x and xp.

I included yp here just for your personal edification. You can see in the code above that we define the sine wave only from the grid points x. Try to create a new Gabor patch by defining the variable th in radians between 0 and 2π. You need to change x to xp when creating the sine wave; do you need to change x to xp in the Gaussian? Why or why not? (Hint: The answer is no, but it's up to you to come up with the reason for that answer.)

Gabor patches (even without animations) are often used in neuroscience, particularly as stimuli in visual psychophysics experiments. For example, many vision experiments involve having research participants discriminate the rotation or the spatial frequency of various Gabor patches. Gabor patches are also used as filters when modeling how neurons in the early visual cortex respond to different spatial frequencies in retinal inputs. Plus Gabors look really cool. You can print them out on T-shirts to impress people at neuroscience conferences.

Anyway, let's breathe some life into our Gabor patch. We begin, as always, by setting the scenery. Notice that we are extending the concept of defining and updating plot handles to defining and updating image handles.

```
subplot(221)
gaus_h = imagesc(randn(size(x)));
subplot(222)
sine_h = imagesc(randn(size(x)));
subplot(212)
gabr_h = imagesc(randn(size(x)));
```

Notice that I set all of these images to be random noise. Does this mean our movie will display random noise? Maybe just in the first frame? Probably you are guessing that the answer is no. Inside the loop over movie frames, we will update the data using the image handles created above. We just need to initialize the axes and handles; it doesn't matter what we put in those images during this initialization.

What shall we animate in the movie? There are three key parameters of the Gabor patch; let's animate all of them. We will have 20 frames, and we first need to specify the ranges of parameters to change at each frame (the phase of the Gabor, the rotation of the sine wave, and the width of the Gaussian).

```
nFrames = 20;
phases = linspace(0,pi,nFrames);
rotates = linspace(0,pi/2,nFrames);
widths = [ linspace(5,12,nFrames/2) ...
           linspace(12,5,nFrames/2) ];
```

Notice how I set the widths. Figure 33.3 shows one frame of this movie. But before watching the entire movie on your computer, try to guess what the Gabor patch will do over time based on inspecting the code.

Now we're ready to loop over frames and update the Gabor patch and the image data. In the previous example with the line movie, we modified the properties XData and YData. With image handles, the main field that we want to update is called CData, which refers to the color data. The

Figure 33.3
One frame of the Gabor patch movie.

following code takes place inside the loop (variable `framei` is the looping index).

```
width = widths(framei);
rotphase = rotates(framei);
xp = x*cos(rotphase) + y*sin(rotphase);
yp = y*cos(rotphase) - x*sin(rotphase);
```

Notice we redefine `xp` and `yp` from the original `x` and `y`. This is a good example of a situation in which overwriting original variables can cause confusion or at least the need for redundant code. More generally, we want to have as little code inside the loop as possible. Especially code that can be time-consuming or that involves command statements should, whenever possible, be precomputed outside the loop. Perhaps it is not so critical in this toy example, but if you are programming a psychophysics experiment in which precise hardware timing is crucial, delays or uncertainties of even a few milliseconds might be unacceptable. Next, we recompute the three functions (sine wave, Gaussian, Gabor) and update the axis data.

```
% define and display the Gaussian
gaus2d = exp(-(xp.^2 + yp.^2) / (2*width^2));
set(gaus_h,'CData',gaus2d);
% define and display the sine wave
sine2d = sin(2*pi*.05*xp + phases(framei));
set(sine_h,'CData',sine2d);
% define and display the gabor patch
set(gabr_h,'CData',sine2d .* gaus2d);
```

Don't forget to include either `drawnow` or `pause` before the end of the loop. Now we have a movie in MATLAB. Before moving forward, try changing the sine and Gaussian parameters, the number of frames, the movie update speed, and other features. The purpose of this movie is to help you gain an intuitive feel for Gabor patches and how they are parameterized, and intuition is built on understanding effects of parameter selections.

What if you want to send this movie to Hollywood for a screening? Or show it to your grandmother? Not everyone has the time to master 33 chapters of MATLAB programming; it would be easier to export the movie to a file that anyone can view. For this, we use the MATLAB `writeVideo` function. (Users of older MATLAB versions might be familiar with the function-combo `getframe` and `movie`, but these are being phased out for the new-and-improved `writeVideo`.)

Using `writeVideo` is very easy and involves three steps: initialize the video object and file name, capture each frame of the movie while inside the loop, and write the object to disk as an .avi file. Step 1 happens before the loop and involves two lines of code, one to create a video object, and a second to open that object.

```
vidObj = VideoWriter('gaborFilm.avi');
open(vidObj);
```

Then, inside the loop, the following line of code will grab the entire figure and place it as a frame inside the video object.

```
writeVideo(vidObj,getframe(gcf));
```

The `gcf` input to the `getframe` function can be replaced with a handle to a specified figure or axis. For example, if you were to replace `gcf` with `gca`, what would be shown in the movie, and would it depend on *where* in the loop that line appears? Guess the answer, then confirm in MATLAB.

Last but certainly not least: After the loop is finished, you need to close the object. Without running the next line, MATLAB will not complete the file, and your grandmother will be disappointed. You don't want to disappoint Gram-Gram.

```
close(vidObj);
```

Now you can open the movie outside of MATLAB using your video player of choice, such as Windows Media Player or VLC.

33.3 Spinning Heads

Let's make one more movie that incorporates some more advanced plotting tools. I had a dream once about two disembodied heads that were spinning around in an empty space, with fluctuating colors that illustrated changes in frequency-band-specific power in the theta and alpha bands (6 Hz and 10 Hz). Let's see if we can reconstruct this dream. The first step of making this movie is to create an electrode-by-time-by-frequencies matrix using wavelet convolution, as you learned in chapter 19. The online code does most of the work but leaves out a few key lines so you can test your knowledge.

Time acts funny in dreams, so let's have our movie play forward and then backward. The movie will span 100 frames between 0 and 800 milliseconds and then play those same 100 frames backward. Unlike with the previous movies, here we are making images from existing data. We'll have to figure out which time points in the data best coincide with our requested time points, which is exactly the kind of problem for which `dsearchn` or min-abs is designed.

```
nFrames = 100;
tbnd = [0 800]; % time boundaries
tidx = dsearchn(EEG.times', ...
       linspace(tbnd(1),tbnd(2),nFrames)');
```

To make time go forward and then backward, we just need to concatenate a backward version of the time vector `tidx` after the forward version.

```
tidx = [tidx(1:end-1); tidx(end:-1:1)]; % use this
tidx = [tidx; tidx(end:-1:1)]; % don't use this one
```

Why is the first line of code better than the second? What happens in the middle of the second vector?

Now we have time going forward and backward; next we need to figure out how to make the heads spin. You'll see below that we are going to make 3D renderings; we can make them spin by changing the azimuth value of the axes on each frame of the movie. In other words, the heads themselves won't rotate; our viewpoint will rotate.

```
azs = round(linspace(0,360,length(tidx)))+90;
```

I added 90° here to make the heads initially facing each other. You can try changing that starting value offset. Now we're ready to set up the plot. We want two axes, one for each rotating head. Usually, we use the function `subplot` to access parts of a figure, but in this case we want to have more control over the locations and sizes of the axes. Thus, we will place axes on the figure and use the `set` function to specify their size and position.

```
ax1_h = axes;
a1pos = [.1 .1 .3 .8];
set(ax1_h,'Position',a1pos,'CameraViewAngle',6)
ax2_h = axes;
a2pos = [.5 .1 .3 .8];
set(ax2_h,'Position',a2pos,'CameraViewAngle',6)
```

The parameter `CameraViewAngle` needs to be initially fixed to a specific value, otherwise it will be recomputed automatically on each frame. After completing this video, you can try testing what happens when the `CameraViewAngle` is not specified.

And now we're ready to begin filming the movie. At each step inside the loop, we specify the axis we want to use, call a function to create the head plot, and then adjust the azimuth component of the viewing angle. The online code includes a function called headplotIndie, which is modified from a function called headplot that comes with the eeglab toolbox. As with the topoplotIndie function that you saw in earlier chapters, I made a few modifications to the eeglab code so it works independent of the toolbox. All credit for the creative work building this function goes to eeglab.

The following code is inside the loop over time points (variable ti is the looping variable).

```
axes(ax1_h);
headplotIndie(squeeze(tf(:,1,tidx(ti))),...
              '3dSpline.spl',[-3 3]);
view(ax1_h,[-azs(ti) 20])
```

This code says that we want the axis indicated with handle ax1_h to be active, then we call the headplotIndie function using the time-frequency matrix at a single time point as the first input, a spline matrix as the second input, and the color limits as the third input. This spline matrix defines the points of the head surface. It must be uniquely defined for each EEG electrode montage, meaning that this spline file is valid only for these EEG data; you cannot apply this spline matrix to your EEG data. If you are working with EEG data, you can create a unique spline matrix in eeglab. The last line of code specifies the viewing angle.

There is another set of three lines to produce the other head plot. The differences are that (1) the other axis is specified, (2) frequency "2" is used (second dimension of the tf matrix), and (3) the viewing angle azimuth (variable azs) is positive instead of negative.

And that's about it for our film (see figure 33.4 for one frame of the movie). The online MATLAB code contains a few extra lines to grab each figure as a frame in a movie and write it to disk in .avi format (the same code as used previously in this chapter). I encourage you to play around with the code to get a feel for the effects of different parameter settings on the resulting plots.

A quick note about the head plot: It is simply a patch, just like the patches you learned about in chapter 9. Rather than calling the function headplotIndie on each iteration of the loop, it would be possible to update the CData property of the patch, similar to how we updated the CDdata property in the Gabor movie (using the handle p1 on line 85 of headplotIndie.m). But this would require moving a lot of overhead code into our script, and I decided against it.

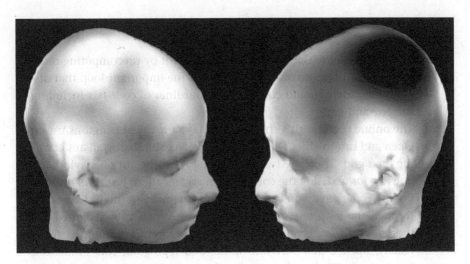

Figure 33.4
One frame of the spinning-heads movie, which shows time-varying changes in time-frequency power at two different frequencies.

33.4 Exercises

1. In the movie in section 33.1, add another subpanel to plot the time series of phase values.

2. Create another movie in section 33.1 that shows only the complex signal in three dimensions (the three dimensions are real part, imaginary part, and time). Make sure all axes limits remain the same throughout the movie. Have the axis slowly rotating in three dimensions as the function "snakes" through time.

3. Rewrite the following line of code using the `cat` function. Make sure the output is identical.

```
tidx = [tidx(1:end-1); tidx(end:-1:1)];
```

4. What is the effect of changing the sine wave phases in the Gabor movie? It's a bit tricky to see when there is also rotation. Turn off the rotation so you can see the effect of the phase changes. This is a general principle in science: When possible, it's best to isolate a single feature of a system in order to understand what that feature does. However, the very valid counterargument is that biology typically involves so many complex interactions that isolating a single feature might be so unnatural as to produce misleading findings that have nothing to do with

the real world. Fortunately for this exercise, Gabors are not biological phenomena.

5.   You can make the Gabor movie more efficient by precomputing more matrices. It might involve two loops, but the important loop that displays the movie should contain only four lines of code (six including the for and end).

6.   The online code for this chapter includes a file called LangtonsAnt.m. Open and run the file. This program involves simulating an ant crawling on a grid. There are two rules that govern the ant's behavior, and one rule for updating the color of each grid location. Just from inspecting the code, see if you can figure out those rules (it might help to go through the loop one iteration at a time). When you've figured out the rules (or have given up), search the Internet for "Langton's Ant." Finally, open a new MATLAB m-file and reprogram this simulation from scratch. I'll close this exercise, this chapter, and this book with the most important piece of advice for scientific programming: There are many correct solutions to a problem. It doesn't matter if your solution looks different from someone else's solution; what matters is that your code works and is accurate.

References

Adams, D. (1989). *The More Than Complete Hitchhiker's Guide: Complete & Unabridged.* New York: Bonanza Books. https://books.google.com/books/about/The_More_Than _Complete_Hitchhiker_s_Guid.html?id=jItwNQAACAAJ&pgis=1.

Bruns, A. (2004). Fourier-, Hilbert- and Wavelet-Based Signal Analysis: Are They Really Different Approaches? *Journal of Neuroscience Methods*, 137(2), 321–332. doi:10.1016/j.jneumeth.2004.03.002.

Burkitt, A. N. (2006). A Review of the Integrate-and-Fire Neuron Model: I. Homogeneous Synaptic Input. *Biological Cybernetics*, 95(1), 1–19. doi:10.1007/s00422-006-0068-6.

Buzsáki, G., & Moser, El. (2013). Memory, Navigation, and Theta Rhythm in the Hippocampal-Entorhinal System. *Nature Neuroscience*, 16(2), 130–138. doi:10.1038/nn.3304.

Cardoso, J. F. (1999). High-Order Contrasts for Independent Component Analysis. *Neural Computation*, 11(1), 157–192. http://www.ncbi.nlm.nih.gov/pubmed/?term =cardoso+high-order+contrasts+for+independent.

Carp, J. (2012). On the Plurality of (Methodological) Worlds: Estimating the Analytic Flexibility of FMRI Experiments. *Frontiers in Neuroscience*, 6(January), 149. doi:10.3389/fnins.2012.00149.

Chang, C.-C., and Lin, C.-J. (2011). LIBSVM. *ACM Transactions on Intelligent Systems and Technology*, 2(3), 1–27. doi:10.1145/1961189.1961199.

Chu, C. C. J., Chien, P. F., and Hung, C. P. (2014). Tuning Dissimilarity Explains Short Distance Decline of Spontaneous Spike Correlation in Macaque V1. *Vision Research*, 96(March), 113–132. doi:10.1016/j.visres.2014.01.008.

Cohen, M. X. (2014). *Analyzing Neural Time Series Data: Theory and Practice.* Cambridge, MA: MIT Press.

Cohen, M. X. (2015). Comparison of Different Spatial Transformations Applied to EEG Data: A Case Study of Error Processing. *International Journal of Psychophysiology:*

Official Journal of the International Organization of Psychophysiology, *97*(3), 245–257. doi:10.1016/j.ijpsycho.2014.09.013.

Cohen, M. X. (2016). Midfrontal theta tracks action monitoring over multiple interactive time scales. *NeuroImage*, *141*, 262–272. doi:10.1016/j.neuroimage .2016.07.054.

Cohen, M. X., & van Gaal, S. (2013). Dynamic Interactions between Large-Scale Brain Networks Predict Behavioral Adaptation after Perceptual Errors. *Cerebral Cortex*, *23*(5), 1061–1072. doi:10.1093/cercor/bhs069.

Delorme, A., and Makeig, S. (2004). EEGLAB: An Open Source Toolbox for Analysis of Single-Trial EEG Dynamics Including Independent Component Analysis. *Journal of Neuroscience Methods*, *134*(1), 9–21. doi:10.1016/j.jneumeth.2003.10.009.

Delorme, A., Palmer, J., Onton, J., Oostenveld, R., and Makeig, S. (2012). Independent EEG Sources Are Dipolar. *PLoS One*, *7*(2), e30135. doi:10.1371/journal.pone .0030135.

Ericsson, K. A., Krampe, R. T., and Tesch-Römer, C. (1993). The Role of Deliberate Practice in the Acquisition of Expert Performance. *Psychological Review*, *100*(3), 363–406.

Faul, F., Erdfelder, E., Lang, A.-G., and Buchner, A. (2007). G*Power 3: A Flexible Statistical Power Analysis Program for the Social, Behavioral, and Biomedical Sciences. *Behavior Research Methods*, *39*(2), 175–191. http://www.ncbi.nlm.nih.gov/ pubmed/17695343.

Haegens, S., Cousijn, H., Wallis, G., Harrison, P. J., and Nobre, A. C. (2014). Inter- and Intra-Individual Variability in Alpha Peak Frequency. *NeuroImage*, *92*, 46–55. doi:10.1016/j.neuroimage.2014.01.049.

Izhikevich, E. M. (2003). Simple Model of Spiking Neurons. *IEEE Transactions on Neural Networks*, *14*(6), 1569–1572. doi:10.1109/TNN.2003.820440.

Jaeger, D., and Jung, R. (Eds.). (2015). *Encyclopedia of Computational Neuroscience*. New York: Springer.

King, J.-R., and Dehaene, S. (2014). Characterizing the Dynamics of Mental Representations: The Temporal Generalization Method. *Trends in Cognitive Sciences*, *18*(4), 203–210. doi:10.1016/j.tics.2014.01.002.

Kriegeskorte, N., Simmons, W. K., Bellgowan, P. S. F., and Baker, C. I. (2009). Circular Analysis in Systems Neuroscience: The Dangers of Double Dipping. *Nature Neuroscience*, *12*(5), 535–540. doi:10.1038/nn.2303.

Le Van Quyen, M. (2011). The Brainweb of Cross-Scale Interactions. *New Ideas in Psychology*, *29*, 57–63.

Lehmann, D., Pascual-Marqui, R., and Michel, C. (2009). EEG Microstates. *Scholarpedia*, *4*(3), 7632. doi:10.4249/scholarpedia.7632.

Li, N., Chen, T.-W., Guo, Z. V., Gerfen, C. R., and Svoboda, K. (2015). A Motor Cortex Circuit for Motor Planning and Movement. *Nature*, *519*(7541), 51–56. doi:10.1038/nature14178.

Luce, R. D. (1963). Detection and Recognition. In E. Galanter, R. D. Luce, and R. R. Bush (Eds.), *Handbook of Mathematical Psychology* (1st ed., pp. 103–189). New York: Wiley.

Macmillan, N. A., and Douglas Creelman, C. (2004). *Detection Theory: A User's Guide*. Hove, England: Psychology Press. https://books.google.com/books?id=2_V5AgAAQBAJ&pgis=1.

Makeig, S., Debener, J., Onton, J., & Delorme, A. (2004). Mining Event-Related Brain Dynamics. *Trends in Cognitive Sciences*, *8*(5), 204–210. doi:10.1016/j.tics.2004.03.008.

Mandelbrot, B. (1967). How Long Is the Coast of Britain? Statistical Self-Similarity and Fractional Dimension. *Science*, *156*(3775), 636–638. doi:10.1126/science.156.3775.636.

Maris, E., and Oostenveld, R. (2007). Nonparametric Statistical Testing of EEG- and MEG-Data. *Journal of Neuroscience Methods*, *164*(1), 177–190. doi:10.1016/j.jneumeth.2007.03.024.

Mizuseki, K., Sirota, A., Pastalkova, E., and Buzsáki, G. (2009). Theta Oscillations Provide Temporal Windows for Local Circuit Computation in the Entorhinal-Hippocampal Loop. *Neuron*, *64*(2), 267–280. doi:10.1016/j.neuron.2009.08.037.

Muir, D. R., and Kampa, B. M. (2014). FocusStack and StimServer: A New Open Source MATLAB Toolchain for Visual Stimulation and Analysis of Two-Photon Calcium Neuronal Imaging Data. *Frontiers in Neuroinformatics*, *8*(January), 85. doi:10.3389/fninf.2014.00085.

Nichols, T. E., and Holmes, A. P. 2002. Nonparametric Permutation Tests for Functional Neuroimaging: A Primer with Examples. *Human Brain Mapping*, 15(1), 1–25. http://www.ncbi.nlm.nih.gov/pubmed/11747097.

Norcia, A. M., Appelbaum, L. G., Ales, J. M., Cottereau, B. R., and Rossion, B. (2015). The Steady-State Visual Evoked Potential in Vision Research : A Review. *Journal of Vision (Charlottesville, Va.)*, *15*(6), 1–46. doi:10.1167/15.6.4.doi.

Norman, K. A., Polyn, S. M., Detre, G. J., and Haxby, J. V. (2006). Beyond Mind-Reading: Multi-Voxel Pattern Analysis of fMRI Data. *Trends in Cognitive Sciences*, *10*(9), 424–430. doi:10.1016/j.tics.2006.07.005.

Onton, J., Westerfield, M., Townsend, J., & Makeig, S. (2006). Imaging Human EEG Dynamics Using Independent Component Analysis. *Neuroscience and Behavioral Reviews*, *30*(6), 808–822. doi:10.1016/j.neurobiorev.2006.06.007.

Palva, J. M., Zhigalov, A., Hirvonen, J., Korhonen, O., Linkenkaer-Hansen, K., and Palva, S. 2013. Neuronal Long-Range Temporal Correlations and Avalanche Dynamics Are Correlated with Behavioral Scaling Laws. *Proceedings of the National Academy of Sciences of the United States of America*, 110(9), 3585–3590. doi:10.1073/pnas.1216855110.

Rescorla, R. A., and Wagner, A. R. (1972). A Theory of Pavlovian Conditioning: Variations in the Effectiveness of Reinforcement and Nonreinforcement. In A. H. Black and W. F. Prokasy (Eds.), *Classical Conditioning II: Current Research and Theory* (pp. 64–99). New York: Appleton-Century-Crofts.

Rey, H. G., Pedreira, C., and Quiroga, R. Q. 2015. Past, Present and Future of Spike Sorting Techniques. *Brain Research Bulletin*, *119*(Pt B), 106–117. doi:10.1016/j.brainresbull.2015.04.007.

Schultz, W., Dayan, P., and Montague, P. R. (1997). A Neural Substrate of Prediction and Reward. *Science*, *275*(5306), 1593–1599. http://www.ncbi.nlm.nih.gov/pubmed/9054347.

Shao, Y.-H., Gu, G.-F., Jiang, Z.-Q., Zhou, W.-X., and Sornette, D. (2012). Comparing the Performance of FA, DFA and DMA Using Different Synthetic Long-Range Correlated Time Series. *Scientific Reports*, *2*(January), 835. doi:10.1038/srep00835.

Shiffman, D. 2012. *The Nature of Code: Simulating Natural Systems with Processing* (1st ed.). New York: The Nature of Code.

Singh, K. D. (2012). Which 'Neural Activity' Do You Mean? fMRI, MEG, Oscillations and Neurotransmitters. *NeuroImage*, *62*(2), 1121–1130. doi:10.1016/j.neuroimage.2012.01.028.

Smith, S. M., Jenkinson, M., Woolrich, M. W., Beckmann, C. F., Behrens, T. E. J., Johansen-Berg, H., Bannister, P. R., et al. (2004). Advances in Functional and Structural MR Image Analysis and Implementation as FSL. *NeuroImage*, *23*(January; Suppl 1), S208–S219. doi:10.1016/j.neuroimage.2004.07.051.

Thorndike, E. L. (1898). Animal Intelligence: An Experimental Study of the Associative Processes in Animals. *Psychological Monographs*, *2*(4), i–109.

van Drongelen, W. (2006). *Signal Processing for Neuroscientists*. Cambridge, MA: Academic Press.

Wells, H. G. (1895). *The Time Machine*. London, England: William Heinemann.

Index